Intelligent Systems Reference Library

Volume 198

The aim of this series is to publish a Reference Library, including novel advances and developments in all aspects of Intelligent Systems in an easily accessible and well structured form. The series includes reference works, handbooks, compendia, textbooks, well-structured monographs, dictionaries, and encyclopedias. It contains well integrated knowledge and current information in the field of Intelligent Systems. The series covers the theory, applications, and design methods of Intelligent Systems. Virtually all disciplines such as engineering, computer science, avionics, business, e-commerce, environment, healthcare, physics and life science are included. The list of topics spans all the areas of modern intelligent systems such as: Ambient intelligence, Computational intelligence, Social intelligence, Computational neuroscience, Artificial life, Virtual society, Cognitive systems, DNA and immunity-based systems, e-Learning and teaching, Human-centred computing and Machine ethics, Intelligent control, Intelligent data analysis, Knowledge-based paradigms, Knowledge management, Intelligent agents, Intelligent decision making, Intelligent network security, Interactive entertainment, Learning paradigms, Recommender systems, Robotics and Mechatronics including human-machine teaming, Self-organizing and adaptive systems, Soft computing including Neural systems, Fuzzy systems, Evolutionary computing and the Fusion of these paradigms, Perception and Vision, Web intelligence and Multimedia.

Indexed by SCOPUS, DBLP, zbMATH, SCImago.

All books published in the series are submitted for consideration in Web of Science.

More information about this series at http://www.springer.com/series/8578

Selby Coxon · Robbie Napper
Editors

Advancing a Design Approach to Enriching Public Mobility

Springer

Editors
Selby Coxon
Mobility Design Lab
Department of Design
Faculty of Art Design and Architecture
Monash University
Melbourne, VIC, Australia

Robbie Napper
Mobility Design Lab
Department of Design
Faculty of Art Design and Architecture
Monash University
Melbourne, VIC, Australia

ISSN 1868-4394 ISSN 1868-4408 (electronic)
Intelligent Systems Reference Library
ISBN 978-3-030-64721-6 ISBN 978-3-030-64722-3 (eBook)
https://doi.org/10.1007/978-3-030-64722-3

This Springer imprint is published by the registered company Springer Nature Switzerland AG
The registered company address is: Gewerbestrasse 11, 6330 Cham, Switzerland

To anyone who has ever travelled and appreciated the difference Design can make to their journey.

Preface I

Our lives are bound by the countless practices that compose life in the city. Few daily routines reveal as much about the rhythm of the city with all its pleasures and disappointments than our daily commute. The discordant sounds of transportation, engines revving, brakes squealing and pedal rotations whispering through the air, the accompaniment to our thoughts of the day ahead or what has been. Travelling anywhere immerses us in a range of emotional and psychological needs responded to by visceral interactions with material objects and the thoughtful or thoughtless inclination behind their creation. Human beings draw pleasure from aesthetic objects that can inflate our moods, underwrite our happiness or diminish our misery.

At the time of preparing this manuscript, the world is experiencing the COVID-19 pandemic. The chapters concern work created Before COVID-19 (BC), but our application and interpretations are inevitably coloured by the experiences of moving into After the Disease (AD). Public mobility of any kind faces complex challenges involving interconnected relationships between various system elements and stakeholders such as passengers, operators, manufacturers, local communities and governments. Design and designers will be central to creating the innovative responses a new urban mobility can offer citizens. The outlook is largely positive. Land-based, low-carbon, transport may flourish. Active travel and the increase in cycle lanes in inner city areas along with fewer commuters in the first place may change the lifeblood of personal mobility. Re-establishing trust in those modes in which we need to share space with others will stimulate new innovative approaches to the management of geometry, materials and our relationship with mobility. Many people might permanently, or at least regularly, work from home, opening up vast vacant office spaces. New delivery-focused mobility will grow and flourish. The pandemic has brought into sharper focus existing social issues such as how much space we give to cars and the role of high-speed rail in a post- or reduced-air travel world.

This book provides transport researchers and practitioners with a rich array of design approaches and examples. Our aim was to provide a volume which highlighted the way in which design approaches to transport are important and compatible with more established schools of thought. It would appear that these approaches are going to be more necessary than ever to sift through a transport system which was still looking very twentieth century and realise the potential for positive change, from the wreckage of one culture to the creation of a new one.

Melbourne, Australia

Selby Coxon
Robbie Napper

Preface II

Mobility is considered an essential part of social inclusion and civil liberty. Our physical geographical limits are bounded by accessibility to a variety of forms of transport. Engaging with the variety of ways we go about our business is far more than timetables and economic expediency. It is experiential and either depleted or enriched by the design of the vehicles, spaces and infrastructure. Advancing a Design Approach to Enriching Public Mobility is a collected discourse on the impact of design approaches upon the travelling public from a collection of the leading design thinkers on the future of urban mobility. There is no shortage of commentators on transport planning and engineering and vital work it is too, for any possibility of moving far beyond our own house would otherwise be quite impossible. As human beings with emotions and sensitivities, our contact with mobility across the urban landscape is highly conditioned on the visceral and behavioural qualities of that experience. Design is the mediator of experience.

There is a panoply of challenges designers face to accommodate our expectations and create a good travel experience. There is an unspoken expectation that we should arrive safely and feel safe at all times in the journey, and yet for many sectors of our community, spaces around transport are unfriendly, inequitable and threatening. Women experience public mobility in different ways to men. In advanced societies, the right to universal accessibility is enshrined in law. This is leading to transport operators scurrying to find design solutions that overcome the legacy of inadequate and less enlightened attitudes to personal disability. The triumph of the car to hold a place at the pinnacle of personal freedom has led to congested and polluted cities that are anything but free. Sensitivity to designing these objects and spaces to mitigate these issues goes a long way to creating a better and more mobile environment for society.

The authors of this book come from a wide range of design backgrounds and approaches, unified by an urgency to embed good design practice into the physical and organisational structures that connect us to one another. Their work spans the psychological and cultural, the pragmatic and the poetic and serves as a primer for transport and mobility professionals to engage design resources to address the prescient problems of urban mobility.

This book has been organised into fifteen chapters that more or less segue to one another, taking the reader through the mobility design research landscape. Leading researchers and practitioners from around the world share with us their insights into new ideas and approaches to enriching the transport use experience. From the digital connectivity and freedoms of emerging Mobility as a Service through new concepts for active travel, rail, automotive and public art amongst others. It is hoped and anticipated that this book will stimulate the professional to question practices and contemporary orthodoxy in the light of a rapidly changing transport environment.

Chapter 1. Mobility as a Service Through Design

There is a considerable debate within the design community over the extent to which the consumer, patron or more widely termed 'user' sits in the design process. Designing for someone other than oneself in the industrial context has largely been driven by needs of manufacturing. However, with the limitations of what industry can make diminishing, there is a greater acknowledgement that it is the needs of the consumer that should take preference. After all, the efficacy of an object is driven by the consumer's experience of using it. Human-centred approaches are not as straightforward as they may seem. Rapid advances in technology are not fully understood as quickly as they emerge. There seems to be an appetite for discovery before we really know what something is for or what it can do.

In Chap. 1, the authors examine research methods by which people can be more actively engaged in the shaping of their technological systems. In particular, the authors have demonstrated how an examination of everyday life contexts can inform the future of intelligent transport. Their approach drawn from design anthropology enables a better understanding of how people experience the existing mobility systems and technologies. The particular example discussed in this chapter focuses upon the relationship between MAAS and autonomous vehicles and their implications upon the models of private car ownership. Workshops with a wide variety of stakeholders actively co-design future scenarios and shape ideas in a collaborative structure. History and research have consistently shown that people do not always use technologies as intended: people will abandon technologies that are not designed to adequately fulfil their everyday needs, or they appropriate technologies to fit their own routines and to accomplish their everyday objectives.

High-density populations and traffic congestion caused by ubiquitous private car ownership are unsustainable. An ethnographic perspective to designing mobility in the urban environment can disrupt orthodox thinking by understanding and taking the real-life situations of what people do, think and experience when moving through their everyday urban spaces and nudges the discussion towards a shared agenda and perhaps more sustainable and realistic mobility solutions.

Chapter 2. Visual Conflict Framing in Public Transport

Public transport embraces a wide range of physical infrastructure, organisational and human-centric challenges. Often, difficult issues contain ambiguity, change, contradictions and potentially insurmountable barriers of culture and understanding. Successfully solving or at least trying to manage these sorts of problems requires a reassessment of some of the traditional approaches to analysing and solving complex and sometimes contrary interrelationships. Since such challenges are multi-faceted, they call for thinking that is capable of grasping the wider picture, including the interrelationships of a full range of causal factors. They require broader, more collaborative and innovative approaches.

While the practice of design thinking has been with us for some time, its value is being increasingly recognised in a wider variety of industries and organisations. Public transport is one such example within which social, cultural, technological and economic factors intersect with often conflicting demands. The authors of this chapter discuss their research in which they developed visualisation tools to enable better analysis, synthesis and assessment of complex systems. This is especially useful in communicating with a wide variety of stakeholders where innovation may emerge and the repercussions for the wider system therein.

Chapter 3. Unbounded Ergonomics—Addressing Gender Transport Poverty

The neglect of embracing a truly human-wide equity to mobility needs, especially with regard to women, has been to the detriment of humanity. Bias, discrimination and inaccessibility have blighted lives for centuries. In unbounded ergonomics, the authors have examined confronting data on gender poverty in Karachi, Pakistan. Social mores and rigorously enforced cultural laws have meant that women have very constrained and even dangerous mobility experiences.

The chapter explores whether transport ergonomics, the use of the hexagon-spindle (HS) model, can add any value to the debate by systematically unpicking the problem and mapping issues into a solution space to identify social and cultural factors which inhibit gender transport equality and develop culturally appropriate solutions. The previous research has applied the HS model to map out key performance indicators for passenger satisfaction; however, little exploration has been made of factors falling into the 'personal sector'. It may be argued in this and other cases that these along with external factors have a significant and sometimes overwhelming effect on mobility and need greater consideration in the design of transport systems.

Although Pakistan provides an extreme example, women across the globe are denied the freedom to travel freely and without fear of harassment on public transport. Putting the female traveller at the centre of a concentric ring (or the HS) model enables a clearer view of the transport system from a user's perspective and may show where small inroads can be made. At a wider level, it is hoped that transport and mobility can be used to illustrate and reduce the inequalities women in the global south experience in their everyday lives.

Chapter 4. Unwanted Sexual Behavior and Public Transport

Following the hexagon-spindle analysis of gender transport poverty in Karachi, the authors of Chap. 4 examine how gender-sensitive policy and design is critical to the future of public transportation. Not only does this approach positively impact women's safety and their access to and use of public transport (and therefore public life), it makes significant contributions to the increasingly important sustainability and viability targets of public transport. The chapter considers the complex issue of safety and the feminist scholarship that has influenced the everyday challenges and risks faced by women internationally. By examining the literature and data around women's safety in public transport spaces, the research draws on case studies and policy actions in this arena. Advocating for a feminist framework, this chapter concludes by making a case for the critical and central role a gender lens and gender-sensitive co-design have in breaking the hold that technical concerns have over transport.

Chapter 5. Speeds Deeds and Needs on Two Wheels (or Three)

Continuing the thread on a personal, human approach in contrast to a technical one is this presentation of a model representing what people do on bicycles. Set in the context that the bicycle is back. For the last decade, progressive cities across the globe have been turning to the bicycle to add to the array of transportation options for citizens. In an attempt at developing infrastructure that helps to legitimise the bicycle as a transport form in our urban centres, designers require a robust framework within which they can design appropriate facilities for bicycle users. Practitioners need to deeply understand the transport sphere that is occupied by the simple push bicycle in order to design effective solutions for its users. A design framework that has been adopted in a growing number of cities around the world today focuses on 'AAA' infrastructure, or bicycle facilities for 'All Ages and Abilities'. While this aims at an inclusionary design approach for bicycle infrastructure, it is not often broken down into its component parts and investigated from a tangible and ground-level perspective. Who fits into this 'All Ages and Abilities' model? How fast do they move? What kind of contraption are they riding? Where are they going and for what purpose? What do they need in order to feel comfortable enough to ride in the space? This deeper framework can help urban designers align form and function when developing new bicycle facilities for this AAA model. Whether the design user is riding in the peloton at 35 km/h, or riding to the corner store at 15 km/h, this makes a significant impact on the design choices of the texture treatments, signal phasing and bends along the bicycle corridor. Whether the user is transporting themselves on a carbon fibre bullet or their two-year-old in a Danish Bullitt, this helps determine appropriate levels of protection of the facility from vehicle exposure. Whether the user has been riding since they were a toddler or just learned to ride in a straight line last month at the age of 30, this informs the design of a bicycle facility width for both to use the space efficiently and comfortably. It is through the examination of these speeds, deeds and needs that the urban designer can truly create spaces that are attractive, accessible and understandable to bicycle users of all ages and abilities.

Chapter 6. Safety, Design and Law: A New Interdisciplinary Approach to Bicycle Rider Road Safety

This chapter demonstrates a novel problem-solving approach to study the interaction at an intersection between a driver (turning left) and a cyclist (continuing straight). Road rules and the built environment were not designed in harmony, and there is a high level of confusion about this road rule in the community. Further to these community concerns are professional ones. Previously, road safety and engineering pointed to the law (road rules), as being the source of collisions and injuries from this negotiation. From the position of the law, problems arose concerning how the road rules were applied in a road safety and engineering context.

The road space and the road rules are two manifestations of controls and guidance. Safety science has identified that legal considerations are made externally to road engineering, if at all. In order to develop solutions to the problem of left turn confusion and conflict, the research team hypothesised that a design approach could bring these fields together.

The overarching method is the implementation of convivial design techniques to bring together safety science, legal investigation of road rules and design. Three specific design techniques were used. First, two-dimensional drawing was tested by the research team. Next, two-dimensional drawings and photographs of key intersection types were combined with three-dimensional scale modelling using LEGO® elements in facilitated discussions with road safety practitioners and experts. The third technique brought road rule considerations to the foreground by encouraging road rule annotation on two-dimensional drawings.

Chapter 7. Innovation Opportunities of Compact Vehicles for Urban Transportation

The rapid motorisation of China has led to widespread traffic issues. While solutions exist, transport providers can often create new problems while solving another. Currently, two choices dominate the public debate; firstly, the automobile as a comfortable transport mode that offers the idea of good mobility for the individual, an offering degraded by traffic congestion; and secondly, the alternative of high-efficiency public transport which offers the consumer-less door-to-door comfort but moves more people. This chapter investigates opportunities in a third choice of smaller vehicles. Such vehicles take into account efficiencies, comfort and door-to-door access, but require both design innovation and regulatory innovation to be a more attractive option for urban commuters in China. We conceive of a traffic system which can balance the high efficiency of public transport with the mobility and comfort of personal vehicles.

Chapter 8. ReinCARnation: Re-conceptualising the Idea of the Privately-Owned Car

Is the one-hundred-year love affair with the privately-owned motor car drawing to an end? Are we entering a period where unfettered enthusiasm for the peak of personal prestige and perceived freedom is tempered by a world of climate anxiety, economic challenges, densely populated cities and shifting cultural priorities? Within the design industry, automotive design has long been the popular and jealously guarded elite of industrial design practice, from the genesis of mass production on the assembly lines of Henry Ford to the elite robot populated super factories of Volkswagen. The push to appeal to consumers through the exceptionalism of developing form and surface has been the backbone of the automotive industry's business model.

However, as we move into the third decade of the twenty-first century, the supremacy of the privately-owned car is under attack. Developing nations are seeing a decline in young people taking up licences, and while former developing countries such as China and India are embracing the motor car, sales in other parts of the world are in decline. More of the world's population live in cities than at any other time in human history, and corresponding challenges for mobility have not been kind to the car. The motor car's expansive footprint relative to occupant capacity has congested streets and choked arterial roads making the former tactile pleasure of 'motoring' seem like a distant fantasy. All the while, an accompanying shift away from fossil fuels with their incumbent problems of pollution, expense and finite life.

In reincarnation, the author discusses a road map for changing our relationship with the personal ownership of cars, citing pressures upon the automotive manufacturing industry to respond to a decline in interest, certainly in developed countries for personal ownership, and looking at time-related sharing. That the notion of hiring and 'borrowing' a vehicle to respond to particular moment by moment needs is a direction that has implications upon the future direction of automotive design. The emerging power of artificial intelligence and big data along with an ecological imperative is slowly driving a new business model and relationship between the consumer and their addiction for the car.

Chapter 9. Implementing a User Experience Design Approach in Melbourne's Public Transport System: The Case of Wayfinding and Customer Information Design at Flinders Street Station

Passenger information design and delivery is a key driver of use and satisfaction with the system. In other words, from the passenger perspective, a usable public transport system is one that provides services that are easily understood and internalised. When networks start to grow from a few lines and services to a comprehensive and extended network, a user experience approach is a key to design, development and delivery of public transport's information to drive use and experience, as demonstrated by the case study. This chapter documents the development and application of a colour wayfinding system, significantly improving the usability and comprehension of a system for all user types, reducing the amount of working memory resources required to find their way.

Chapter 10. The Challenges of Designing the Rail Network Map/Diagram as the Critical, Missing, Transit Infrastructure for Mumbai

In contrast to the previous chapter, it is intriguing how a population of 7 million passengers travelling daily on the Mumbai rail network manage to navigate in the absence of a well-designed public information system. The authors inquire, why is it necessary to design a consolidated map/diagram for this high-density rail network (suburban, metro, monorail) controlled by distinct public and private entities, running across multiple districts. Can this map be understood within a multicultural, multilingual context of Mumbai, with varying degrees of literacy and exposure among people? In response to these questions, authors designed a network map/diagram for Mumbai Rail, as an alternative to the current method of 'asking around' to access information. The designers of the map have based their design on the DNA of the famous London underground diagram drafted by Harry Beck in 1933. They have introduced new features in the map to enhance its utility. Intercept study (guerrilla usability testing with tasks, pre- and post-task questions) method was employed with convenience sampling to test the designed map with 102 users in commute context. The insights, challenges, failures and design decisions about the use of station index and grid, the effect of perceived visual complexity, advantages of pseudo-geographic representation and appropriate use of colours or the colour-blind are captured in this chapter.

Chapter 11. Developing a Research and Development Approach to Inter-city Rail Carriage Interiors

Historically, rail travel had been the first great arbiter of travel over significant distances that could be experienced by the common person. A nineteenth-century technology overcame distances in journey times unobtainable by the contemporary means of that century. Since that time, the travelling public has fallen out of love with the mode—contrast the romanticism of travelling by luxury sleeping coach to the crowded standing room only of a commuter service. The late twentieth century has been dominated by increased accessibility to air travel and the associated culture that manifests itself in airport termini, aircraft interiors and associated accessories.

However, high-speed rail (HSR) is enjoying a resurgent interest in global national transport planning agendas. Economic and sustainability metrics indicate benefits over alternative transportation methods over comparable geographical distances. Recent developments in rail carriage interior design both leverage the 'flavour' of the airline aesthetic with a new confident form language that celebrates the space to move about, amenity, comfort and psychological wellbeing that places the mode at a competitive advantage. This chapter examines the critical issues and contemporary approach as exemplified by the French National Railway (SNCF) to create a human-centred resurgently confident approach to the future of rail.

Chapter 12. Data Visualisation Prototypes for Problem Definition in the Design of Accessible Transport Solutions

Public transport is a challenging environment for both passenger and operator. The operation of large networks carries expectation of timely delivery of passenger-oriented services. With numerous stakeholders such as government, private industry and the general public, the expectation around the primary concept of mobility within transport networks often differs. This chapter examines how data visualisation has been engaged to better understand an enduring mobility challenge of Australian metropolitan rail networks, the universally accepted accessibility issue widely known as 'the gap'. With increasing global requirements for disability compliance and anti-discrimination within public transport, data visualisation is applied within a broader design project. Engaged at the problem-definition stage this data-driven approach provides investigators greater insight and visual transparency towards understanding, the impact this issue has on mobility for all users of public transport.

Chapter 13. Mitigating Rail Vandalism Through Design Intervention

This chapter describes a research project undertaken to explore how the design of vulnerable rail assets might be better adapted to reduce incidents of vandalism within the context of Melbourne's rail network. The design approach utilised the novel application of linking rail assets, users and operators through both physical and system-based interventions. Ultimately, the project's outcomes demonstrate how the proactive re-design of vandalism-affected assets can assist operators in developing more targeted, proactive and effective solutions.

Chapter 14. Creating the Circuit Breakers: An Examination of the Sociotechnical System Factors Which Impede and Enable the Delivery of Safe and Healthy Neighbourhood Street Design in Aotearoa New Zealand

Taking a sociotechnical systems approach, the aim of this chapter is to describe the barriers and enablers to innovative street projects that promote wellbeing. We explore these barriers and enablers through the lens of five proposed, current or delivered niche street re-design projects or programmes in Aotearoa New Zealand. Through a thematic analysis of project and programme information, the key themes of leadership, funding, policies and procedures, organisational norms, community and delivery tensions, and social environment emerged. These themes were used to analyse the extent to which the projects and programmes succeeded as niches and influenced the wider system. While there was varying success across the projects and programmes in influencing the wider regime and social landscape, the analysis found that niches need to be supported within government planning systems as a way of managing investment risk and testing future scenarios. The lessons provide direction for those seeking to expedite transport system change so that positive health, safety, environmental and social outcomes can be realised.

Chapter 15. Practicing the Art of Wayfinding Design and Finding the Way to Art Materials: Designing for Los Angeles' Transit Future: Applying Lessons Learned for New Project Delivery Methods for Metro Signage and Artwork

This chapter presents the selected case studies of Los Angeles County Metropolitan Transportation Authority (Metro) Arts and Design through the lens of 'lessons learned'. During nearly 30 years of project delivery amidst tectonic shifts in the construction environment, refined frameworks have been developed using time-tested materials and processes that optimise the customer experience with the aim of creating a world-class transit system. As mediators of that experience, station architecture, environmental graphic design and artwork operate in synergistic harmony. Coordinating the methods of wayfinding and artwork integration allows for maximal visual impact while ensuring adaptability for future growth and ease of maintenance. Through this combined effort, it is possible to create navigable landmarks and a unique sense of place within the complexities of a massive urban transit network. This study includes a description of a new subset of typologies in wayfinding and environmental graphic design, as well as the integration of contemporary visual art specific to the requirements of a transit system in transition—where expansion represents design challenges and consequent opportunities. The discussion deals both with the specifics of implementation in design build contracting as well as the broader conceptual framework.

Melbourne, Australia

Selby Coxon
selby.coxon@monash.edu
Robbie Napper

Acknowledgements

The editors would like to recognise and acknowledge the contribution of the following organisations in the preparation of this anthology.

Monash University.
Coventry University.
SNCF.
Copenhagenize Design Co.
Los Angeles Metro.
Tsinghua University.
Griffith University.
Metro Trains Melbourne.
Volvo Cars.
Ford Motor Company.
Indian Institute of Technology Bombay.
Mackie Research.

Contents

Chapter 1
Mobility as a Service Through Design: A Human Approach

Sarah Pink, Rachel Charlotte Smith, Vaike Fors, Jesper Lund, Kaspar Raats, Katalin Osz, Thomas Lindgren, and Robert Broström

Abstract This chapter explains how designing for future intelligent mobility systems is advanced by a human-centered approach, based in design anthropology. It provides an accessible introduction the theory and methodology of this approach, the production of ethnographic insights, and their translation into design probes for use workshops tailored to enable stakeholders to actively co-design future mobility and autonomous vehicle services and outlines the potentials and challenges of engaging

S. Pink (✉)
Emerging Technologies Research Lab, Faculty of Art, Design and Architecture, Faculty of Information Technology, Monash University, 900 Dandenong Road, Caulfield East, VIC 3145, Australia
e-mail: sarah.pink@monash.edu

S. Pink · R. C. Smith · V. Fors · J. Lund · K. Raats · K. Osz · T. Lindgren · R. Broström
School of Information Technology, Halmstad University, Box 823, 301 18 Halmstad, Sweden
e-mail: rsmith@cc.au.dk

V. Fors
e-mail: vaike.fors@hh.se

J. Lund
e-mail: jesper.lund@hh.se

K. Raats
e-mail: kaspar.raats@hh.se

K. Osz
e-mail: katalin.osz@volvocars.com

T. Lindgren
e-mail: thomas.lindgren@hh.se

R. Broström
e-mail: robert.brostrom@volvocars.com

R. C. Smith
School of Communication and Culture, Aarhus University, Langelandsgade 139, 8000 Aarhus C, Denmark

K. Raats · K. Osz · T. Lindgren · R. Broström
Volvo Car Corporation, 91200 Customer Experience Center, PV3A 405 31 Torslanda, Göteborg, Sweden

S. Coxon and R. Napper (eds.), *Advancing a Design Approach to Enriching Public Mobility*, Intelligent Systems Reference Library 198,
https://doi.org/10.1007/978-3-030-64722-3_1

diverse stakeholders—from industry and policy to the people who will use future technologies and services—in the development of future mobility.

Keywords Mobility as a service · Autonomous driving vehicles · Design anthropology · Human-centered design · Co-design · Urban planning · Design ethnography · Futures

1.1 Introduction

Mobility as a Service (MaaS)—future intelligent transport systems, which will integrate different modes of transport, seeking to create potentially seamless, effective and sustainable mobility solutions—are increasingly regarded as a new way to confront the opportunities and challenges of future mobilities in an era of Artificial Intelligence (AI) and Automated Decision Making (ADM). Autonomous Driving (or self-driving) cars, the most hyped emerging technology in 2015 [1], are increasingly conceptualised as part of the ecology of future MaaS systems. However, we know little about how the relationships between Autonomous Driving Vehicles (ADVs) and MaaS will evolve, either from the perspective of how key policy and industry stakeholders in their design and delivery will collaborate, or in terms of how people will engage with them in their future everyday lives. Large-scale initiatives towards the global roll out of ADVs and MaaS (e.g. the World Economic Forum's *Shaping the Future of Mobility Systems* initiative; and European H2020 research investments) tend to be informed by attempts to resolve regulatory, ethical and safety issues and to achieve their 'public acceptance' and user 'trust', through User Experience (UX), Human-Computer-Interaction (HCI) and Safety research [2], in order to accrue their projected sustainability and quality of life benefits. While these visions of smart mobility 'offer optimistic visions of a society in which technological advances have delivered a benign mobility system that all users can access seamlessly and on demand, avoiding much of the waste, pollution and environmental degradation of today's conditions', there is however, 'no guarantee that such a vision will come to pass' [3].

Critical social science and humanities scholars argue that the search for technological solutions to societal problems fails to acknowledge two key points: first that societal engagement with and acceptance of emerging technologies and services (such as ADVs and MaaS) involves significant processes of social learning which policy makers should see as opportunities for 'more active engagement in the shaping of technological systems' [4]; and second, that future mobilities will also be constituted by what people will do with emerging technologies in everyday life contexts [5]. These critical points call for a re-thinking of how our futures with ADVs and MaaS will play out, and of the types of interventions required to ensure that this happens in equitable, responsible and ethical ways. We argue that this requires combining Science and Technology Studies (STS) and planning scholars critiques of policy and governance of ADVs with design anthropological critiques which are advanced through design focused ethnographic engagements with people in everyday worlds. It also necessitates bringing together different stakeholders in the future of intelligent transport in order to refigure the technological and systemic possibilities that they have under consideration. Design anthropology in particular, has a pivotal role to play in such a process because it is able to: broker the relationship between the ethnographic everyday and those who seek to intervene in it through technology design and policy; create new relationships between policy makers and planners and technology design organisations; and, as described here, to use ethnography as an interventional device that refigures the different ambitions of automotive design and policy into a shared agenda shaped around people and human futures.

This chapter is an accessible introduction to our approach, written for an interdisciplinary readership. It draws on our ethnographic and design workshop research and practice, developed in the project 'Co-designing future smart urban mobility services: A Human Approach project' (AHA) undertaken in Sweden in 2018–19. This project was a collaboration between Halmstad University, Volvo Cars and the Cities of Gothenburg and Helsingborg (Sweden) with Monash University (Australia) and Aarhus University (Denmark).

Thus our work brought together these three academic, industry and public sector stakeholders and in doing so, university researchers (ethnographers, interaction designers, design anthropologists), user experience designers and city planners. Volvo cars' existing focus on people, was significant to our collaboration, as outlined on Volvo's web site:

> As a human-centric car company, we are a brand for people who care about other people and the world in which we live. This is at the core of our business and of our approach to sustainability. Our commitment is about re-thinking sustainability, and goes beyond our operations and our cars, and into society. It makes us think again every time we take decisions that affect the world and the lives of people (https://group.volvocars.com/sustainability)

This, along with our design anthropological approach and the ethnographic insights this brought into our discussions, created a focus on people that brought about new modes of visioning MaaS futures. According to Smith et al. [6], Sweden and Finland 'have arguably acted as global pioneers of MaaS thus far', therefore making Sweden an ideal example through which to explore additional advances through anthropological and co-design approaches. This, brought together with Volvo's approach which also puts people at the centre enabled us to develop a unique approach.

In what follows, we first draw on selected recent academic contributions to this field to contextualise our discussion. Then we describe our design anthropological theory and methodology, how the ethnographic insights were produced and how they were translated into design workshop probes. We next demonstrate how workshops were tailored to enable stakeholders to actively co-design future Maas and ADV services, by: aligning interests and setting the scope; collaborative creation and shaping of ideas; and producing tangible design input for future MaaS and ADV services. Finally, we discuss the potentials and challenges of a human-centred approach to engaging diverse stakeholders in the development of future mobility.

1.2 Visions of Intelligent Mobility Futures

Contemporary discussions about the participation of ADVs and MaaS in future mobility systems have highlighted the question of how policy, planning and governance interface with commercial interests. Stilgoe [4] has proposed that 'Rather than seeing the technology as fixed and looking to plug the deficits of law or public understanding that are imagined around it, policy makers should instead see self-driving cars as an opportunity for more active engagement in the shaping of technological systems, prioritizing social learning' and Cohen et al. [7] call for 'more active form of governance for responsible innovation' in this area. Doherty et al. [8] suggest that a 'Smart Mobility transition' will involve tensions between state-imposed obligations, the interests of producers, public value and citizens' rights and benefits. Legacy et al. [9] highlight a 'tension between the private sector's ambition for creating new platforms for mobility centred around the commodification of the individual journey' on the one hand, with 'transport planning that positions public transport, well integrated with land-use planning, at the centre of public-purpose planning' on the other. Their review of the urban geography, sociology and political theory literature outlines how these two different stakeholder groups represent 'competing visions for the future of AVs in cities'. In one vision AVs participate in maintaining the existing model of individual car ownership. In the other future sharing economies present ways to confront future urban challenges [10]. They warn that private companies may commercialise individual travel through AVs and MaaS if planning and governance matters relating to MaaS are not engaged with [11]. Doherty et al. [12] likewise call for 'governance intervention' and management of the transition to smart mobilities in order to maximise benefits of smart mobility for individuals, ensure equality and social justice and adhere to internationally agreed environmental questions. And Smith et al. reflecting on the Swedish context, argue that we need to align societal and key MaaS stakeholder goals, involving governmental organisations and public and private transport companies including start-ups [13].

These recent reviews and commentaries, across STS, planning and policy research, emphasise the need to bring together commercial industry and public planning and policy stakeholders over the governance and implementation of fair, just and equal future intelligent mobility systems. There is a concern for 'individuals' [12] and for 'who will ultimately benefit [from AVs] and how these benefits are experienced, spatially and over time, across the city' [14]. Yet, as these works show, in the Australian and UK contexts, there is little knowledge about the people whose lives ADVs and MaaS is set to intervene in. Likewise in Sweden, a study based on workshops and interviews with stakeholders from the public and private sectors involved with MaaS, revealed that they did not profess 'much knowledge about end-users' and 'saw the need for further pilots to learn more about users' attitudes, preferences and behavior' [15]. They note 'uncertainty regarding the size of the MaaS market and its potential is the result of a lack of knowledge regarding users and their willingness to adopt MaaS as a genuine alternative to private vehicle use' [15]. These literatures take a significant step in calling for public and private sectors to be brought together to plan for MaaS, and for further knowledge about future users. Our agenda advances this further, by asking how future users, or *people*, can become likewise engaged and represented as stakeholders in planning processes for future MaaS. The *Co-designing future smart urban mobility services—A Human Approach* (AHA) project, which we now outline exemplifies how this might be developed through design anthropological research and collaboration.

AHA was a proof of concept project, designed to tailor and demonstrate a methodology that combined experimental prototyping and testing of intelligent services for urban development and autonomous vehicles, with participatory and human centred research for city development and sustainable lifestyles. It brought together diverse stakeholders—technology designers, urban planners, citizens and social researchers—to create a transferable and scalable design anthropological approach to research, co-design and evaluation for future projects. AHA was developed as a critical response to existing agendas behind Autonomous Driving Vehicles and Smart City technologies which often take the view that technologically driven change will create improvements which will benefit society, individuals, business and government. Such solutionist [16] approaches tend to depend on the underlying assumption that, as long as people accept the technologies and use them as intended, the predicted benefits will come about. However, as social science research has consistently shown, people do not always use technologies as intended: people abandon technologies that are not designed to adequately fulfill their everyday needs; or they appropriate technologies to fit their own routines and to accomplish their everyday objectives [17]. Our design ethnographic research into experiences and future imaginations of commuting, human experience of simulated ADV tests, and people's experiences of semi-autonomous cars on the roads [5, 18] undertaken within the Human Expectations and Experiences of Autonomous Driving (HEAD) project (a collaboration between Halmstad University and Volvo Cars), demonstrates that such activity is integral to how people use cars (see below). Hence, to be sustainable, a shift is needed, to design Smart City technologies *with* rather than *for* their user communities and future users.

In the AHA project City and Automotive industry partners collaborated to learn how Autonomous Driving (AD) technology design and urban planning can work together to account for human needs, anxieties and expectations. This was undertaken with a commitment to go beyond simply what is effective or possible from an urban management or system development perspective, to instead develop a viable means of accounting for how different people and perspectives can contribute to developing visions of desirable and sustainable futures. AHA's innovative design anthropological methodology, was based in rigorous in depth research and design processes that brought together citizens, industry and City stakeholders in collaborative research about possible AD scenarios and services. Thus creating interfaces between the experiences, actions and knowledge of citizens, urban planners and AD technology design to propose routes towards a smooth transition towards effective and possibly alternative digital services and its implications for industry, policy and city development. The objective of AHA was not to develop a new MaaS system, but to create the methodology needed to achieve this in future projects.

When we developed the AHA project, the stakeholders (Volvo Cars, Drive Sweden, City of Gothenburg, and City of Helsingborg) identified a need to align the development of ADV services in relation to new understandings of multi-modal commuting, car-pooling and ride-sharing as a priority ADV service for research and development. Therefore, we focused on the development of future intelligent services within everyday commuting and the new configurations of transportation that are developing in relation to emerging sharing economies. Everyday commuting was the ideal starting point for the agenda to understand and engender a human perspective on AD because: (1) the spaces and digital platforms and services through which the transitions between different modes of commuting and the different technologies they use (car, train, cycling, pedestrian) have implications for urban planning and for the automotive industry; (2) everyday commuting is already becoming connected to elements of the new digital economy through e.g. ridesharing, carpooling, and through services such as route planning and location aware apps.

1.3 Design Anthropology

Design anthropology approaches are increasingly significant internationally in the development of new understandings of automobilities, in our work with Volvo Cars in Sweden [5, 18, 19] and Samsung Institute in Brazil [20, 21], and at Nissan in the USA [22]. Our approach to Design Anthropology aligns with recent developments that draw on the phenomenological anthropology of Tim Ingold and reposition anthropology's relation to design research and practice [23, 24]. Here we engage with design anthropological concepts of: *emergence*, which emphasises the ongoingly changing circumstances of everyday life, technology design and use; *improvisation,* to understand how people use technologies, often in unintended ways in order to accomplish their everyday goals in the face of the contingent circumstances they often face; and the sensory and embodied elements of *knowing* and *learning* that the use of technology within everyday life circumstances entails. In taking this approach design anthropology is critical of quantifying, objectifying and predictive approaches to futures. It offers a theoretical critique of the product-oriented innovation paradigm, and is coherent with the STS and policy critiques outlined above. Methodologically design anthropology brings together approaches from anthropology and design, to create a blended interdisciplinary practice which is rigorous and in depth in its investigations of how people live out their everyday lives and imagine their futures with technologies, and is future focused and collaborative in its co-creative work with stakeholders. It uses innovative future-focused video-based, sensory and digital ethnographic research techniques. These produce knowledge that can be translated into effective materials and probes for participatory design processes. It considers those who participate in research and design to be people and communities rather than users, and situates their experiences of existing and possible future technologies within plausible everyday life circumstances, where they are doing more than simply 'using' technologies. In our work we seek to engage stakeholders with these everyday life experiences. We use approaches from participatory design and co-design to engage differently positioned stakeholders in processes of exploring, negotiating and forming concerns and common sustainable directions [25, 26]. Such focus on developing dialogues and capabilities, also called infrastructuring, is building strategies for long-term engagement of diverse groups in developing new practices, spaces, artefacts or systems [27, 28]. However, in contrast to co-design approaches of bringing 'users' into design workshops to prototype technologies, we undertake in depth research and analysis with people in the places where they live and bring the knowledge created through this research into workshops and ongoing research processes. Bringing such perspectives and approaches together allows for an exploratory and human-centred research process in which multiple stakeholders are engaged in co-designing possible futures and imaginations. We next demonstrate this with examples from the AHA project.

1.4 Producing Ethnographic Insights and Probes

The first step in creating a human-centred approach to co-designing new MaaS systems involves addressing the knowledge gap evidenced by the literature discussed above, to understand how people experience existing mobility systems and technologies, and how they imagine future mobilities. Smith et al. [15] indicate an assumption amongst private and public sector stakeholders in Sweden that this gap can be filled by research about 'users' attitudes, preferences and behaviour' and Legacy et al. [32] note that 'Much of the research assessed under both the neoclassical and sustainable development frames is technical and quantitative: reporting on the findings generated by employing surveys, scenarios and modelling techniques' which 'allows very little space for critical engagement with the AV future'. To create this critical engagement we need to go beyond psychologically informed surveys, for two reasons. Such quantitative methods tell us 'what' people answer when asked certain questions in controlled environments, but they cannot tell us 'why' everyday life situations play out as they do. Ethnographic research methods are ideally equipped to get under the surface of everyday life to produce the insights required to understand the social, cultural and experiential dimensions of future mobility systems [19].

The ethnographic insights used in the AHA project were produced through a series of subprojects that composed the Human Expectations and Experiences of Autonomous Driving (HEAD) project and a small commuting study that focused specifically on future mobilities. Our research explored how, where and why people make particular decisions relating to transport (where meaning and benefits lie for them) and how they imagined futures. This enables us to create a basis from which to understand what we expect people will do in future scenarios, in ways that are continuous with what they already do. The HEAD ethnographies focused specifically on AD cars, and therefore provided materials through which we could critically address, and unpick tensions concerning how and where ADVs might participate in future MaaS systems, with automotive and city partners.

Here we outline three methods used to produce ethnographic materials and insights. These are all discussed in detail in existing publications which we refer to as relevant. Our approach to ethnography differs from the ethnomethodological approach that has dominated in HCI research, and is based in a collaborative visual and sensory approach to ethnography [30, 31], where researchers collaborate with participants during research encounters to perform, discuss and record aspects of their embodied and verbalised experiences.

In-car video ethnographies [19] involved undertaking video re-enactments of morning routines with participants in their homes, before they drove us along their normal commuting routes to the city (Fig. 1.1). During video performances in their homes and the drive-alongs we asked them questions and discussed their experiences with them, using a checklist, but allowing the discussion to roam across areas of interest that arose. For example, a key unplanned outcome of this study was that

Fig. 1.1 **a** A participant drives his car along his normal commute, while discussing these experiences with the rese-archer. **b** The researcher video records while discussing with the driver. Photos by Vaike Fors

Fig. 1.2 An example of the experimental setup in the autonomous driving Wizard of Oz car. Photo by Volvo

we learned about the significance of the car-smartphone relationship, as well as developing insights into the feeling of driving.

Wizard of Oz studies [32] involve experimental testing of AD prototypes in which participants 'drive' simulated ADVs, (Fig. 1.2) and therefore experience possible design concepts of future vehicles that do not as yet exist on the roads. Participants have the opportunity to test their ideas of freedom in autonomous cars, such as having breakfast, reading a book or watching a movie while doing their morning commute. Our ethnographic research focused on participants' experiences of these test scenarios, and the embodied and emotional feelings that they had when 'driving' an ADV.

The Drive-Me study was a trial led by Volvo Cars in which 5 families were given research cars with autonomous features for a period of one and a half year. These autonomous features can currently be used on the roads in Sweden. Our research team undertook research with the families during this period, including home visits, drive and commute alongs, speculative future scenarios, discussions and mapping activities over a shared take-away meal at the family home. This included mapping the technologies and 'stuff' kept in the car (Fig. 1.3b) which when put together with findings from the in-car ethnographies (Fig. 1.3a), and video recorded looks into cars, raised the question of the implications of this for future car sharing schemes in MaaS scenarios.

1.5 Translating Ethnographic Insights

The ethnographic materials were translated into probes to be used in workshops. These included: the *AD Futures Design Cards*, which are designed to disrupt some concepts that are commonly taken for granted in user experience (UX) and interaction design research (such as trust and sharing); *video drive-along clips* of illustrative moments; storyboards and pain point maps, and *Commute Journey Maps* (Fig. 1.4) developed from the idea of journey mapping to not only focus on how the car was used but instead centre how the car became part of people's daily routines on their way to and from work.

Fig. 1.3 **a** An example of a box with 'stuff' in it in an in-car ethnography participant's car. Photo by Vaike Fors. **b** A map of the different kinds of stuff a DriveMe family kept in their usual car. Photo by Thomas Lindgren

The ethnographic research methods discussed above were developed and tailored to respond to the needs of particular projects while also producing new and unexpected knowledge, which can sometimes be pivotal to research insights. We do not recommend that others replicate our methods of research or translation, but that they

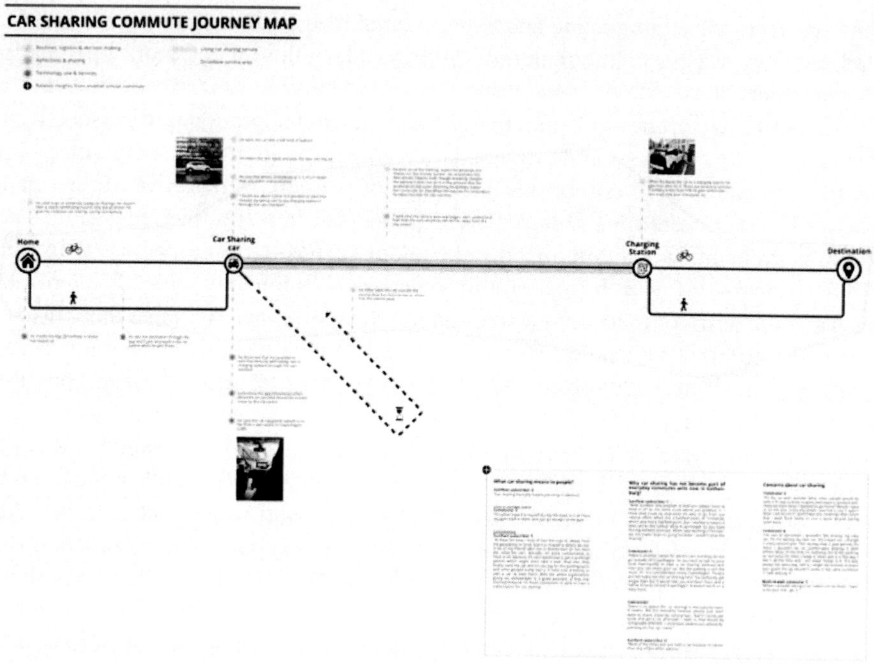

Fig. 1.4 An example of a how to map a person's commute routine and how different modes of transportation become part of it. Map created by Kaspar Raats and Katalin Osz

are used as starting points and inspiration in the design of specific methods and materials suited to the project being undertaken.

1.6 Workshops

In total the project conducted six collaborative workshops, the first was a start-up workshop in which the different stakeholders shared their perspectives, challenges and opportunities to be able to align and prioritise the goals and expectations of the project. The following four were led by Halmstad University, the two Cities, and Volvo Cars respectively, and focused on developing a shared agenda for future urban mobility with AD. The final workshop brought together the stakeholders to analyse and evaluate the process and outcomes, and set directions for a following larger scale research and innovation project. All the workshops were designed to bring together, as well as to share and communicate stakeholder perspectives, rather than to serve the pursuits of the organisers. Each workshop was organised and led by shifting partners in order to co-develop ownership and dialogue between the different perspectives (and sectors) in the project. The materials presented as examples here were produced for the second workshop held at Halmstad University which focused on turning findings from the ethnographic and experimental investigations into useful insights, and enabling access to the materials in ways that allowed for more open-ended engagements.

The first stage of the workshop brought the human people-focused perspective to the centre of the discussion. Participants were presented with talks and video clips of specific examples of two people's commutes, from the perspective of their daily routines, which were next represented as visualised in a 'routine map' (Fig. 1.4). Next, participants worked in mixed stakeholder groups to discuss points of interest (POI) in the routine maps by pinpointing situations they found specifically interesting or critical in terms of their perspectives on mobility solutions. Through these discussions, the different viewpoints became visible and tangible, and created a concrete starting point from which to undertake the next exercise, which targeted possible commute innovations.

Participants were asked to complete tables (Fig. 1.5) which encouraged them to write the perspectives of the different stakeholders alongside each other and to bring these together in terms of their commercial and societal values, interest for other stakeholders, and feasibility. Subsequently, the main exercise of the workshop was to *learn how to create mobility solutions together*, under the following two conditions: (a) all stakeholders were needed to develop the solution, and (b) the solutions should address actual real-life situations based in ethnographic insights and materials.

The collaborative activities directed and inspired by these materials and frameworks enabled the groups to discuss and create future mobilities scenarios that accounted for the perspectives of city planning, the automotive industry, and also everyday commuters with different resources and priorities (Fig. 1.6). For example, one of the groups re-developed the everyday commuting journeys of people presented

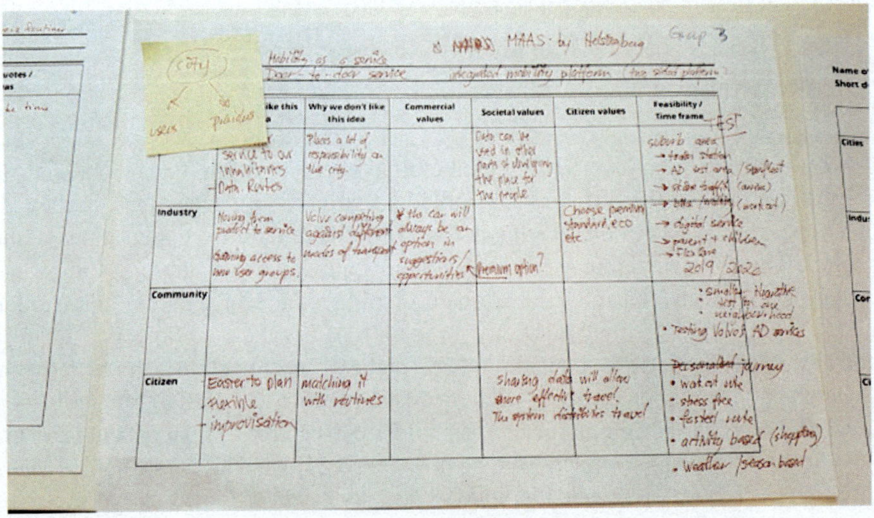

Fig. 1.5 Map of different stakeholder perspectives in creating a joint mobility technology. Photograph by Thomas Lindgren

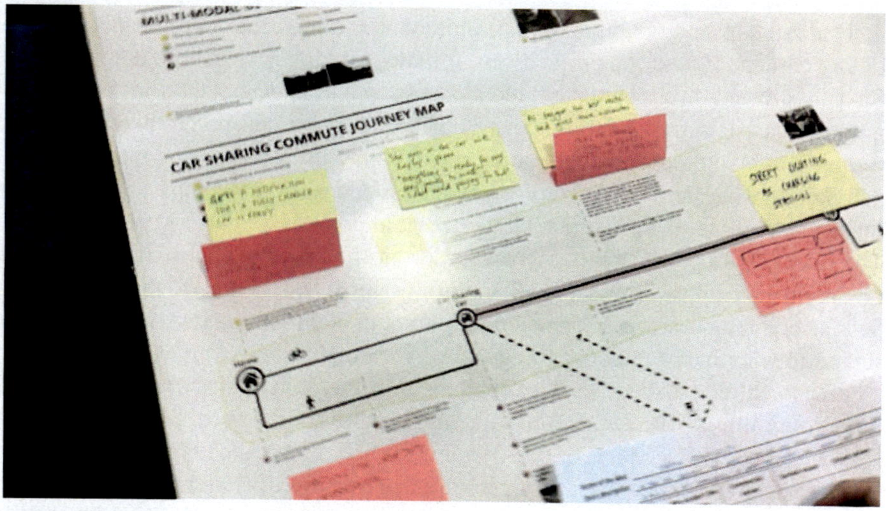

Fig. 1.6 The groups collaborated to develop future mobilities scenarios. Video still by Kaspar Raats

to them in journey maps with a focus on the concept of a transition hub, while another group thought about an app-based mobility service. While these solutions may not appear innovative at first glance, the specific solutions themselves are less important than the process of dialogue engagement creating them. These were later described by the participants as highly unusual in their home organisations, because they were based on non-competitive interaction points between industry and the municipality and were based on real-life ethnographic investigations.

Each of the examples developed in the workshop suggested modes of harnessing the creativity of human improvisation to shape ways of designing and using MaaS applications. They show how the ambitions, emotions and needs of particular citizens might be translated, understood, and accommodated by anticipatory AI in such a way that has personal, social and environmental benefits for individuals, cities and service providers through the activity of engaged users. For example, in the POI exercise one group emphasized data security and privacy, making commuting meaningful, and making services tailored. They emphasized multimodal transport and creating efficient ways of shifting between different modes of transport that work for citizens and are based on efficient infrastructures (so bringing together citizen needs and infrastructures). Their notes also indicate the complexity of stakeholders involved including platform providers, contractors, and others. The group also considered motivation for citizens and possible ways that citizens would become engaged ethically in cities by participating to increase sustainability, through energy credits and sharing schemes; how this could be part of brand building for industry; questions of community, trust and ownership; and how citizens could benefit in other ways—e.g. currency, credits, free tickets etc. All of these latter possibilities suggest scenarios that could be developed in further workshops, or could be explored ethnographically.

Under the heading of 'Imagining Future Commuting', the framing and structure of two cornerstones of the project were initiated and interwoven into the workshop. First, it allowed different stakeholders access to and engagement with rich ethnographic materials from people's everyday experiences and commuting practices, setting a human perspective at center stage of the discussions. The materials triggered discussions around citizens' everyday practices and helped participants focus on concrete, rather than abstract, challenges and potentials of everyday commuting. The intentional openness in the presentation of the materials allowed industry and city partners to apply both personal and professional experiences to the discussions. Moreover, while searching for potential points of common interest, the discussions served to widen their understandings of what future mobility might look like from a human-centred perspective. The challenge for the participants was to grasp/cope with the wealth of diverse materials, and the insights they carried, in short workshop sessions. We had placed one or two members of the research team in each workshop group to deal with this issue, and make sure that rich discussions based on the ethnographies would continue into the group work, but still the participants felt that the wealth of materials was too much. As one of the City partners expressed it, the materials were really inspiring, but there were so many of them that it was difficult to take it all in. Another challenge relating to the citizens' perspectives was that their voices as stakeholders were only represented by proxy through the research and materials. As seen in the joint template above (Fig. 1.3), the 'citizen' and 'community' interest and values were added based on the stakeholders' interpretations and not as a result of direct citizen engagement in the workshop. The possible engagement of urban commuters in the workshop was discussed at length in the planning of the workshop, but ultimately we decided to instead engage with the ethnographic materials and the project partners.

Secondly, the workshop scaffolded dialogues and understandings between stakeholders with diverse economic and societal interests and organisational cultures around future urban mobility, in developing possible points of common interest while forming capacities and capabilities for a longer-term engagement. Discussion of human-centered perspectives on urban mobility was not present in the organisational strategies and discourses, and cross-sector stakeholder engagement on these topics were non-existent. This meant that each workshop in the proof of concept project, was central to creating the foundations, capacities and infrastructures for a longer-term engagement around such topics. The focus, therefore, was not on the concrete services and concepts developed through the workshop, but rather on the dialogues and new understandings that could come out of the common engagements around them. It was a way of opening up boundary spanning dialogic spaces of engagement around diverse perspectives and understandings. Hence, the wealth of ethnographic materials worked to ensure the presence of multiple voices and real-world complexity in the debates, while triggering exchanges of concrete interests and perspectives among the stakeholders. Importantly, the researchers' roles as facilitators in each group together with the workshop frameworks, was crucial for directing the conversation and brokering between different perspectives. Thus, the conversations moved quickly between different levels of abstraction, for instance from day-to-day experiences and commuting practices of citizens, to 5, 10, and 50 year perspectives of urban planning and infrastructures, to commercial interests in in-car and shared technology solutions. The human, social and societal agendas and values, new to the stakeholders, were continually kept on the table by the probes and material used in the workshop as well as by the researchers facilitating the different sessions. Hence, the workshop ended with a common discussion about 'what new insights and understandings about commuting the workshop created', and 'what would be the first steps in collaborating around these issues'. The intention was to develop ways of working together and align perspectives and interests, as well as building a continuity between the workshops by scoping the project and focus together.

1.7 Potentials and Challenges

Finally, we discuss the potentials and challenges of a human-centred approach to engaging diverse stakeholders in the development of future mobility across local and global contexts. Existing research in the potential of AVs and MaaS in cities has emphasised the need to develop the critical capacity of transport planning to make visible 'the political and economic forces shaping policy and innovation', since 'public and private sectors will together shape the rollout and the penetration of the AV into our future city' [33]. Our design anthropological approach to this question proposes not only to do this, but also to introduce everyday life into these discussions, and offers a framework for doing so.

Co-design was activated by the ethnographic studies presented as probes to fuel discussion and perspective sharing and co-creative activities between stakeholders. As such, stakeholders collaborated to develop plausible understandings of how their different interests might intersect and come together in the design of future MaaS systems, that put people and everyday life concerns at their centre. As emphasised, the AHA project did not aim to create future mobility solutions, rather it aimed to prototype a methodology through which this could potentially be achieved. By centring human experience, everyday life improvisation and the ongoingly emergent nature of life as it happens, the possibilities offered by and visions proffered by city planning and the automotive industry could be explored contextually. However, this cannot be done without working through the meaning of 'human-centred' that different stakeholders bring to the table, and also doing long-term design-oriented engagements with citizens in urban environments. End-user, consumer and citizen are different perspectives, and the reasons why specific organisations bringing them into the discussions vary. However, it serves a common goal to use these human-centred perspectives to achieve different organisations interests in a socially sustainable way. An ethnographic perspective can disrupt these views by understanding and taking the real-life situations of what people do, think, and experience when moving through their everyday urban spaces seriously, and nudges the discussion into a shared agenda:

to create realistic mobility solutions. To achieve this, future research would need to be larger scale and to encompass in depth ethnographic studies with future MaaS user communities and concrete prototyping in such contexts.

Acknowledgements Our research was funded by the following two projects: 'Human Expectations and Experiences of Autonomous Driving' Vinnova, Sweden; and 'Co-designing future smart urban mobility services: a human approach' (AHA), Drive Sweden. We thank everyone who generously participated in these projects.

References

1. Fortune.: The most hyped emerging technology of 2015 (2015). http://fortune.com/2015/08/20/self-driving-car-hype/
2. Mok, B.K.J., Sirkin, D., Sibi, S., Miller, D.B., Ju, W.: Understanding driver-automated vehicle interactions through Wizard of Oz design improvisation. In: Proceedings of the International Driving Symposium on Human Factors in Driver Assessment, Training and Vehicle Design, pp. 386–392 (2015)
3. Docherty, I., Marsden, G., Anable, J.: The governance of smart mobility. Transp. Res. Part A: Pol. Pract. **115**, 114–125 (2018)
4. Stilgoe, J.: Machine learning, social learning and the governance of self-driving cars. Soc. Stud. Sci. **48**(1), 25–56 (2018)
5. Pink, S., Fors, V., Glöss, M.: The contingent futures of the mobile present: beyond automation as innovation. Mobilities **13**(5), 615–631 (2018)
6. Smith, G., Sochor, J., Sarasini, S.: Mobility as a service: comparing developments in Sweden and Finland. Res. Transp. Bus. Manag. **27**, 36–45 (2018)
7. Cohen, T., Stilgoe, J., Cavoli, C.: Reframing the governance of automotive automation: insights from UK stakeholder workshops. J. Responsib. Innov. **5**(3), 257–279 (2018)
8. Doherty, et al.: p. 222
9. Legacy, C., Ashmore, D., Scheurer, J., Stone, J., Curtis, C.: Planning the driverless city. Transp. Rev. 3 (2018)
10. Legacy, et al. p. 9
11. Legacy, et al. p. 14
12. Doherty, et al. p. 116
13. Smith, G., et al. p. 43
14. Legacy, et al. p. 13
15. Smith, G., et al. p. 42
16. Morozov, E.: To Save Everything, Click Here: Technology, Solutionism, and the Urge to Fix Problems that Don't Exist. Penguin Books, London (2013)
17. Miller, D., Horst, H. (eds.): Digital Anthropology. Bloomsbury, London (2012)
18. Osz, K., Rydström, A., Pink, S., Fors, V., Broström, R.: Building collaborative testing practices: doing ethnography with WOz in Autonomous Driving research. IxD&A J. 37, 12–20 (2018)
19. Pink, S., Fors, V., Glöss, M.: Automated futures and the mobile present: in-car video ethnographies. Ethnography **20**(1), 88–107 (2019)
20. Pink, S., Lucena, R., Pinto, J., Porto, A., Caminha, C., Maria de Siqueira, G., Duarte de Oliveira, M., Gomes, A., Zilse, R.: Location and awareness: emerging technologies, knowing and mobility in the Global South. In: Wilken, R., Goggin, G., Horst, H. (eds.) Location Technologies in International Context. Routledge, London (2019)
21. Pink, S., Gomes, A., Zilse, R., Lucena, R., Pinto, J., Porto, A., Caminha, C., de Siqueira, G.M., Duarte de Oliveira, M.: Automated and connected?: Smartphones and automobility through the Global South. Appl. Mobilities (2018). Online first at https://doi.org/10.1080/23800127.2018.1505263

22. Stayton, E., Cefik, M., Zhang, J.: Autonomous individuals in autonomous vehicles: the multiple autonomies of self-driving cars. In: Proceedings of the Ethnographic Praxis in Industry Conference, EPIC'17, vol. 1, pp. 92–110 (2017)
23. Smith, R.C., Kjærsgaard, M., Vangkilde, K., Otto, T., Halse, J., Binder, T. (eds.): Design Anthropological Futures. Bloomsbury Academic, London (2016)
24. Gunn, W., Otto, T., Smith, R.C. (eds.): Design Anthropology: Theory and Practice. Bloomsbury Academic, London (2013)
25. Simonsen, J., Robertsen, T. (eds.): Routledge International Handbook of Participatory Design. Routledge, New York (2012)
26. Smith, R.C., Iversen, O.S.: Participatory design for sustainable social change, special issue on participatory design. Des. Stud. **59**, 9–36 (2018)
27. DiSalvo, C., Clement, A., Pipek, V.: Communities: participatory design for, with and by communities. In: Simonsen, J., Robertson, T. (eds.) Routledge International Handbook of Participatory Design, pp. 182–210. Routledge, New York (2012)
28. Pink, S.: Doing Visual Ethnography. Sage, London (2013)
29. Broström, R., Rydström, A., Kopp, C.: Drivers quickly trust autonomous cars. In: Ahram, T., Falcão, C. (eds.) Advances in Usability, User Experience and Assistive Technology. AHFE 2018. Advances in Intelligent Systems and Computing, vol. 794. Springer, Cham (2019)
30. Pink, S.: Doing Sensory Ethnography. Sage, London (2015)
31. Huybrechts, L., Dreessen, K., Hagenaars, B.: Building capabilities through democratic dialogues. Des. Stud. **34**(4), 80–95 (2018)
32. Legacy, et al. p. 7
33. Legacy, et al. p. 15

Chapter 2
Visual Conflict Framing in Public Transport Innovation

Ilya Fridman and Selby Coxon

Abstract With the rise in popularity of design thinking and other innovation practices, designers are being asked to contribute their skill sets and creative methods to emerging and often complex problem spaces. Public transport provides one such example of a space for design, where social, cultural, technological, and economic factors intersect with often conflicting requirements. These form complex challenges which must be understood before they may be addressed.This chapter considers how conflicting factors manifest within public transport and how they may be identified for greater understanding during design development stages. It describes how designers visually frame conflicting factors to establish platforms that allow development of innovative outcomes. Two public transport design research projects–one focusing on trains and the other on busses–are drawn upon to provide examples of work in this space. The idea of visual conflict framing and reframing is then discussed as a tool for analysis, synthesis and assessment that may also be used for communication between design teams and other stakeholders during public transport development stages.

Keywords Visual conflict framing · Frame innovation · Public product service systems

2.1 Complex Problems in Public Transport

Dorst suggests that today's designers are tasked with working on challenges that are open, complex, dynamic, and networked [1]. This is to say that they have unclear boundaries, are comprised of many interconnected elements, change over time, and are coordinated across multiple organisations. It is difficult today to think of products that sit in isolation outside of the services offered within broader systems. From household washing machines that are leased based on use, to carshare services that are shared amongst community members to reduce their collective vehicle ownership

I. Fridman (✉) · S. Coxon
Department of Design, Monash University, 900 Dandenong Road, Caulfield East 3145, Australia
e-mail: ilya.fridman@monash.edu

S. Coxon and R. Napper (eds.), *Advancing a Design Approach to Enriching Public Mobility*, Intelligent Systems Reference Library 198,
https://doi.org/10.1007/978-3-030-64722-3_2

[2]. These products need to be considered for how they relate to other elements within a system and how these combine to provide satisfactory customer experiences. Tukker describes this type of design through eight categories of Product-Service Systems (PSS) [3], suggesting that although isolated products still do exist, they are at the very narrow end of a much broader spectrum. Public transport fits the PSS categorisation as designers are required to consider products and services that must work together within a broader system to deliver safe, viable, reliable, and enjoyable transport experiences. They cannot simply solve problems in isolation and must consider relationships between various system elements that cross-over between different organisations and stakeholder groups including passengers, governments, operators, manufacturers, and local communities. Individual services within these systems require coordination as part of the greater travel experience that takes people from one destination to another. In order to develop effective services, designers must be skilled at recognising and managing conflicting factors which are commonplace in these complex systems. As services are funded by and implemented for tax-paying people, an extra layer of accountability is created for designers working on public transport. To capture the added layer of complexity, this type of design activity has been referred to as a Public Product-Service System (PPSS) [4]. Acknowledging the term public within this classification is particularly important, as it differentiates these public transport projects from privately-owned commercial PSS, or consumer products and services.

Designers working in PPSS development must consider social, cultural, technological, and economic factors that can often result in conflicting ideas for how a successful outcome should be achieved. These projects need to acknowledge many stakeholders with varying interests and expectations. Transport operators, for example, seek to generate revenue by operating services for fee paying passengers. The more passengers they can move on each vehicle, the greater their revenue and the better their outcome. On the other side of this transaction are passengers who seek services which are accessible, safe, reliable, frequent, comfortable, and affordable. Some expectations are aligned between operators and passengers. Safety, for example, is something that both stakeholder groups strive for. While other expectations, such as comfort, can differ. Passengers may want more personal space onboard a vehicle, while operators seek to accommodate the greatest number of people possible while attempting to maintain a safe travel environment. In most cases, this conflict is invisible as transport services typically operate at partial capacity for most of the day. Where we can clearly see, and feel, this tension is during peak commuting hours as passengers travel on crowded vehicles that stretch levels of comfort, safety, and frustration to their limits bringing the greatest challenge to designers working on public transport.

To extend the example of interconnected complexity, another stakeholder group can be taken into consideration. Bus drivers follow a strict schedule, which involves a break period for them at the end of a route. Completing a route quickly will result in more time for them to rest before they need to resume driving. They have certain checkpoint stops which they have to arrive at by a particular time, though any stops outside of these are a matter of flexibility and time allocation. During operation, someone may hail the bus incorrectly or take a few extra minutes to ask the bus driver about the service. These common occurrences can result in the bus driver falling behind in their schedule and encouraging them to hurry in order to reach their next stop on time. In this situation, the driver may see someone running for the bus while they are pulled over at a stop. They need to decide: do they continue to delay the timetable further by waiting for this passenger or do they pull away with the reasoning that a subsequent service will eventually come? This situation is all too relatable to anyone who has relied on buses for their daily commute.

These two examples help to build our concept of public transport complexity by showing how interactions between just three of the stakeholders can lead to conflict. The operator who seeks to get a return on investment by generating revenue, the bus driver who wishes to stay on schedule set by the operator and get a much-deserved rest break for themselves, and the passenger who relies on a bus for their commute. Although these two common examples raise all sorts of questions, they are just the tip of the proverbial iceberg when it comes to public transport design. Challenges in this field extend far beyond three stakeholders, often also considering local councils, state governments, and communities in the areas where service is provided. Designers must understand this web of connections before they can attempt to develop innovative outcomes within this space.

To add to the challenge of designing public transport, these social, cultural, technological, and economic factors are dynamic and constantly evolving. As populations continue to grow in cities, they bring with them greater urban sprawl, a higher number of commuters and greater commuting distances. These place greater strain on existing public transport services raising issues around demand and infrastructure capacity. Societal expectations of transport service also evolve through technological advancements, such as vehicle electrification and digital connectivity. Mobility as a Service (MaaS) provides an example of this type of evolution in recent years as digital platforms such as mobile devices allow people to plan, monitor, and coordinate their daily trips across different transport modes [5]. People gain access to information from different services that encourages them to consider mobility as a connected system comprised of various travel modes including walking, cycling, driving, ride-sharing, buses, trams, trains, and other micro-mobility share schemes. These advancements bring exciting potential to improve passenger experience. At the same time, they have the potential to disrupt the industry through significant changes to revenue generation, service management and planning.

Dorst [1] proposes that framing is one of the most important tools of a contemporary designer as it allows them to address these open, complex, dynamic and networked challenges. This chapter elaborates on how public transport designers create visual frames of focus that connect various elements within a PPSS through a particular project vision in order to understand problem contexts and achieve innovative outcomes. It shows how visual frames allow designers to creatively explore conflicting factors and better understand their relationships—the push and pull between different stakeholder's interests.

Within this chapter, the concept of visual conflict framing is reviewed in relationship to PPSS development. A process for acknowledging and considering conflicting factors is presented through two examples of designers working on projects in different public transport areas—trains and buses. Although projects are focused on different tasks with their own unique factors, both designers use the same visual process to analyse, communicate and work at managing factors which exist in each situation. Conflict framing is then discussed as a tool that can be used for communication between designers, project partners and other external stakeholders. Following this process also creates potential for factors to be analysed and creatively reconsidered by reframing them to achieve innovation. Visual conflict framing is proposed as a valuable process for designers working on PPSS development projects such as those found in the field of public transport.

2.2 Frame Creation When Factors Conflict

A frame can be described as someone's perspective or understanding of a situation or a problem to be solved [6]. While working on a project, designers may create various frames of reference through a variety of methods including metaphors, objects or visuals which may be used to build a shared understanding of specific project themes. The act of visualisation is one of the most fundamental methods that designers have to externalise their ideas and capture a vision for how they imagine something to be. Visualisations, such as drawings and diagrams, can capture existing situations in order to develop a better understanding of them or to explore visions for what potential outcomes may look like in the future. By visually framing their understanding of the context, designers discover new ways of seeing the problem at hand and as a result also see new pathways for how to resolve it.

Dorst's [1] theory of frame innovation has expanded our understanding of how designers develop frames for innovative thinking in complex contexts. The process of framing a creative challenge is described as the way in which designers consider products, processes and outcomes simultaneously within a given context. This is because problems and solutions constantly co-evolve throughout the design process as they are closely linked with each other [7]. Recognising how this is done and how combinations of certain products and processes result in specific outcomes allows designers to create new ways of approaching the problem space and generate innovative solutions that work around seemingly paradoxical situations.

The term paradox in this understanding is used to describe a complex statement that consists of two or more conflicting statements [8]. In the field of public transport, conflicting statements are commonly numerous and intertwined. The tension between designing vehicles to accommodate more seating or standing space is just one such example. Passengers enjoy the ability to easily enter a train, tram or bus and move throughout its interior. At the same time, they also like having access to seating so they can travel in comfort. Though these two individual statements may not seem contradictory at first, they do in fact present a tension when considered together. This is because seating creates bottlenecks which block the free-flowing movement of passengers onboard a vehicle. By slowing down passenger flow, seating inadvertently also delays the service as passengers take longer to board and alight. As services are delayed, extra passengers queue up on the platform also expecting to get on, thus contributing to an undesirable cycle of delays and discomfort.

Dorst [1] suggests that there are typically three ways to deal with paradoxical statements. They are to either pick one side of the conflict and design for it, find a compromise between the two conflicting sides, or to redefine the problem situation in which these conflicts exist to remove them or make them complementary. Finding a compromise between conflicting factors is often what happens in public transport, as one cannot simply select a single side of a conflict, like that between seating and standing passengers. This is because it is difficult to understand which side is the right one to pick and, in most cases, a single correct solution simply doesn't exist. Selecting a solution also depends on which stakeholder groups is making the decision and what outcomes they want. Redefining how a problem is understood provides the greatest chance for developing novel and innovative outcomes; however, in public transport there is often a reluctance to explore truly novel solutions for fear of them being costly or potentially dangerous if poorly considered. It is therefore important for public transport designers to build a robust understanding of a given situation, which can be used by themselves and others involved in a project. Relevant factors within a PPSS should be visualised by identifying conflicts during the frame creation process in order to develop a thorough understanding that informs innovative thinking.

Communication on PPSS projects is often required across different stakeholders who often only see their side of the system. The creation of frames can provide a common ground for participants to discuss the problem and possible solutions [1]. By visually framing existing conflicts, designers develop a communication tool that can enable a shared understanding which allows ideas to be discussed, analysed and addressed collectively. This process makes them more tangible by adding a level of concreteness to the issues and allows them to be approached creatively with a view that they may be manipulated.

This chapter proposes that visualising conflicting factors during the frame creation process is a useful tool to develop knowledge and creativity in a design. Using two examples of PPSS projects, it showcases how designers working on PPSS can visually frame conflicting factors to establish a ground for innovative thinking. The projects have been taken from studio-based Ph.D. design studies conducted at Monash University, Australia. Both examples are from the area of public transport with the first focusing on managing crowding on trains, and the second exploring the transition to electric vehicle technologies for bus services.

2.3 Example 1: Train Carriage Passenger Crowding

The first project example has been referenced from Coxon's [9] Ph.D. work, which was focused in the area of metropolitan trains and explored how passenger dispersal could be controlled to alleviate crowding through novel train carriage designs including interior layouts and furniture.

Metropolitan train services are vital to the movement of people and goods in many of the world's large cities. In 1863 when the world's first underground railway opened in London, only 10% of the world's population lived in cities. Now, in the early twenty-first century, over 50% of the world's population resides in city centres [10]. As a result of this urban densification, train systems have become regarded as a highly efficient way to move large numbers of people relative to their land use. In Tokyo, for example, 80% of the population uses the subway train system, which is one of the highest levels of patronage anywhere in the world.

The relatively long life–around 30 years–of the train rolling stock, the near permanence of track, and the high level of patronage make rail the lowest operating cost per passenger kilometre of any mode of transport [11]. In addition to this economic benefit, transport planners also see the appeal in the mode's peak-hour carrying capacity [12]. This capacity is determined by carriage size, number of carriages per train, and maximum number of trains circulating through the network. However, rail networks in many cities around the world struggle to be punctual. The most significant variable in a train's journey is the length of time it must stop at each station. This dwell time depends on how long it takes passengers to board, alight and disperse within the train carriage and across the platform. At peak periods, dwell times can become extended as passengers jostle to board or alight the train carriage. Extended dwell times reduce the headways between services, therefore affecting network capacity, ultimately impacting on an operator's revenue and contributing to poor passenger perceptions of their experience due to service delays and potential for crowding. From a physical design perspective, a number of factors conspire to impact upon the efficient dispersal of train passengers: carriage dimensions, number of doors, and seating arrangements.

Within Coxon's [9] Ph.D. work, a visual conflict frame (see Fig. 2.1) was constructed to understand and communicate relationships between all of the separate elements. It was created early in the study, following a literature review, and efficiently captured the dilemma and challenge. This activity created a communication tool that allowed research to be conveyed to others to develop a shared understanding of issues existing within this space. It was also used by the designer to explore potential solutions and was subsequently referred to frequently during the project as a check and measure on emerging design solutions. The subsequent research in this project centred on a studio-based design practice methodology that collated evidence from a wide range of literature on the topic to inform an empirical design activity. This type of synthesis through studio-based practice has been discussed in detail by Farias and Wilkie [13], who suggest that design prototyping brings together a heterogeneous variety of elements as a novel and coherent whole. In order to synthesise information, this research used design activities such as drawing, computer aided design, physical models, and computer simulations to develop concepts and to evaluate their efficacy. The outcome was a designed train proposal that consisted of a series of innovations pertinent to the problem context.

To begin with were the spatial limitations of the carriage itself. The dimensions of this box emerged from a series of gauges particular to the network in which the trains travelled. Three different gauges were considered: the train track gauge,

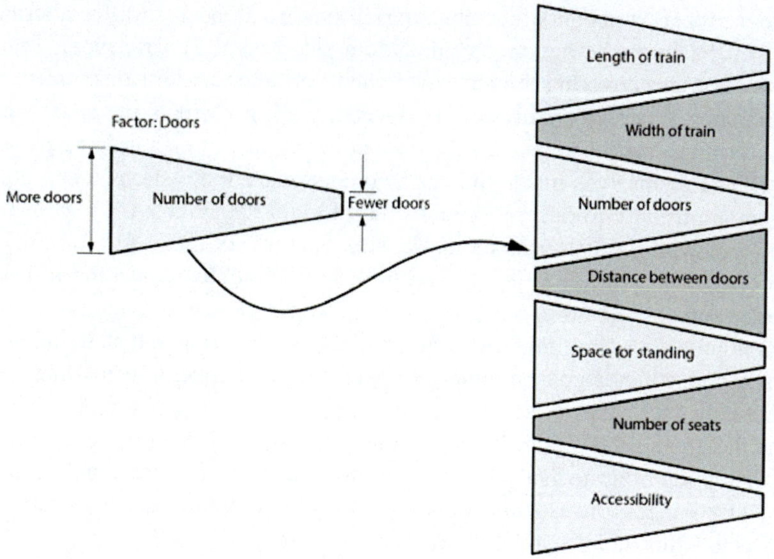

Fig. 2.1 Visual conflict framing of train factors. Diagram referenced from Coxon [9]

defined as the distance between rails; the loading gauge, which described the outer extremities of the train coachwork; and the structural gauge, which determined the closest external point of the carriage to any hard-fixed infrastructure. The overall train carriage length was determined by two factors: the prevailing length of station platforms, and the clearance dimension required for travelling around bends in the track. These determinants of the carriage envelope were largely finite; however, the impedimenta of the interior and how it is accessed by passengers was subject to a variety of complex trade-offs.

The doors were identified as the most significant points of bottleneck in this research because ingress and egress determined platform flows and, as seen in Hirsch and Thompson [14] could influence how passengers dispersed within the carriage. Increasing the number of doors would intuitively assist passengers to board and alight. However, the more doors and wider doors available the less space there would be for seating. Having somewhere to sit is rated very highly by passengers [14]; Therefore, the removal of seats to accommodate more standing passengers, especially during peak patronage, would not be a decision taken lightly by operating companies.

There are two fundamental orientations of seating on train carriages. The first is longitudinal, running along the windows and facing towards the centre of the carriage, and is commonly the choice by train designers when the track gauge is Standard or narrower. This arrangement is commonly referred to as metro-style seating. On networks of traditionally high patronage the wider central floor space makes passenger flow quicker and less obstructed [15]. The second arrangement is transverse, with seating at right angles to the windows. Typically, this composition of seating is used in wider carriages, which tend to be regional, and intercity trains. There is evidence that passengers favour sitting in the direction of travel or alternatively with their backs to the direction, but seldom sideways [15]. Transverse seating is used for services covering longer trips, where the tolerance for discomfort is less than on short, inner-city journeys. Transverse seating narrows the space between doors, forming corridors, which is the least effective way of encouraging passengers to distribute themselves inside the carriage. Passenger motivations when boarding the train are usually to occupy the most comfortable positions, often perceived as those areas with the least passenger density. Passengers move further away from the doors if their journey is longer and they do not anticipate needing to leave the carriage soon after boarding [16].

Through this research investigation, parameters of seating and standing arrangements, carriage dimensions, number of doors, and door spacing were discovered to conflict with each other in terms of achieving an optimum performance. The more of one the greater the diminishment of the other. As with many design challenges even the option of a median point compromise only left the whole design in some way reduced. In response to this, the design study therefore focused upon negotiating conflicts to better stabilise or shorten train dwell times.

A range of factors were identified through a review of literature and design precedents. Knowledge was then synthesised into a graphical interpretation that framed conflicting factors and provided a good enabler to capture the essence of this problem space. This helped to build an understanding of what the actual issue was before speculating upon a solution could begin. Figure 2.1 shows how distinct conflicting factors were visually framed by highlighting tensions between them as part of the broader picture. The representation was constructed through a visual nomenclature of parallel lines of different lengths to demonstrate magnitude. Longer lines and field edges represented large dimensions while the shorter lines and edges represent small dimensions. A title to denote the specification being considered was written in the middle. The boxes formed angles as they were read alternately from left to right as either large or small in number. The relationship between these elements was read by looking at the diagram as a whole. These angular boxes, almost arrow heads, interlocked conveying that the diminishment of one specification was the gain of the other. For example, the greater the area for standing, the fewer number of seats. This in turn enabled more doors which shortens the distance to those doors and so on. The left to right and right to left, greater than and less than interpretation was referenced from mathematical symbolism. The vertical reading of the diagram enabled the interrelationship of elements to be understood at a glance. Each factor held a relation to the overall composition and not only the adjacent box. For example, the vehicle length would have a greater influence on the number of doors than the vehicle width; however, all of these factors must be considered for their influence on standing space, number of seats and accessibility.

A number of innovations were subsequently explored and assessed with the assistance of this visual conflict frame. They embraced three key features to affect an improvement to dispersal and passenger ingress and egress: (1) extra doors that operate only during peak periods for increased passenger exchange. During off-peak periods, the space behind these temporary doors is occupied by seating. Accompanying the extra doors are physical and graphical devices to encourage patrons to move to the left to facilitate simultaneous boarding and alighting. (2) a central aisle of seating clusters with both longitudinal and transverse seating, creating two corridors down the length of the carriage. (3) folding seats that can be locked into an upright perch position to create more standing positions during peak periods and released to form conventional seating during less crowded periods.

2.4 Example 2: A Platform for Electric Bus Design

The second project example has been referenced from Fridman's [17] Ph.D. work, which explored how Electric Vehicle (EV) technologies could disrupt products and services within the Australian public transport bus system. It was specifically focused on bus services in Melbourne, Victoria.

Buses have long formed an important part of the transport systems of cities. Since the early Omnibuses of Europe in the 1820s, commuters have relied on these services as an affordable alternative to the automobile. Particularly as populations have grown and sprawled further away from their original town centres. Buses have helped to complete the public transport network of cities by serving areas where rail infrastructure is lacking, under repair, or difficult to access. As a result, vehicles and operational strategies have had to adopt a high level of flexibility in order to work around roadworks or act as rail replacement services when the train network has been disrupted. This capability has also been their main advantage when compared to trains and trams in competing with the mobility offered by automobiles [18, 19].

The diesel internal combustion engine has been a vital factor in the flexibility of busses, allowing them to operate for long periods in one stretch whilst manoeuvring around detours or changing routes. Due to their maturity, reliability and market position, diesel engines have continued to outnumber all other fuel technologies on buses in Australia, as well as many other countries. However, enhancements in diesel engines have reached a development ceiling. The Intergovernmental Panel on Climate Change has previously predicted that even with further system refinements, an internal combustion engine could never reach the emission reduction requirements for a fuel system in the year 2050 [20]. These concerns have encouraged a shift in the bus industry away from diesel engines to the exploration of EV systems.

Fridman's [17] Ph.D. work considered how these advancements may be incorporated to enhance services through novel vehicle designs and suitable operational strategies that leveraged EV technologies. This research was conducted through financial sponsorship from an Australian bus builder who wanted to be prepared for the upcoming technological shift within their industry. It was, therefore, important for the project to develop a broader resource that others working in the bus industry could use to gain an understanding of how their approaches to vehicle design and service provision may need to adapt.

Fridman's [17] project began by analysing the current bus systems in Melbourne, Australia to develop an understanding of how they functioned and who were the relevant stakeholders. A global review of emerging EV technologies was then conducted to identify which ones could be, or had already been applied to buses. This was done through a combination of a literature review, as well as six interviews with industry professionals from various companies in Australia and Europe such as component manufacturers, vehicle builders and bus operators. This process helped the researcher to develop a base level of project-relevant knowledge within two areas. The first area was focused on Australia-specific requirements and considerations including how vehicles were specified, developed, purchased, and operated. The second area was focused on EV technologies considering which technologies had been tested and how they performed. These two areas were then cross-referenced against each other to see which EV technologies had potential for application within Australia.

It was discovered that EV technologies would bring new tensions into the bus system due to their inherent characteristics conflicting with current operational paradigms. A key point of contention was immediately noticed as the result of a vehicle's energy source changing from diesel fuel to lithium-ion batteries, which were far less efficient at storing energy. Vehicles on Australian roads must conform to a maximum weight allowance stipulated by the National Heavy Vehicle Regulator. Standard buses were already operating at this maximum allowance during times of morning and evening work commutes. Adding extra battery weight, therefore, directly conflicted with weight allocation reserved for passengers, which in turn reduced a bus's capability for generating revenue–its primary purpose in the eyes of an operator. Two research pathways were established to further investigate this topic of energy storage. The first was to explore new vehicle designs that incorporated large battery packs that matched current operational expectations. The second considered how operational approaches may be changed in response to new technologies and how this would influence factors within the system. Information developed through these two experiments would inform the development of a visual frame for this research project.

Taking large batteries as a given, the first design experiment tested how battery packs may be incorporated within the vehicle and what this would mean for accessibility, weight distribution and passenger accommodation. This process discovered that incorporating large battery packs could introduce numerous new conflicts which had not been considered in the literature. These included vehicle-relevant factors such as weight distribution and increased height, as well as passenger-relevant factors such as obstructions to accessibility. The second experiment further built on this understanding by testing smaller batteries with a change in operational strategy. The main point of difference was that rather than being recharged once a day–as was the current approach with refuelling diesel buses–they would recharge frequently throughout operation in order to maintain a service with a small battery pack. This allowed them to continue accommodating a full passenger load comparable to today's diesel buses, which meant that services could continue generating revenue as expected by operators. Upfront investment was, however, discovered to be higher in this scenario as charging infrastructure would need to be installed along routes to support vehicles during service. The time required for charging was also a further consideration as it had the potential to delay service schedules.

A visual frame (see Fig. 2.2) was created to summarise conflicting factors that had been discovered through the two experiments. Both large and small battery options were seen to be potential solutions at either end of a spectrum that considered design around energy storage. Both presented their own associated benefits and drawbacks, which were captured in this single visual diagram. No one-size-fits-all solution could be identified and the visual did not claim to capture every consideration that would present itself during actual vehicle development or operation. It was, instead, presented as a diagram that bus designers could use as a launching place for their creative thinking processes and one that they could add to as new conflicting factors were discovered.

In this example, conflicting factors were only visually framed towards the project's end as a retrospective activity to make sense of the myriad of factors at play and capture a coherent snapshot of the problem context. Inspiration for representing conflicting factors as a single visual diagram was taken from the prior work of Coxon [9], which is the previous example provided in this chapter. Project stages were analysed chronologically with information at each stage being captured as part of a growing and changing diagram.

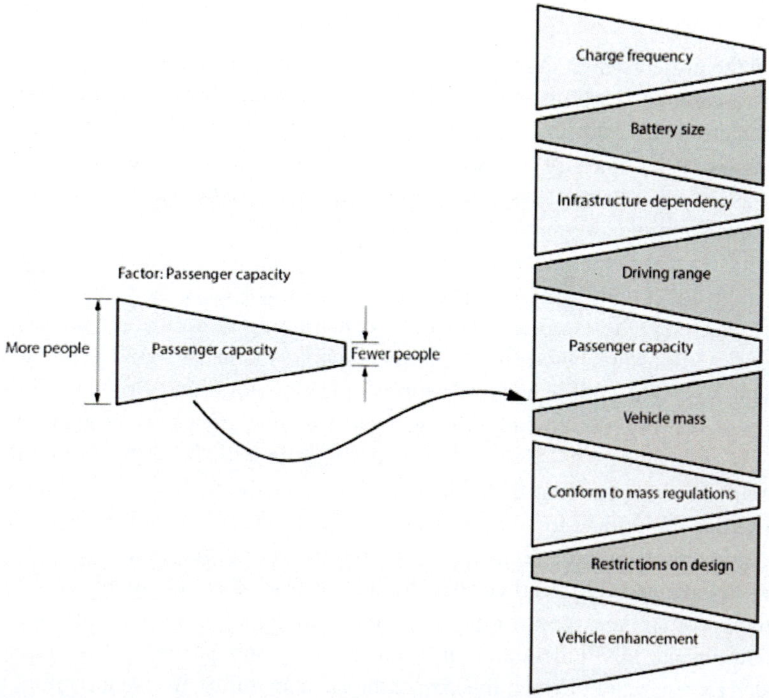

Fig. 2.2 Visual conflict framing of EV bus factors. Diagram referenced from Fridman [18]

Initially, this was done as an exercise to help clarify the design researcher's own understanding by making ambiguous concepts tangible and giving them a visual value through illustration. Prior to this exercise, the factors were known, though were all considered at the same level. They were concepts of equal weighting, floating around within the problem space without an anchor to position them in relation to each other. Considering every factor and analysing its influence over others helped to develop an understanding of the connections within this space. Influencing factors were identified for their capacity to make change within the system. Though these ideas had already surfaced throughout the project's design experiments, it was the first time they could be articulated clearly within a single frame of reference.

Once this diagram was created, it was soon recognised that it was itself a key knowledge contribution from this project by forming a platform that others could use when working in the field. This visual conflict frame became the main method for communicating project findings to the research sponsor. It allowed a variety of seemingly isolated factors to be presented as a single understanding of the PPSS which could be discussed and manipulated by designers and other departments within the company.

2.5 Platforms for Communication and Creativity

Both of the projects have showcased how visual conflict framing has been applied by design researchers working on PPSS challenges. Although these examples focused on different problems, both identified and synthesised conflicting factors into visuals that could assist with creativity, assessment, and communication throughout the project. This was a significant process within both projects as challenges were comprised of many open, complex, dynamic and networked elements.

On the train project, visual conflict framing allowed the designer to see their challenge more holistically and gain a greater idea of the space they were working within. Their creative synthesis process began by constructing a single coherent visual that captured relationships between seemingly disparate elements found in the literature. This starting point established a platform for the exploration of potential solutions that could address the myriad existing factors at play. Once developed, proposed concepts were then assessed against the conflict frame to check their suitability and highlight any areas for improvement. At the project's completion, the diagram was also used as a platform to communicate designed outcomes to a broader audience who could understand the context for which they were developed.

On the bus project, visual conflict framing provided the designer with a greater and more comprehensive awareness of the challenge. These diagrams were used to communicate this understanding to the industry sponsor who was interested in knowledge generated through this research. The capability of visual conflict frames to act as a communication tool was highly important to this research as it had to establish a platform for future work in this field. This diagram was able to present design insights to other stakeholders and develop a shared understanding of the issues that may have otherwise been considered in isolation. In this sense, it was able to stimulate further strategic discussions between various people within the industry.

2.6 A Potential to Reframe

Once challenges are understood, visual conflict frames may be utilised for their generative power to assist with exploring new possibilities and potential solutions. By making issues more tangible, they create potential to redefine the situation and explore alternative scenarios where these factors are no longer conflicting but complementary. This relates to the third approach for dealing with paradoxical statements as described by Dorst, who suggests that these situations may be redefined by creating analogous connection to other observed phenomena. The following example is given for how such problems may be reframed using a simple statement that creates an alternate point of view: "If the problem of… was approached as if it was a problem of…, then the solution should be…" [1, p. 78].

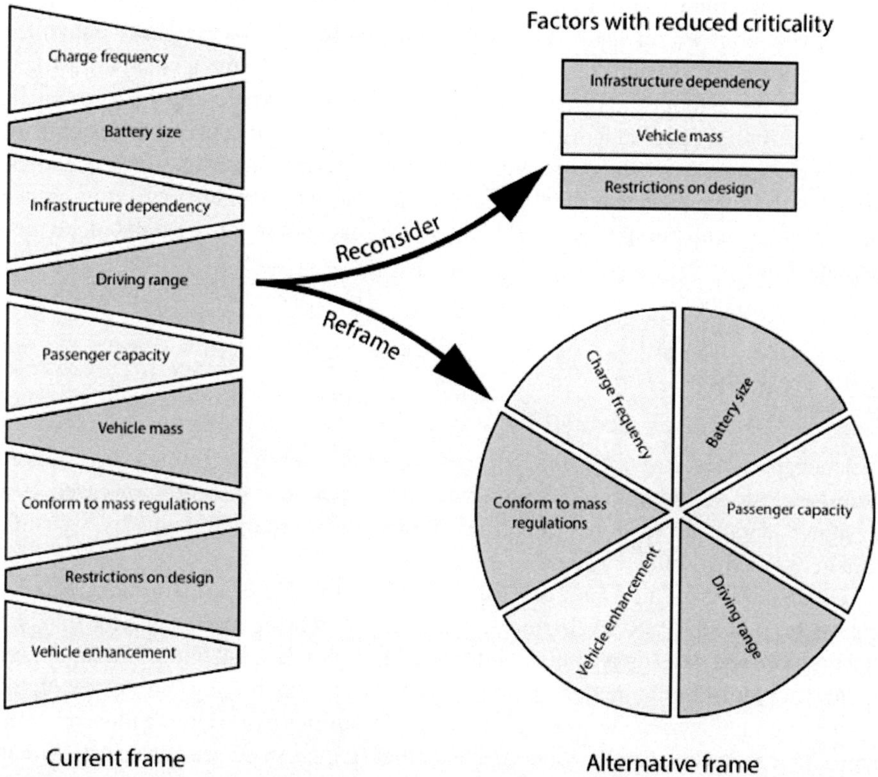

Fig. 2.3 Visual conflict re-framing of bus factors

The process of reframing conflicting factors is summarised in Fig. 2.3. In order to achieve this type of outcome, designers would first need to visually frame existing conflicts by identifying individual factors and understanding why they are at odds with one another. Once this initial visual is created, the conflicting factors may be re-framed to explore new possibilities. As each possibility is tested, individual factors would need to be reassessed to understand if the same problems still exist, or if this new scenario allows them to be complementary. This process would require iteration, with new frames being tested until a suitable one is identified that reduces conflicts and creates a more ideal relationship between a system's factors. It is therefore important that these visual frames remain dynamic and flexible as they would need to be updated throughout PPSS development when new information is introduced.

As an example, in the case of the bus project, if the problem of battery weight was approached as if it was a problem of demand responsiveness, then the solution should be to design using modularity. This way of looking at the problem would be in contrast to the current industry standard of buying a one-size-fits-all bus that is neither the biggest nor the smallest option available. Rather, vehicles could be developed as a smaller module which can accommodate passengers and has a battery. These modules could then be connected and disconnected depending on demand forming a type of road train. With each module, the bus would get more energy as freshly charged batteries would also be attached. In this frame, more passengers would mean more batteries in a system that is demand responsive. With just a small adjustment in how we think about the EV bus system, a scenario can be created that is mutually beneficial for both passengers and operators. This alternative way of approaching the challenge would avoid the current battle between having larger batteries that allow vehicles to travel further but with less carrying capacity, versus or having greater passenger capacity but with a smaller battery that may not reliably get people to their destination.

2.7 Conclusion

This chapter has reviewed the concept of visual conflict framing in relationship to PPSS development. Two examples from public transport showcased how designers creatively explore conflicting factors which are open, complex, dynamic and networked within a system.

Conflict framing can act as a tool for analysis, synthesis, assessment and communication throughout PPSS development. Public transport designers can use it to analyse the problem and develop a greater confidence through a detailed understanding of the factors at play. After an understanding has been established, they can work with the factors within this frame to synthesise potential solutions by reshaping their relationships. Once concepts are developed, the conflict frame can be used as an assessment tool to recheck relationships between the factors and determine a concept's suitability. Throughout this entire process, conflict frames can also support communication between designers and other relevant stakeholders through their visual nature.

Challenges in public transport are complex involving interconnected relationships between various system elements and stakeholders such as passengers, operators, manufacturers, local communities and governments. The risk of failure can be costly if solutions are under considered or poorly implemented; therefore, designers must develop a robust understanding of the various factors within a given problem context before they can attempt to propose innovative solutions in response. The proposition here is that visual conflict framing may assist with this on multiple levels and is a valuable process for designers working on PPSS development.

References

1. Dorst, K.: Frame Innovation: Create New Thinking By Design. MIT Press, Massachusetts (2015)
2. Beuren, F.H., Ferreira, M.G.G., Miguel, P.A.C.: Product-service systems: a literature review on integrated products and services. J. Clean. Prod. **47**, 222–231 (2013)
3. Tukker, A.: Eight types of product-service system: eight ways to sustainability? Experiences from SusProNet. Bus. Strategy Environ. **13**(4), 246–260 (2004)
4. Fridman, I., Napper, R., Roberts, S.: Data co-synthesis in developing pubic product service systems. CoDesign (2018). https://doi.org/10.1080/15710882.2018.1546322
5. Hensher, D.: Future bus transport contracts under a mobility as a service (MaaS) regime in the digital age: are they likely to change? Transp. Res. A: Policy Pract. **98**, 86–96 (2017)
6. Schon, D.A.: Problems, frames and perspectives on designing. Des. Stud. **5**(3), 132–136 (1984)
7. Dorst, K., Cross, N.: Creativity in the design process: co-evolution of problem-solution. Des. Stud. **22**(5), 425–437 (2001)
8. Dorst, K.: Design problems and design paradoxes. Des. Stud. **22**(3), 4–17 (2006)
9. Coxon, S.: A Design Study of Metropolitan Rail Carriage Interior Configuration to Improve Boarding, Alighting, Passenger Dispersal and Dwell Time Stability (Doctoral Dissertation). Monash University, Australia (2015)
10. Burdett, R., Sudjic, D.: The Endless City—The Urban Age Project, London School of Economics and Herrhausen Society. Phaidon, London (2009)
11. Vuchic, V.: Urban Transit: Operations, Planning and Economics. John Wiley & Son, New Jersey (2005)
12. Costa, L., Costa, F.: The evolution of the Rio de Janeiro subway system. In: Paper Presented to the 12th World Congress of Transport Research, 11–15 July, Lisbon, Portugal (2010)
13. Farias, I., Wilkie, A.: Studio Studies: Operations, Topologies and Displacements. Routledge, London (2015)
14. Hirsch, L., Thompson, K.: I can sit but I'd rather stand: commuter's experience of crowdedness and fellow passenger behaviour in carriages on Australian metropolitan trains. In: Paper Presented to Australasian Transport Research Forum (ATRF), 28–30 Sept, Adelaide, Australia (2011)
15. Ruger, B., Tuna, D.: Influence of railway interiors on dwell time and punctuality. Railway Interiors International, UKIP Media Events, Dorking, pp. 32–36 (2008)
16. Regirer, S., Shapovalov, D.: Filling space in public transport by passengers. Autom. Remote Control Arch. **64**(8), 1301–1309 (2003) (Plenum Press, New York)
17. Fridman, I.: Battery-Electric Route Bus: A Platform for Vehicle Design (Doctoral Dissertation). Monash University, Australia (2016)

18. Beirao, G., Sarsfield Cabral, J.A.: Understanding attitudes towards public transport and private car: a qualitative study. Transp. Policy **14**(6), 478–489 (2007)
19. Bunting, M.: Making Public Transport Work. McGill-Queen's University Press, Montreal (2004)
20. Ueda, M., Hirota, T., Hatano, A.: Challenges of widespread marketplace acceptance of electric vehicles—towards a zero-emission mobility society. SAE International Online (2011)

Chapter 3
Unbounded Ergonomics: Addressing the Wicked Problem of Gender Transport Poverty in Karachi Through the Application of the Hexagon-Spindle Model

A. Woodcock, S. Iqbal, and J. Osmond

Abstract Gender transport poverty is truly a wicked problem. It is an extreme example of transport poverty, bounded with social and cultural practices, which needs to be addressed if women can enjoy safe mobility and rights as equal citizens. This chapter explores whether transport ergonomics, in particular the use of the hexagon-spindle (HS) model, can add value to the debate by systematically unpicking the problem and mapping issues into a solution space to highlight and identify social and cultural factors which inhibit gender transport equality and develop culturally appropriate solutions to. Previous research has applied the HS model to map out Key Performance Indicators for passenger satisfaction, however, little exploration has been made of factors falling into the 'personal sector'. It may be argued, in this and other cases, that these, along with external factors have a significant and sometimes overwhelming effect on mobility and need greater consideration in the design of transport systems. Although Pakistan provides an extreme example, women across the globe are denied the freedom to travel freely and without fear of harassment on public transport. Putting the female traveller at the centre of a concentric ring (or the HS) model enables a clearer view of the transport system from a user's perspective and may show where small inroads can be made. At a wider level, it is hoped that transport and mobility can be used to illustrate and reduce the inequalities women in the global south experience in their everyday lives.

Keywords Gender · Transport poverty · Hexagon-spindle model

A. Woodcock (✉) · S. Iqbal · J. Osmond
Future Transport Research Institute and Centre for Arts, Memory and Community, Coventry University, Coventry, UK
e-mail: adx974@coventry.ac.uk

S. Coxon and R. Napper (eds.), *Advancing a Design Approach to Enriching Public Mobility*, Intelligent Systems Reference Library 198,
https://doi.org/10.1007/978-3-030-64722-3_3

3.1 Context

Pakistan's population in 2017 was 207.7 million with 106 million (51%) men, 101 million (49%) women, and 10,418 transgender persons [1]. Karachi (where the fieldwork took place) is the most populous city, with more than 20 million [2]. Pakistan ranks 128th out of 182 on Human Development Index (2010), 124th out of 155 on Gender Development Index (2009) and 132nd out of 134 on the Global Gender Gap Report (2009). It has been stated that:

> '…transport remains a neglected area among gender specialists and transport specialists are still reluctant to take on gender issues. Until this is done, the prospects for many women who live in areas characterized by poor physical accessibility and inadequate transport will remain poor' [3, p. 51].

Karachi has an estimated population of 20 million [4], of which 96% are Muslim [5]. It is the economic hub of Pakistan and typical of expanding megacities in the Global South. The transport system not only fails to support the needs of the city and its population, in some ways it has worsened since independence, in the year 1947.

The unregulated bus service is hazardous, of poor quality, run by untrained staff and private operators, with just 6000 buses operating on 111 routes—a decrease from 329 routes that were operational in the 1970s [6]. Despite increases in demand for mobility and no other available forms of mass transit [7], the passenger to seat ratio remains 1:40 (in comparable cities of Mumbai and Hong Kong, the ratios are 1:12 and 1:8 respectively).

With almost 50% of the city's population dependent on public transport [8], and no immediate plans to develop other modes of transport, three-wheeler paratransit vehicles (namely auto-rickshaws and qingqi) fill the lack of regulated service provision. This is typical of other south Asian and African cities, where informal (and sometimes expensive) services fill the gaps in transport service provision for those unable to access services or afford private modes of transport [9].

As a predominantly Muslim city, gender-segregation in public spaces is considered culturally appropriate and legally enforced. Women do not enjoy or experience the same rights of men. The segregation carries forward into transport. For example, in public buses, men and women sit in separate compartments [7], and women are allocated less than one-third of the seats. This maintains the on-going lack of acceptance of women's occupation of public space in the wider context such as on the streets and general public realm [10]. Men hold positions of power in most transport organizations, where they make little attempt to acknowledge or understand the rights of female travellers. Interviews with transport stakeholders revealed an almost total lack of understanding, or willingness to consider the rights and needs of women to access transport or move freely without gaining permission of their male family members [11]. Denied access to transport, women are deprived of opportunities to become full citizens—they cannot take up employment, education or health care services, let alone be free to meet their friends. When they do travel, they have to pay a higher percentage of their income to travel in safety and without fear of harassment from other road users, passengers and bus drivers [12].

The Centre of Economic Research in Pakistan's survey found nearly 30% of respondents considered it "extremely unsafe" for women to walk in their neighbourhood, with 70% of male respondents discouraging "female family members from taking public wagon services" [13]. Walking on poorly maintained roads or footpaths and avoiding vehicles [14] in urban areas is stressful and tiring owing to environmental effects—overcrowding, temperature, humidity and pollution.

With very real and ever-present threats to female safety (e.g. kidnappings, abductions, rape and sexual and harassment) on the streets and on public transport, women and their families live in perpetual fear when travelling alone, late at night or with strangers [11].

3.2 Gender Transport Poverty

Mobility issues in Lower Middle-Income Cities (LMICs) are wicked problems, systemically linked to many socio-political and cultural problems [15, 16]. Although transportation in LMICs is 'a challenge for everyone at the same time' with under-investment in public transport, transport infrastructure, rapid urbanization and rises in car ownership etc., Anand and Tiwari argued that inadequate pedestrian infrastructure, non-existent or badly designed pedestrian crossings, poor location of bus stops and shelters, and high steps on buses contribute to a system that is hostile to women [17]. Gender transport poverty research [16] acknowledges disproportionate burdens on women—they earn less, have multiple roles and their mobility is restricted by social, cultural and religious norms. Lastly, the lived experience of women reveals that they face more harassment on transport and public spaces [18, 19] and are more affected by road rage [20]. Anand and Tiwari commented that "the nature of the entire transportation system of the city is then not only insensitive to the needs of women, but also actively disables accessibility and induces poverty" [17, p. 78]. 'Work arounds' include gender-segregated transport [21, 22], zero tolerance of harassment on public transport, gender aware planning [23] and products increasing attention on the aggressor [24]. It is not clear whether these on their own can have an impact on the wider issues underlying gender transport poverty. Faiz et al. identified cases where women were either not aware of transport interventions designed to help them, were not told how to use them, did not think they had permission to use them or were intimidated and threatened if they did use them [1].

Transport-related social exclusion recognizes that mobility is socially produced and is unevenly distributed along ethnic, class, gender, and religious lines [25, p. 15]. Lucas et al. developed a theoretical framework of transport poverty which stressed that transport-related social exclusion should not be limited to the inaccessibility of transport services but should also encompass social disadvantage caused by poor health (e.g., disabilities) and land use [26]. The four conditions which can contribute to transport poverty are:

- Mobility poverty; Non-availability or poor design of motorized transport appropriate to the needs of an individual.
- Accessibility poverty; The inability of transport to reach destinations which ensure a decent quality of life.
- Transport affordability; not being able to afford the cost of transport.
- Exposure to transport externalities; include the harmful effects of exposure to the transport system itself, such as traffic congestion, pollution or accidents.

This model conceptualises transport poverty as encompassing not only the lack of physical and financial resources but also the social and spatial deprivations which can influence an individual's mobility. Hence, socioeconomic status, marital status, gender and age can also contribute to social disadvantage. These factors can lead to reduced mobility and marginalisation of groups [27].

Additionally, in some cases social and cultural issues inhibit women's equality and agency, adding significantly to their burdens. Here, women's mobility is not only influenced directly (e.g. through poor bus design making it difficult to enter or egress buses) but also indirectly through social and cultural norms which deny them agency and freedom to travel.

The chapter argues that there is a need to go beyond the conventional measures of understanding transport poverty and to identify gender-relevant themes. It is essential to analyse women's (im)mobility and understand whether it is imposed or chosen [28], or a by-product of poor practices. This necessitates investigating the barriers that exert power on women's independence such as traditional gender roles [29]. There is a need to emphasize the importance of decision-making regarding journeys and not assume that mobility is a right everyone enjoys. For example, women in western cultures do not have to ask permission to go for a walk or need to be chaperoned when they go out. Such freedoms are not universal. Freedom of movement is subjective, and gender has a significant role to play in determining one's mobility choices. The impact of gender norms dictated by the social and cultural milieu, also needs to be explored. This requires looking at the micro-level and macro-level factors such as behaviour, that affects one's mobility to fully understand the impact of gender and intersectionality on transport poverty. The hexagon-spindle model may be useful to help widen the perspective of these engaged with usability and inclusion.

3.3 Ergonomics and Design

Looking at this from a transport ergonomics perspective, traditionally only task, workspace and some organisational factors lie within its remit (see Fig. 3.1). However, Woodcock [30, 31] recognised the potential for a wider role for design and ergonomics in this (and other) arenas following the argument of Wilson that: "*Human-machine interaction does not take place in a vacuum…ergonomic techniques are needed to predict, investigate or develop each of the possible*

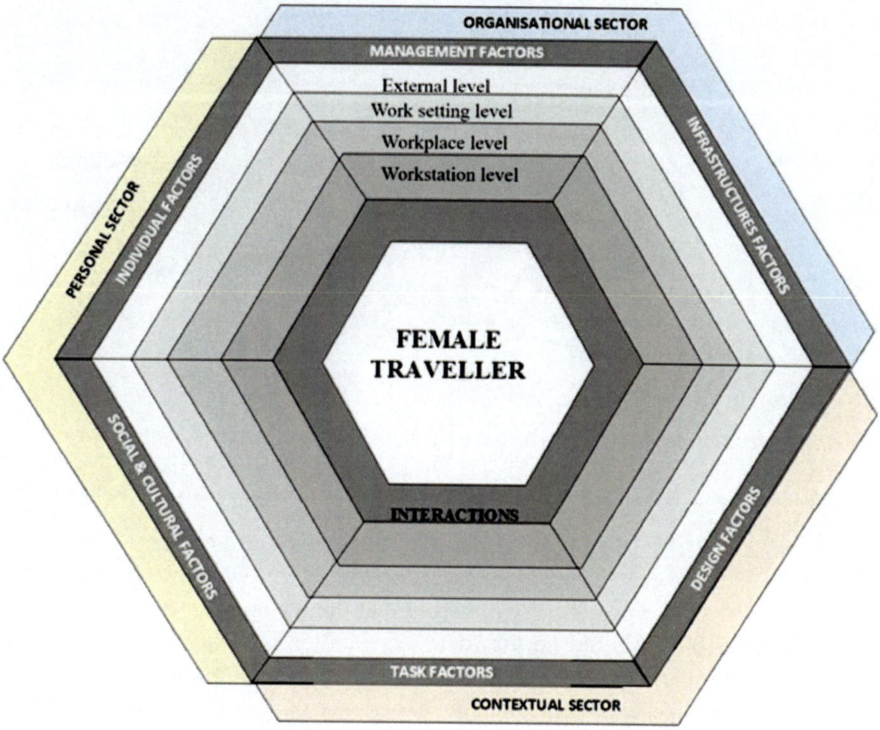

Fig. 3.1 Generic hexagon spindle model

interactions, between person-task, person-process, person-environment, person-job, person-person, person-organisation and person-outside world" [32, p. 10].

The Hexagon Spindle (HS) Model [33, 34] provides a systematic division of factors which influence task completion, between the constant characteristics of the user and individual task-dependent variables. Although developed to elucidate factors which effect task interactions in learning environments, its generic nature means that it has been applied by Woodcock to transport design [31, 35] with only slight adaptations.

The HS model starts in the centre, emphasizing the user's interaction with different parts of the system, for transport this could be the design of seats, bus stops, transport infrastructure, fellow passengers etc. The overall traveller's task may be simply determined as the need to get from a to b, but each part of the journey comprises a set of tasks which take place using different modes requiring different interactions—only some items remaining constant across time (such as fixed factors in the environment (e.g. heat and humidity) or the social cultural milieu). Traditionally, transport operators have simply concentrated on that part of the journey for which they are responsible—such as the design of the bus service or ticket system. In a heavily fragmented system, the responsibility for station design or the pedestrian infrastructure

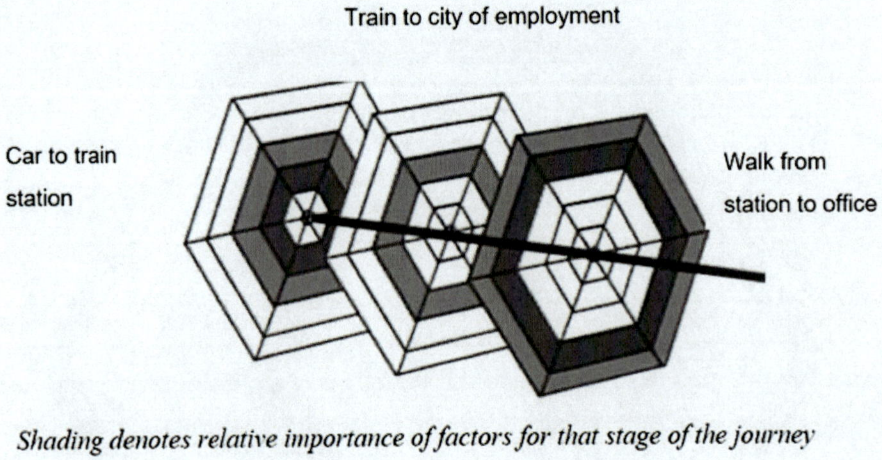

Fig. 3.2 Stages of a commuter's journey. Shading denotes relative importance of factors for that stage of the journey

lies with other service providers. So, saying that the whole journey, from origin—to—destination (OD) should be the focus of attention has only come to the fore recently e.g. through smart mobility as exemplified by multimodal ticketing. The FP7 METPEX project explored passenger experience of multimodal whole journeys [36].

In taking a person-centred perspective to the design of transport systems the HS model recognizes;

- The need to support a person's entire journey, not just the design of one particular aspect, or interaction. For example, a commuter's journey may be composed of different elements. Therefore, an effective user centred transport system must support the entire, door-to-door journey (Fig. 3.2).
- The changing nature of human characteristics across the day and task. Tailorable systems, including transport ones, need to acknowledge and adapt to changes in human characteristics over time such as length of journey, motivation, reactions of others. Clearly negative experiences change behaviour. For example, if a woman feels unsafe on public transport, or walking along certain roads, she will change her travel behaviour.

With reference to Fig. 3.1, the personal sector considers:

1. Individual factors which may vary in relation to a specific task, from individual to individual. This includes social, psychological and physiological factors such as intelligence, motivation, attitudes, socio economic group, earnings, peer group, age, gender and sensory, physical and cognitive abilities and the implications of these on all levels of activity from purchasing decisions, through to choice of transport and driving performance.

2. Social and group factors, such as cultural and social norms, peer pressure, behaviour of others and the influence of this on one's behaviour.

The contextual sector is similarly broken down into two factors:

1. Task design factors which relate to the way in which transport has been operationalised and how it influences behaviour. In terms of new mobility paradigms there has been a shift in the way transport is organized, with more integrated services covering the end to end journey. This in turn influences travellers' behaviour, for example in relation to buying mobility services rather than single transport tickets.
2. Product design factors. This sector deals specifically with the way in which the design of the product influences user interaction. The product in question could refer to a visual display through to the design of the vehicle or transport system. Historically this is where a lot of ergonomics effort has been expended.

The organizational sector relates to:

1. Infrastructure and the wider environment transport system i.e. the network over which the vehicles operate and its related technology, through to the implementation of national rules and regulations governing flow and speed of transport and the design of the network as it affects drivers and passengers. Additionally, the integration and interoperability of different transport services is now given attention with a new focus on the whole journey experience (whether for freight or passengers).
2. Operational and organizational issues (such as organizational ethos, culture and regulations) can affect the operators of public transport, and hence the safety and wellbeing of not only the drivers but the passengers and other transport users as well.

The hexagon spindle model was used as a basis for the FP7 funded METPEX [36] developing passenger experience measurement instruments. The Key Performance Indicators (KPIs) established from a pan European survey undertaken by the project have been transposed on to the model, revealing points at which interventions are needed to increase passenger satisfaction (summarised in Fig. 3.3). Notably improvements are needed across all areas of public transport, to increase passenger satisfaction and attract ridership.

In the following sections, the chapter explores how such an approach could be used to differentiate the many factors which can inhibit women's mobility using a case study of Karachi.

3.4 Discovering How Transport Poverty Effects the Everyday Lives of Women in Karachi

A phenomenological approach using mixed methods (survey, interviews, focus groups, autoethnography and conversations with purpose) was used to understand

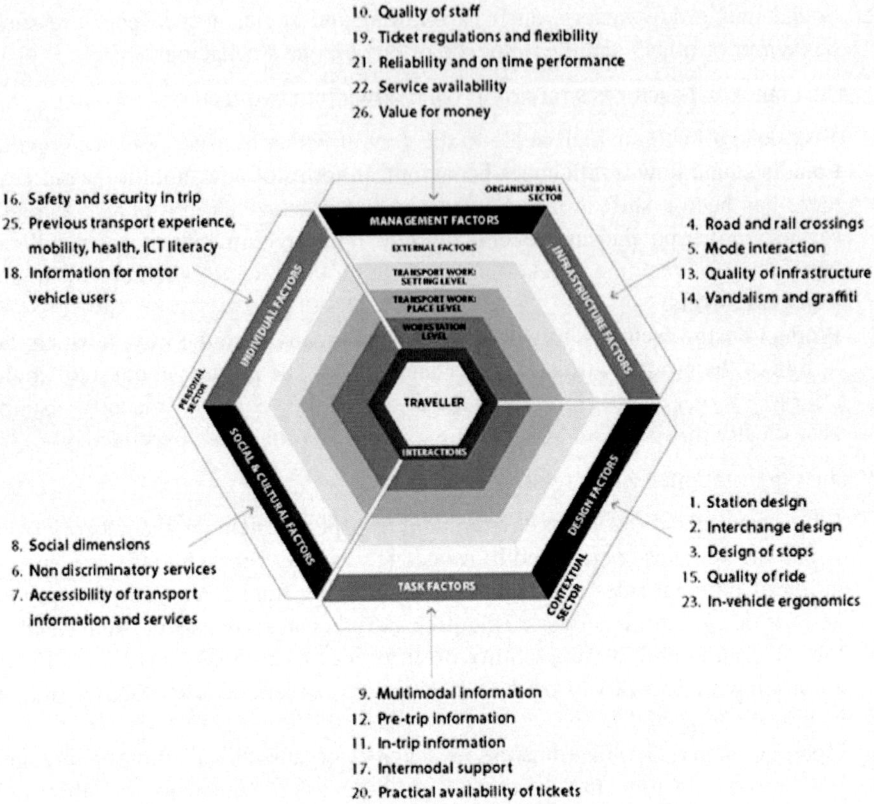

Fig. 3.3 HS model for METPEX key performance indicators

the experiences of young working women in Karachi and the attitudes of representatives of the wider transport business ecosystem to women's travel. The methods and number of participants in each phase of the research are shown in Table 3.1.

The thematic analysis started with reading the transcripts and coding the discussion. Coding was followed by organizing and classifying the data into meaningful parts or categories/themes, which are consists of several codes [37]. The process of coding involves aggregating the text or visual data into small categories of information, seeking evidence for the code from different databases being used in a study, and then assigning a label to the code' [37 p. 184]. Emergent themes, in order of relative importance, are shown in Table 3.2.

The survey and interviews revealed that most women spent between 10 and 15% of their daily salary on transport, men spend less than 10%. These differences are explained because women need safer, more direct forms of transport than that provided by public transport, so they have to use private and semi-private modes of transport. On average 24% of women spent about one-third of their monthly incomes on rickshaw journeys (having unregulated fares) providing door-to-door services, while only 5% of men spend a similar amount. Men also enjoyed greater

Table 3.1 Participant breakdown

Methods	Participants	Duration
On line survey	185 responses from 16 to 40 years old (94 women and 91 men)	Over 4 months
One-to-one interviews	28 interviews with 21–35-year olds (18 women and 10 men)	90 min
Participant observations	Researcher	Extensive observations (at 4 locations)
Focus group discussion (FGD)	12 participants aged between 21 and 35 years (6 women and 6 men)	60 min conducted in 2 groups
Expert interviews	10 men 1 woman	30 min

freedom and mobility as car and motorbike owners. Men's commuting time can be shortened by 50% if they use motorbikes, compared to commuting by bus. These allow greater flexibility and time savings during congestion. Women are culturally discouraged to ride motorbikes even as passengers. Therefore, women make more complicated journeys if they are allowed to work.

Substantial gender-based differences were found related to temporal availability. Most married women preferred or were required to seek employment and education close to their homes to avoid spending time travelling across the city and being distant from their families. Women may simply be denied an opportunity to work by their families. Even if they do, they are still required to do all household related tasks. Our survey showed over half of the employed women spent and additional 8–12 h or more each day on household chores (while most men spend less than 4 h). This gendered division of household labour means that women are time poor. They have to squeeze into their day housework, employment, and travel time (up to 2 h each way in tiresome and stressful conditions). Poor public transport, social and culturally imposed travel restrictions and expensive on demand options increase inequality and gender-based poverty. In line with this, women were also found to undertake more journeys related to unpaid care work. As expected, the survey also revealed that omen made fewer leisure trips, such as those to see friends or for simple relaxation, such as window shopping. Such trips are regarded as non-essential for women, requiring permission from the male members of the household. This permission is often denied, or trips which are taken are 'chaperoned'. Although decision-making can foster power and confidence related to mobility, it was quite common for both single and married ladies to feel questioned about their choices.

The survey is replete with accounts of gender differences related to safety on public buses, which leads women to select door-to-door services if they can afford them. Although poor maintenance and dirty transport affects both genders equally, differential societal expectations for men and women shape mobility. An obvious

Table 3.2 Summary of thematic analysis

Theme highlighting gender transport poverty	Description	Example	Method
Increased cost of travel	Women spend more money on transport, due to their dependence on door-to-door transport service that are more expensive	Women spend 5 times more each month on their travel	Survey and interviews
Lack of seats due to the partition in the bus	Unequal gender segregation, with only 1/3rd seats allocated to women	Men can sit on the roofs of buses too	Survey and interviews
Lack of enforcement of traffic regulations	Poor maintenance of vehicles; no central ownership and hence no accountability	Dirty seats affect women as they are supposed to look clean and tidy at work, or not have the stigma of travelling on public transport	Expert interviews
Temporal restrictions	Women cannot travel alone unaccompanied, especially at night	Cultural norms operate on young women requiring them to either stay at home or undertake accompanied journeys. In both cases, these norms limit women's mobility to 'approved journeys'	Survey and interviews
Time pressures	Time available for these women is prioritised for their role as caregivers	Most of the women spend 8–12 h versus men who spend less than 4 h	Interviews, survey
Increased risk on transport	Lack of personal safety No mechanism to complain	Women are subjected to harassment and need to depend on door to door transport service to avoid harassment	Interviews, survey, FGD
Lack of coordination between civic agencies	Federal, provincial and city government bodies have overlapping roles	No consensus about the number of buses in Karachi	Expert interviews
Poor quality of infrastructure	Lack of basic shelter at bus stops, buses are outdated and lack of	Rudimentary bus stops, lack of ramps	Interviews
Lack of gender representation	No female bus conductors or drivers	All transport staff and policymakers are males	Expert interviews

(continued)

Table 3.2 (continued)

Theme highlighting gender transport poverty	Description	Example	Method
Lack of accountability	Corruption at different levels	Bus drivers pay bribes to traffic police	Expert interviews
Lack of information	Basic information related to trips Fares, timetable	People have to wait longer for buses and women complained about inconsistent fares and being harassed	Interviews
Lack of route coverage	Public transport only runs on certain routes	Most of the women had to depend on contract carriages to reach their work since they cannot drive motorbikes	Survey and interviews

example is the expectation that working women should look professional, clean and tidy for work. Journeys on public transport leave them dirty and dishevelled and they are judged inferior by colleagues if they are seen as having to depend on public transport.

Gender norms impact mobility choices of women. Although around 40% of the women surveyed, agreed it was more important to feel accepted in society, most were frustrated with how they had to adjust their lives due to socially constructed notions of gender. This made them feel disempowered to make independent decisions and cared when they were justified in demanding their due rights. As shared by one of the female respondents,

> 'I am dependent on my family to pick and drop me if I am going to a wedding or any event with my friends. They do not allow me to travel alone in a cab at night after 9 pm….' [11]

With little regulation of transport, basic health and safety provision is not enforced. This impacts more on women more than men, because many of the male transport drivers and operators believes that women should not use transport or take part in activities outside the home. This is exacerbated by lack of transport information, poor route coverage and service quality which exposes women to long waits As such they are subjected to harassment on a daily basis. With few buses, unfair seat allocation and poorly maintained bus stops, women complained that long waits on the roadside left them exposed to harassment, '*Out of 100 times, 95 times I have faced harassment*' [12, p. 14]. This not only inhibits women's mobility (for who wants to walk anywhere when they are constantly harassed or need to be aware of potential threats) it also provides justification for gender-related restrictions on mobility—for the streets and public transport are indeed not safe for women. This exacerbates the difficulty of conducting daily tasks and requires detailed planning to ensure safe travel, leading to higher levels of stress.

Thematic analysis of interviews with stakeholder representatives (e.g. NGOs, transport operators and planners) revealed that at governmental level, there was no transport master plan, no clear division of roles and duties, and that different agencies were hostile to each other. Moreover, gender-based assessment of initiatives was simply not a priority. Stakeholders either overlooked the needs of women, possessed a minimal understanding of their issues, or displayed insensitivity towards the issues faced by women—they simply did not count. For example, an engineering expert employed in BRT (bus rapid transport) planning believed there are more important issues to consider. Hence, he commented, 'This gender element might come after 30–40 years and cannot be integrated with current planning'. With few women in the industry, in policy or decision-making positions, or consultation groups, such views remain unchallenged. Commitment to gender mainstreaming was perversely manifest in the assertion of the need to have gender segregation in public buses (for the protection of women), along with the assertion that fewer seats were justified for women as their place was in the home. Thus, the gender pattern of social disadvantage was reproduced by policymakers.

3.5 Mapping the Problems Using the Model

The following tables illustrate how the HS model can be used to map out gender related issues from the primary research, moving in levels out from the centre of the HS model (Key: 1. Workstation level, 2. Transport workplace, 3. Transport work setting and 4. External factors).

Tables 3.3 and 3.4 show that the factors of poor design, management and infrastructure across the system effects all travellers. However, for women these have more serious consequences. Lack of affordable transport means that a higher percentage of their income is spent on other means of transport, lack of transport coverage required them to walk longer distances increasing the likelihood of being subjected to harassment, theft and pollutants, dirty and poorly maintained buses, and poor in vehicle ergonomics causes more damage to women and their clothing.

There are several gender-based differences which are more evident in the social and cultural domains as well as individual factors (as shown in Table 3.5). Women prioritized their safety but also wanted more acceptance of their mobility at the societal level. They wanted to have the agency to decide about their trips and enjoy the liberty of not being dictated to about their choices regarding appearance or the time of journeys. The share of household responsibilities also needs to be balanced along with non-discriminatory social attitude in order to allow women to increase their participation. Since women were more dependent on multiple modes of transport such as paratransit vehicles/rickshaws, they wanted to have an intermodal support system too. Another key indicator for female traveller is a journey free from harassment.

Figure 3.4 when contrasted to Fig. 3.3 (from the European project) shows a higher number of external factors which curtail women's use and satisfaction with transport. It should be noted that these permeate right to the heart of the model. Design factors

Table 3.3 Design sector issues

	Design sector			
	Task		Transport design	
Level	Male	Female	Male	Female
1	Lack of intermodal support	Lack of intermodal support	Poor noise control and ventilation	Poor noise control and ventilation
2	Poor driving practices	Poor driving practices	Poor seat design	Poor seat design, in-vehicle ergonomics interacts with clothing design leading to greater H&S risks
3	Lack of information regarding tickets	Lack of information regarding tickets	Poor design of buses, stops, crossings, etc.	Less seats allocated
4	Inconsistent fares, lack of affordable transport	Inconsistent fares, lack of affordable transport	Poor planning, poor traffic control	Poor planning, poor traffic control

Table 3.4 Organisational sector issues

	Organisational sector			
	Management		Infrastructure	
Level	Male	Female	Male	Female
1		Poor behaviour of passengers and staff		Lack of choices for transport due to gender discrimination
2	Poor quality of public transport staff, poor service	Poor quality of public transport staff, poor service, lack of cleanliness	Lack of capacity, high level of crowding in vehicles	Lack of capacity, high level of crowding in vehicles
3	Lack of traffic management, lack of surveillance	Lack of traffic management, lack of surveillance	Lack of route coverage	Lack of route coverage
4	High level of informality, lack of integration between civic agencies	High level of informality, lack of integration between civic agencies	Poor quality of urban infrastructure roads, streets, etc.	Poor quality of urban infrastructure roads, streets, etc.

are much more centred around the vehicle design (buses are older an not replaced). Although some factors are similar in both cases, such as the quality of the transport staff, design of bus stops and availability of information, there can be no comparison, as the situation and context in Pakistan is so different.

Table 3.5 Personal sector issues

	Personal sector			
	Individual		Social and cultural	
Level	Male	Female	Male	Female
1		Strict dress code	High level of criminality on vehicles	Harassment as a norm, high level of criminality on vehicles
2		Greater household responsibilities, lesser time		Lack of women employees in workplace and decision making bodies
3		Fear of being judged		Lack of gender mainstreaming, gender stereotyping with respect to travel
4		Restricted decision making, lack of agency	Bribery, corruption and lack of accountability	Lack of equity, bribery, corruption and lack of accountability

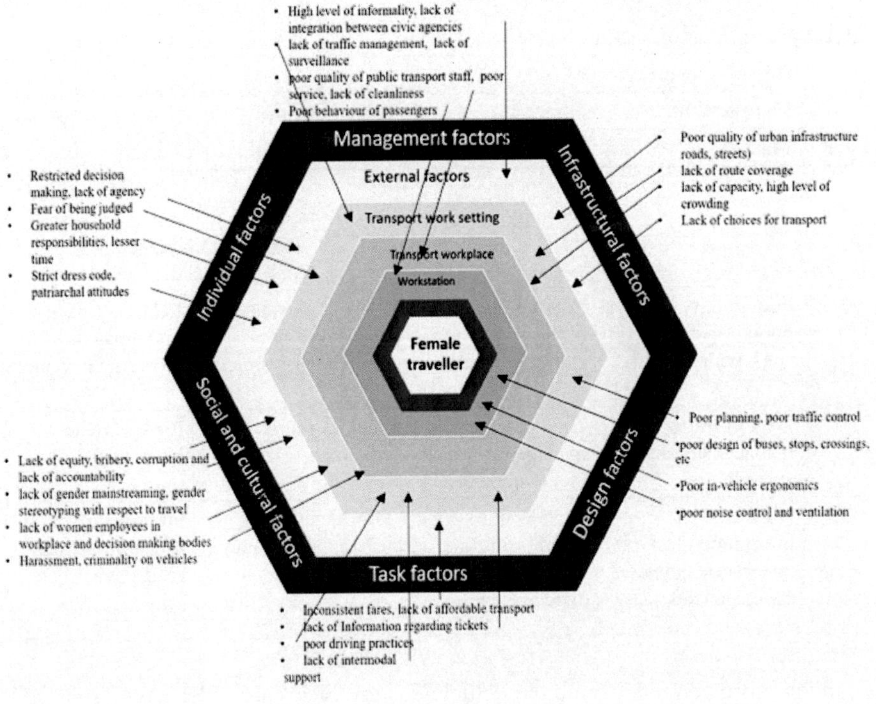

Fig. 3.4 KPIs for female travellers mapped on to the HS model

3.6 Recommendations for Delivering Change Against the KPIs

- There is a need to understand and address macro-level issues that impact on women's mobility decisions since the gender pattern of social disadvantage is not recognized by the policymakers. This understanding may be present in some cases, but a willingness to do something about it, is not evidenced.
- Gender mainstreaming or gender action plans should be developed to integrate women into employment, decision making bodies, and consultation processes as a means of developing gender inclusivity within the transport business ecosystem and its operation. This requires an integrated transport planning process, with reliable stakeholder and user engagement processes (for all citizens).
- At a more operational level, transport infrastructure needs to be improved. Timetabled services may be problematic owing to levels of congestion and disruption, but safer, more frequent formal bus stops/pick up points can be designed that offer shelter from weather conditions, reduce harassment e.g. through cameras or better design, which do not require long walks so that women are confident that they will not be harassed. These could include time tabling and pricing information.
- Replacement of existing bus fleet with safer, more ergonomically designed and accessible vehicles.
- Cameras can be installed for the safety of women at stops, stations and pedestrian bridges, as well as a robust complaints mechanism. This would increase the safety of all travellers.
- Where regulations exist to protect human rights they should be enforced.
- More women employed in the transport and law enforcement areas should be part of a concerted gender action plan. This has been attempted elsewhere in Pakistan, but requires gender neutral working environments and organisational change and commitment to guarantee safety and authority of women moving into such employment.

3.7 What Has This Type of Analysis Shown

The potential role of designers as catalysts in framing problems, bringing disparate parties together and in envisioning solutions in the transport domain has been recognised [35]. Models of transport poverty can be used as a theoretical basis to map out problems on the H-S (or similar) models to show how these relate to micro as well as macro levels, and where solutions may be possible. It allows gender-specific transport issues to be unpacked. The primary research [11] provides clear indications of the impact of social and cultural barriers on women's mobility. When compounded by the problems associated with inadequate, inefficient and unregulated transport, poverty and criminality, women's mobility is severely compromised. Physically tiring, long journeys, disrespect and harassment on the street, by transport staff and passengers, naturally lead to reduced mobility especially when women are also required to spend long hours on household duties. The KPIs acting towards the centre of the model, and in the contextual sectors and infrastructure/organisation sectors are those which may be more solvable and may lead to improvements in mobility for both men and women. However, it is only by improving women's equality in the other sectors will big inroads be made. During the focus group discussion, when the prospects of encouraging women to ride motorcycles were under-discussion, a female participant made a comment which manifests the impact of context on their mobility,

> I think it will be very encouraging to see women riding motorbikes as they can be independent if this can be made possible. Men still consider women their *pair ki jooti* (translation: shoes) and this is visible in public transport too. Such an attitude needs to be changed and women need to start taking drastic steps now.

References

1. Faiz, K., Woodcock, A., McDonagh, D., Faiz, P., Nordin Nikmatwal, A., Shamsul Haruma, Y.: Permeating the barriers between the individual and policy designers in Pakistan: a cross-cultural study of women's mobility. In: Proceedings of RSD7, Relating Systems Thinking and Design 7, Turin, Italy, 23–26 Oct (2018). Available at http://openresearch.oca.du.ca/id/eprint/2677/
2. Khawar, H.: karachi's population-fiction and reality. Express tribune online 4 Sept (2017). Available at https://tribune.com.pk/story/1505657/karachis-population-fiction-reality/
3. ILO: Decent Transport for Working Women: ILO Country Office for Pakistan (2011)
4. Ali, S.A.: The wheels of Karachi buses grind to a halt. The Express Tribune (2017). https://tribune.com.pk/story/1532176/wheels-karachis-buses-grind-halt/
5. Hasan, A., Mohib, M.: The case of Karachi, Pakistan (2003)
6. Hashim, A.: Karachi public transport on the verge of collapse: Report. DAWN Online 23 January (2015). Available from https://www.dawn.com/news/1158772. Accessed 14 June 2017
7. Qureshi, S.: The fast-growing megacity Karachi as a frontier of environmental challenges: urbanization and contemporary urbanism issues. J. Geogr. Reg. Plann. **3**(11), 306–321 (2010)
8. Qureshi, I., Lu, H.: Urban transport and sustainable transport strategies: a case study of Karachi, Pakistan. Tsinghua Sci. Technol. **12**(3), 309–317 (2007)
9. Sen, M.: The problems in public bus transportation system: assessment by passengers and transport authority in Dhaka. Master of Philosophy Thesis in Development Studies (specializing in Geography) Department of Geography Norwegian University of Science and Technology Trondheim, Norway (2016)
10. Harrison, J.: Gender segregation on public transport in South Asia: a critical evaluation of approaches for addressing harassment against women. Dissertation Submitted to the University of London (2012)
11. Iqbal, S.: Understanding the gender dimensions of Transport poverty in Karachi. Unpublished PhD Dissertation, Coventry University (2020)
12. Iqbal, S.: Mobility justice, phenomenology and gender: a case from Karachi. Essays Philos. **20**(2), Article 4 (2019). https://doi.org/10.7710/1526-0569.1634
13. Sajjad, F., Anjum, G.A., Field, E., Vyborny, K.: Overcoming Barriers to Women's Mobility (2103). http://cdpr.org.pk/images/publications/cities/Overcoming-Barriers-to-Womens-Mobility-Extended-Brief.pdf

14. Seedat, M., Mackenzie, S., Mohan, D.: Phenomenology of being a female pedestrian in an African and an Asian city: a qualitative investigation. Transp. Res. Part F Traffic Psychol. Behav. **9**(2), 139–153 (2016)
15. Rittel, H.W.J., Webber, M.M.: Dilemmas in a general theory of planning. Policy Sci. **4**(1973), 155–169 (1973)
16. Lucas, K.: Transport and social exclusion: where are we now? Transp. Policy **20**, 105–113 (2012)
17. Anand, A., Tiwari, G.A.: Gendered perspective of the shelter–transport–livelihood link: the case of poor women in Delhi. Transp. Rev. **26**(1), 63–80 (2006)
18. Osmond, J., Woodcock, A.: Are our streets safe enough for female users? How everyday harassment effects mobility. In: Sharples, S., Shorrock, S., Waterson, P. (eds.) Contemporary Ergonomics, pp. 495–502. Taylor and Francis, London (2015)
19. Madan, M., Nalia, M.K.: Sexual harassment in public spaces: examining gender differences in perceived seriousness and victimization. Int. Crim. Justice Rev. **26**(2), 80–97 (2016)
20. Gil, N.: Road Rage is Sexist and these Female Drivers know it (2018). https://www.refinery29.com/en-gb/2018/11/216815/road-rage-sexist-women
21. Dunckel-Graglia, A.: Women-only transportation: how "pink" public transportation changes public perception of women's mobility. J. Public Transp. **16**(2), 85–105 (2013)
22. McLeod, E.D.: Pink Public Transport: A Necessary Evil? Senior Projects. Spring, (2018). https://digitalcommons.bard.edu/senproj_s2018/188. Accessed 2018
23. Rakodi, C.: Cities and people; towards a gender-aware urban planning process. Public Adm. Dev. **11**(6), 541–559 (1999)
24. Saul, H.: The badge designed to tackle sexual harassment on public transport (2017). https://inews.co.uk/news/badge-designed-tackle-sexual-harassment-public-transport/
25. Lucas, K.: Making the connections between transport disadvantage and the social exclusion of low income populations in the Tshwane Region of South Africa. J. Transp. Geogr. **19**(6), 1320–1334 (2011)
26. Lucas, K., Mattioli, G., Verlinghieri, E., Guzman, A.: Transport poverty and its adverse social consequences. Proc. Inst. Civil Eng. Transp. **169**(6), 353–365 (2016)
27. Uteng, T., Lucas, K.: Urban Mobilities in the Global South. Routledge, London (2018)
28. Song, L., Kirschen, M., Taylor, J.: Women on wheels: gender and cycling in Solo, Indonesia. Singap. J. Trop. Geogr. (2018). https://doi.org/10.1111/sjtg.12257
29. Hoodbhoy, P.: Women on motorbikes-what the problem?. The Express Tribune, 23 Feb (2013) Online. Available at https://tribune.com.pk/story/511107/women-on-motorbikes-whats-the-problem/
30. Woodcock, A.: The opportunity for design led transport futures. In: Design Research Society (DRS) International Conference 2012 "Re:Search" was held at Chulalongkorn University in Bangkok, Thailand, 1–4 July 2012 (2012a)
31. Woodcock, A.: Design for transport: a user-centred approach to vehicle design and travel. In: Tovey, M. (ed.) Ashgate, p. 21–69 (2012b)
32. Wilson, J.: Methods in the understanding of human factors. In: Wilson, J., Corlett, N. (eds.) Evaluation of Human Work (3rd edn.). Taylor and Francis (2005)
33. Benedyk, R., Woodcock, A., Harder, A.: The hexagon-spindle model for educational ergonomics. Work **32**(3), 237–248 (2008)
34. Woodcock, A., Woolner, A., Benedyk, R.: Applying the hexagon-spindle model for educational ergonomics to the design of school environments for children with autistic spectrum disorders. Work **32**(3), 249–260 (2008)
35. Woodcock, A.: Designing Mobility and Transport Services: Developing traveller experience tools. In: Tovey, M., Woodcock, A., Osmond, J. (eds.). (1st ed.) Routledge, Oxon, pp. 177–192 (2017)
36. Tovey, M., Woodcock, A., Osmond, J. (eds.): Designing Mobility and Transport Services: Developing Traveller Experience Tools, pp. 177–192. Routledge, Oxon (2017)
37. Creswell, J.W.: Qualitative Inquiry and Research Design: Choosing Among Five Approaches. Sage Publications, Thousand Oaks (2013). https://doi.org/10.1111/1467-9299.00177

Chapter 4
Unwanted Sexual Behaviour and Public Transport: The Imperative for Gender-Sensitive Co-design

Gill Matthewson and Nicole Kalms

Abstract This chapter examines how gender-sensitive policy and design is critical to the future of public transportation. Not only does this approach positively impact women's safety and their access to and use of public transport (and therefore public life), it makes significant contributions to the increasingly important sustainability and viability targets of public transport. The chapter considers the complex issue of safety and the feminist scholarship that has influenced the everyday challenges and risks faced by women internationally. By examining the literature and data around women's safety in public transport spaces, the research draws on case studies and policy actions in this arena. Advocating for a feminist framework, this chapter concludes by making a case for the critical and central role a gender lens and gender-sensitive codesign have in breaking the hold that technical concerns have over transport.

Keywords Transport planning · Gender lens · Co-design · Women's safety · Unwanted sexual behaviour

4.1 Introduction

This chapter argues that gender-sensitive policy and design is an essential requirement for any public transport planning, strategy and infrastructure. Before any policy is formulated, the perspectives of women[1] and gender-diverse people need to be engaged. Following this, and through every step of decision making and process,

[1] When we use the word Woman or Women, the authors recognise that they are not a homogenous group. Women represent enormous diversity in their cultural background, socioeconomic status, where they live, their sexuality, ability/disability, and age. The term is therefore inclusive of all women, including cis-women, trans-women, and intersex women.

G. Matthewson (✉) · N. Kalms
XYX Lab, Monash University, Melbourne, Australia
e-mail: gillian.matthewson@monash.edu

N. Kalms
e-mail: nicole.kalms@monash.edu

S. Coxon and R. Napper (eds.), *Advancing a Design Approach to Enriching Public Mobility*, Intelligent Systems Reference Library 198,
https://doi.org/10.1007/978-3-030-64722-3_4

a gender lens (making the too-often-invisible actions of gender visible) is critical. Every decision has design implications and every aspect of design has differing impacts for the many different groups affected by and using public transport. The deployment of a gender lens will positively impact women's experiences of public transport, and will maximise the sustainability, viability and equity of access to and use of public transport.

Public transport is critical to the full and productive functioning of a society. It not only enables efficient and sustainable economic connectivity, but the personal mobility it empowers supports the fundamental human right to social inclusion by removing barriers to education, training, leisure, employment, and health and social services [1, p. 29]. In short, mobility is "both an economic development *and* a human rights issue, a matter of individual access to public space *and* an equitable share of collective resources" [2, p. 50, italics added]. More than that, access to and ease of use of public transport is critical for a wide range of complex socio-cultural processes. As urban planner Carolyn Whitzman argues, cities are highly complex and dynamic places where citizens can also find, create, and recreate their identities [2, p. 50].

Currently, however, access to and use of public transport options is unevenly distributed across society, with gender a strongly differentiating factor. Crucially, for women to use public transport, it needs to be safe; moreover, it needs to be perceived as safe. As such, safety for women must be at the forefront of transport planning, its policy and design. But too often there is a substantial gap between women's safety experiences and the concerns of public transport planning. This paper argues that incorporating women's viewpoints via application of a gender lens can counter this gap and have benefits that exceed the direct benefits to women. The benefits are so well recognised that gender equity is one of the seventeen broad-based and interdependent Sustainable Development Goals (SDG) set by the United Nations in 2015 as targets for 2030. The goal clearly states that "Gender equality is not only a fundamental human right, but a necessary foundation for a peaceful, prosperous and sustainable world" [3]. Some of these connections between sustainability and gender equality will become apparent throughout the chapter.

4.2 Women and Public Transport

Travel and the use of different transport options (or lack of options) reflect social and economic activities and positions. These are strongly related to the temporal and spatial aspects of a society's social institutions, the division of labour and paid and unpaid work within each household, and an individual's access to resources [4, p. 193]. Presumed differences between men and women riddle every society and create unequal power differences that, among many other effects, generate differential transport use and experiences between the genders. While grouping a society on the basis of simple binary genders can be problematic, this kind of analysis is extremely useful for identifying structural issues that only surface when large groups are examined. As geographer Gill Valentine put it thirty years ago: "women's inhibited use and occupation of public space is… a spatial expression of patriarchy" [5, p. 389].

Multiple studies confirm again and again that gender impacts on public transport design, access, and usage, creating and compounding differences between men and women (as well as with gender-diverse people) in cities across the world. The spaces, systems, and networks of public transport carry implicit and explicit hierarchies of power [6, 7], meaning that the resultant patterns of movement through cities ("how people move where, how fast, how often" [8, p. 2]) are profoundly gendered. The differences are so marked that travel sociologist Randi Hjorthol suggests that daily mobility patterns might be one of the more astute indicators of gender equality in a society [4]. Lack of easy access to the services and potentials of the city means that women (as a group) are strongly disadvantaged in their everyday lives; without access, cities and city spaces materialise a form of gender inequity that impacts women's whole life access to service, education, health and identity.

Over forty years ago, Genevieve Giuliana and others noted differences in opportunities and alternatives between men and women using public transportation, and asked questions about the impact of the transportation system on women's choices and agency [9]. Since then, feminist scholars have been investigating and discussing how and why women's mobility and consequent access to health, social identities and economic opportunities has been and continues to be curtailed, detrimentally impacting their rights. The continuing difficulties that women and others experience gaining access to safe personal mobility is a direct result of their needs and requirements seldom being accounted for in public transport policy and goals, let alone the subsequent myriad of decisions and actions that directly generate transportation infrastructure design.

Thus, the first and most critical point to recognise is that public transport is far from 'public' [10, p. 34]. In a 2016 summary of the literature around gender and public transport, urban planner Anastasia Loukaitou-Sideris argues that there are cultural, physical, economic, and psychological barriers that curtail women's mobility and inhibit their use of public transport [11]. These barriers often overlap and at times reinforce each other, but are typically overlooked or marginalised in public transport policy and thinking, which tends instead to focus on the technical challenges of the sector.

Cultural barriers encompass the norms, practices and expectations of a particular society about how women 'should' behave. These might be religious (restrictions on women's access to the public realm), but in the Western world are more commonly social. Notably, because of the still persistent traditional household division of labour, women typically have primary responsibility for domestic activities, like care-giving and family maintenance activities, and these strongly impact on their interactions with public transport. For example, they are more likely to be travelling accompanied by people needing care, to execute multiple short trips to accomplish domestic tasks (trip-chaining), and to be carrying items larger than the "rolled-up newspaper" of a male 9–5 commuter [cited in 10, p. 34].

Physical barriers consist of zoned land-use urban planning and development often designed around private car use, or public transport that is geared towards 9–5 commuters—all of which generate a patchy overall public transport infrastructure adversely affecting women's access. Economic barriers include those where women more than men lack the means to use private transport options. These economic barriers in particular overlap with physical ones for those who live in less expensive outlying suburbs or in rural areas where there is often infrequent or erratic service.

The psychological barriers to public transport usage include the fear and experience of sexual harassment, victimisation, and assault involved in accessing and using public transport. These are the barriers that most strongly impact on perceptions of safety and are discussed in more detail in the next section.

4.3 Safety, Women and Public Space

The gender-based unwanted attention in public space that is sexual harassment towards women (and those who do not conform to binary genders) in public spaces is a product of the unequal power relationship between men and women and is ubiquitous. Those who have a lower status (women and 'others') are typically considered 'open persons', meaning that they are available for unsolicited interaction [12]. In this regard, "It is not *how* women behave in the public sphere that makes them vulnerable to street harassment; it is that they have chosen to enter the public sphere at all" [El Nahry cited in 13, p. 56]. Sexual harassment affects women as they travel from their homes and move about their city by whatever means.

There is, of course, a tension between actual and perceived risk of harassment, victimisation, and assault. Women typically consider attack is more likely at certain times, in certain places, and from strangers, even though sexual assault more commonly occurs in domestic spaces and by known men. However, childhood admonishments for young girls, and popular media lessons about 'stranger danger' encourage and reinforce their fear of public spaces, including those of public transport [1, p. 7] (see also Fig. 4.1). Women are, in effect, socialised to fear such spaces [14, p. 259; 15, p. 557]. In addition, women are told that they need to be careful [16, p. 381]; the underlying logic behind such messages is that women bear responsibility for their own safety.

Any kind of sexual harassment in public (including verbal 'compliments' and staring) serves as a reminder to women of their social and physical vulnerability (and inferiority [12, p. 86]) and has a role in increasing fear of more serious, violent attack [17, p. 300; 18, 19]. Women have no way of knowing whether a 'comment' will escalate into something more serious, and there is a tacit understanding that sexual harassment can be a precursor to harm [16, p. 380]. It does need to be noted, however, that different women will respond differently to sexual harassment incidents. Women are not a homogeneous group and there are clearly compounding factors impacting perceptions of safety and fear [11, p. 555; 20, p. 278]. Ethnicity and sexuality in particular tend to act as multiplying factors for the risk of sexual harassment escalating into physical violence. Age, disability, socio-economic background, education, and previous experience also have strong intertwining impacts.

Overall, sexual harassment in public increases general levels of anxiety for women and decreases perceived levels of safety in both busy and isolated public spaces [21, p. 557]. Or as Fatma El Nahry succinctly put it: “despite claims made by men, harassment is not a harmless, direct reaction by men to women but an institutionalised system of violence that functions to police women’s participation, freedom of movement and behaviour in public spaces” [cited in 13, p. 56].

This power of street harassment to constrain women’s mobility surfaced strongly in the crowd-mapping project *Free to Be* run in Melbourne (2016) and Delhi, Kampala, Lima, Madrid, and Sydney (2018) [22, 23]. Cutting across all the cities, cultures and continents, overwhelmingly the number one reason why the young women participating felt unsafe on the streets of their cities was sexual harassment and victimisation. In particular, they clearly identified unsafe hotspots on and around public transport.

Participants in the *Free to Be* survey were able to tag multiple activity/locations, and ‘On the street’ was the most frequent tag for the location of bad experiences. But there were also substantial overlaps between ‘On the street’ and ‘Public transport’ in all of the cities, reflecting the continuum of sexual harassment that is faced and managed by women and girls as they move about cities. In response to harassment incidents, many participants avoided the locale if they were on their own, modifying their travel itineraries and modes of transport; some never went back because of their fear of further victimisation and sexual harassment; and a small proportion stopped their travel to study or work altogether. What the *Free to Be* project clearly demonstrated was the high degree of limitations that young women and girls experienced in moving through their city.

Substantial research on women’s transport needs and experiences confirms the results of *Free to Be*; it also finds that gender-based violence and the fear of it are key barriers that can dramatically reduce women’s transport options [2, p. 35; 7, 24]. The active and sustainable transport options of walking, cycling, and using public transport, in particular, are hampered by fear of sexual harassment [2, 25, p. 92]. One participant in *Free to Be Sydney* bought a car after an incident at a bus interchange in order to avoid a repeat occurrence. The link between public transport use and gender equality have implications for sustainability, as recognised by the UN SDG #5.

4.4 Reasons for Public Transport as a Focus for Sexual Harassment

The spaces on and around public transport have been identified as key sites of sexual harassment [22, 23]. There are a number of reasons for this, which are summarised in Fig. 4.1 and detailed below (Fig. 4.1 also shows the wider societal factors that impinge on perceptions of vulnerability and safety previously discussed).

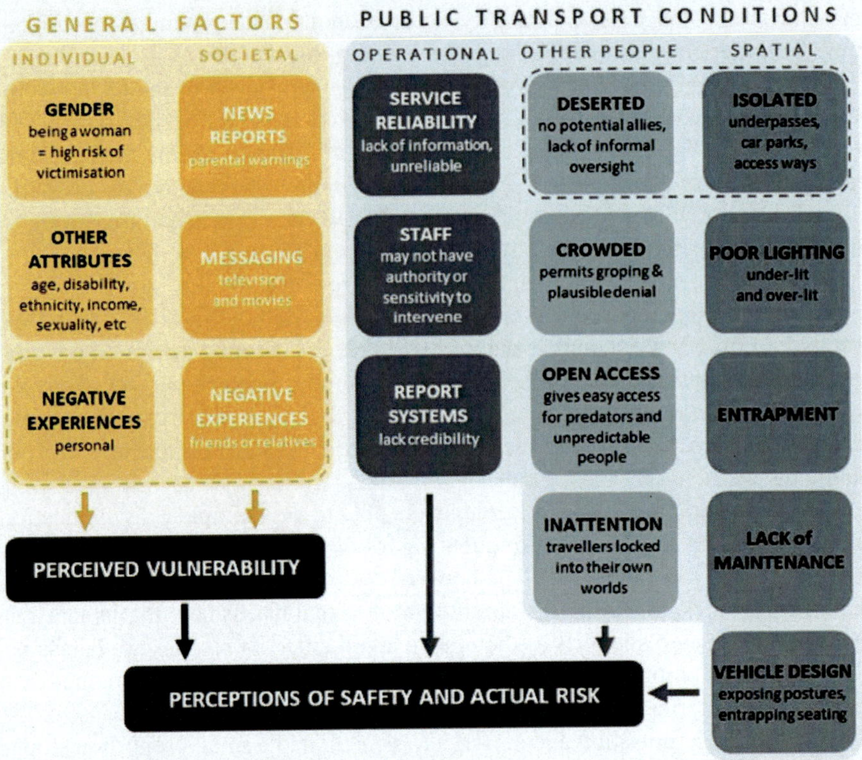

Fig. 4.1 Factors affecting incidence of sexual harassment and perceptions of safety. Diagram adapted and expanded by authors from [26, p. 2495]

The public transport specific reasons can be roughly grouped into three often reinforcing types: behaviours by other users that are part and parcel of the particular features of public transport, operational matters, and critically, for the theme of this book, infrastructure design and conditions that can facilitate harassment behaviours, or the fear of them.

Behaviours of other people

The spaces of public transport are characterised by peak-hour crowding, but off-peak they are often deserted, meaning that they can lack the informal oversight provided by the presence of other people, affecting perceptions of safety. This density of people can also change very quickly.

At peak times, the spaces of public transport concentrate many people into small areas which provide multiple targets or opportunities for predators. Crowded spaces facilitate groping and rubbing, but also give ‘plausible’ deniability for such contact [20, p. 279; 25, p. 4].

As well as predators, transport environs can attract/host beggars, alcohol and drug-affected people, and rowdy crowds because of their publicness and easy accessibility. Such people often exhibit unpredictable behaviour, and their presence increases perceptions of risk [15, p. 556].

The spaces of public transport are very often anonymous spaces—travellers are 'tuned out' and focused on their own journey (their music, podcast, reading, phone, daydreaming, et cetera), and therefore don't tend (or want) to notice what is happening to those around them. The possibility of not having to concentrate on the world around them (in contrast to driving a car) can be one of the attractive aspects of travelling by public transport, but it can also induce this inattention.

Operational factors

Services that are unreliable or infrequent mean often mean unknown waiting times, which is a situation that induces fears for safety [26, p. 2496]. Good information around services is a way to combat this insecurity.

The opportunism of perpetrators can be inadvertently enhanced by public transport staff behaviours. They often lack authority to intervene, and/or union or company rules to protect staff may advise or mandate against such intervention. Staff are not necessarily viewed as allies, but are seen as focused on fare evasion rather than travellers' safety.

Finally, reporting incidents is often difficult. In addition to the usual reluctance of women to report such incidents due to an expectation that it will make no difference [24, p. 4], reporting will generally cause delays to the journey. Because of the strong perception that they will not be believed and/or there would be little support, many women adopt an 'escape and forget' default position. Consequently, there are high rates of non-reporting of sexual harassment on public transport across the world [24, p. 4] Poor reporting infrastructure, such as no means of reporting in the moment, can aggravate the situation. A notable exception to this is Transport for London's 'Report it to stop it' campaign, using a simple text phone system for reporting that has significantly changed awareness of and attitudes towards sexual harassment on the London network [24].

Infrastructure and design spatial factors

The design of stations, stops and surrounding environs can negatively impact perceptions of safety. Access underpasses, tunnels, overbridges, and associated carparks are consistently identified by women as places of fear and trepidation, particularly at night when they might be poorly lit [15, p. 555]. Poor lighting appears regularly in audits of unsafe spaces more generally.

Public transport spaces can also lack informal oversight provided by the presence of other people, an important element for the sense of safety [27]. A sense of entrapment can also be experienced in underpasses and the like—participants in *Free to Be* detailed walking across busy roads rather than use the provided underpasses, especially at night.

When financial pressures lead to lack of maintenance of public transport infrastructure, the consequent proliferation of litter, run-down, empty, and desolate spaces, and graffiti are also associated with increased fear [1, 26, p. 303; 28, p. 2].

Finally, the design of public transport vehicles can force postures and positions that make women vulnerable—for example high hand holds which can be exposing (and uncomfortable) and seating arrangements that might feel like entrapment for women.

4.5 Design Responses to Sexual Harassment

Design responses to the conditions that support sexual harassment on public transport are neither simple nor straightforward. Crime Prevention Through Environmental Design (CPTED) is often invoked when discussing crime and personal safety in public spaces, and especially in complex spaces such as public transport hubs. In their careful review of the current status of CPTED, environmental criminologist Paul Cozens and CPTED specialist Terence Love argue that the connections between crime and design are far more complex than many of the policy makers and designers encouraging the use of CPTED would care to admit [28, p. 406]. They maintain that there are no simple answers, and a checklist of generic CPTED features is no match for the complexity of factors impinging on the occurrence of crime and perceptions of safety. More importantly, for this chapter, there is very little acknowledgement in CPTED literature of the fact that women experience both public space and the fear of crime differently [29, p. 285]. In Cozen and Love's extensive and comprehensive bibliography of over 180 CPTED references, just one acknowledges gender in the title.

Loukaitou-Sideris likewise maintains that there are no easy fixes or single responses to sexual harassment and victimisation in public transport, because such harassment is a phenomenon that is remarkably persistent and pervasive [11, p. 558] (also discussed in Sect. 4.3 of this chapter). For example, in the USA, one judge in a sexual harassment case argued against conviction, because in doing so "the court ran the risk of 'criminalizing generally accepted behaviour'" [cited in 16, p. 383]. Gender inequality—or the idea that men are superior to women and women's needs are less important—is deeply entrenched in society, and consequently seems like a universal truth underwriting 'generally accepted behaviour'. But it is a 'truth' that clouds the "interests it serves and perpetuates the myth that organizational and social arrangements are gender-neutral" [30, p. 115].

To effect any change, a broad-spectrum approach is needed to generate a multi-faceted and coordinated package of complementary initiatives [24, 26]. While no stand-alone single initiative is sufficient to tackle sexual harassment on and around public transport, design can make a difference. Loukaitou-Sideris argues that, because of the scale and nature of changes required, initiatives need to target structural issues and be tailored to particular socio-spatial contexts [11, p. 558]. She also maintains that any approach needs to have at least four prongs: research, design, transport policy, and technology.

The chapter will now discuss case studies that demonstrate the implementation of a gender lens: one to policy and the other to design.

4.6 Transport Planning and the Gender Lens—Policy

While women's mobility in cities is challenged by the physical, economic, cultural, and psychological constraints detailed previously, Loukaitou-Sideris also argues that inadequate transportation policies that too often neglect or disregard women's needs present a further major challenge [11, p. 547]. Increased knowledge accumulated over the last forty years around the gendered nature of mobility and access to public transport has not typically been incorporated into nuanced or effective policy and action responses [15, p. 558].

Transport planning and policy have long been dominated by men [14. p. 258; 25]. Consequently, there is a long history of transport issues being viewed through a purely utilitarian or rational lens, and thus as solvable by 'more and better' technological interventions [31]. Safety provisions for users mainly focus on tripping and falling hazards, and separate the general public from the dangerous machinery of public transport, rather than considering personal safety through a gendered lens. Transport providers tend to claim that they are 'gender-blind' and their measures benefit men and women equally. Given the evidence of the markedly different gender usages and experiences of public transport detailed for more than forty years, the 'blindness' more accurately refers to the refusal to see and acknowledge this difference.

From a planning point of view, compared to the complex technological, economic, and environmental challenges of transport infrastructure, gender appears irrelevant or, at best, inconsequential [31, p. 10]. Standard commuter trips with their high peak travel numbers place high demands and stress on a transport system, and so efficiency and expediency tend to dominate transport planning and thinking. The more erratic and harder-to-quantify needs of female public transport users appear to place less stress on the system and so do not tend to have an impact on planning, despite their complexity. The technical focus of much transport work and policy concentrates on the purely operational and transactional value of public transport, and gender, which tends to be bundled with other matters as 'social', is consistently marginalised from the mainstream focus of transport studies and planning [13, p. 44]. Commuter trips may have generally dominated transport, but the type of caring trips completed by women are also a significant proportion of all public transportation trips. Ines Sanchez de Madariaga calculated for Spain that caring trips accounted for 25% of all public transportation trips, not far behind commuter trips at 30% [32]. To marginalise that volume of trips is highly problematic.

To counter this strong historical emphasis, a gender lens or gender mainstreaming is an effective tactic. A key element of gender mainstreaming is to take seriously the different needs of users of space, and most importantly, not to allow one set of needs to dominate the others. There are a number of tactics that can encourage this equivalency. First, a full review of policies with a gender lens is an important process, although gender sensitive evaluation should *precede* any planning [33], and is a necessary step in setting the best conditions for gender-sensitive design. Evaluating existing policies through a gender lens tackles the assumptions in the sector, 'outing' practices that people did not know were harbouring assumptions.

This kind of determined gender lens can yield some surprising results. A good example and outcome of this process comes from Sweden, and is detailed by Caroline Criado Perez [34, p. 29–32]. Following a directive to evaluate all policies through a gendered lens, one local government official joked that at least snow clearing would not need scrutiny! But he was wrong. The priority had always been to clear the roads first, followed by footpaths and cycleways—a policy that advantaged drivers over footpath users, the former typically male and the latter typically female. An unanticipated but highly beneficial result of a reversal in the order of clearing was a significant drop in pedestrian accidents in icy conditions, saving the city and its residents both hospitalisation and loss of productivity costs. The gender lens in this case also acted as a 'human' lens, prioritising people over machines; a reverse of the norm in transport.

More typically, however, the uptake of evidence-based research and gender-sensitive approaches in public transport planning, policy, typologies, and contextual settings, has been reluctant [33]. Even in Sweden, which is one of the more advanced countries in terms of gender equality, an analysis of key transport policies by transport planner Lena Smidfelt Rosqvist found that the policies she examined did not tackle the gender differential in access and use beyond somewhat generic high-level statements [33]. Moreover, gender issues were trivialised [33, p. 80]. This is why every step of decision making, from the way a problem or issue is defined and the generation of design alternatives, through to the establishment of goals, targets, actual plans, and investments, requires a gendered perspective to ensure that the differences, preferences and experiences of women and men are included [33, p. 83].

There is recorded difficulty in translating high-level goals and ambitions into action [25, p. 96]. Sometimes this is due to a misalignment of administrative or management control systems—administrators and managers are unable to see how gender applies [33, p. 73], as the local government official in the snow-clearing case above demonstrated. And sometimes the default technical priorities of transport exert their strong pull. The operationalisation of goals and policies into shorter term objectives and targets is critical to counter this slippage.

More women at the table as planners, transport policy advisors, and designers, are critical to facilitate this level of operationalisation and gender lens scrutiny. Because of the technical bent to transport, it has traditionally been more difficult to attract women to some of these professions (although this is continually changing). In addition, as gender mainstreaming researcher Lena Levin and impact assessment expert Charlotta Faith-Ell point out, representation needs to exist at *every* level of decision making [25]. Too often the proportion of women drops when peer-selection methods are used to generate committees, as happens with regional and some private sector groups involved in transport [25, p. 95; also 33, p. 82].

Just as a shift in the pattern of snow clearing yielded some unexpected benefits, Smidfelt Rosqvist identifies ongoing research showing a correlation between women's representation in decision-making on municipality transport boards and the sustainability performance of local transport [33, p. 83]. Sustainability has begun to exert a strong influence on transport discussions, planning, and policy, because in order to transition towards a sustainable world and lower carbon emissions, effective public transport systems are vital. It is highly significant that research is indicating a correlation between equal gender representation and improved sustainability; this reinforces the inclusion of gender equality in the UN SDG. Women's mobility patterns tend to be more sustainable than men's. The expectation has been for women's patterns to eventually match those of men [33, p. 79], however, the reverse would be more in line with sustainability targets [33, p. 83].

In practice, there is often a perceived mismatch between sustainability and gender goals in transport [10, p. 37; 33, p. 73], mainly through the relegation of gender to the 'social' dimension, as discussed earlier. Like Smidfelt Rosqvist, urban planner Caren Levy strongly argues that the technical and sustainability challenges of transport will not be successfully solved without full and equal consideration of social matters [13, p. 45]. She identifies three critical issues that are (or should be) absolutely central to transport and transport planning: the different social positions and multiple identities of transport users (encompassing not just gender, but race, age, disability, et cetera); the social construction of space, public and private (who is 'allowed' where and when and why); and the politics of transport in the context of social relations (what institutions, groups and individuals have the decision-making power). Bringing these issues towards an equal standing in transportation is key, and women at the table is an essential tactic towards this end.

Nonetheless, more women at the table should not be viewed as a quick fix either. Due to their common education and socialisation, women transport professionals may not necessarily have a different view than their male peers [25, 35, p. 94]. Women professionals are not a homogenous mass who might think the same and act in concert to promote women's issues and interests. Instead, women are multiply fractured by class allegiances, professional socialisations and education, and personal perspectives derived from individual experiences of life. This means they do not necessarily have different views to those of their immediate male colleagues. Equal gender representation in policy and process deliberations is an important start (and necessary for a democratic process), but not sufficient.

To institute a clearer gendered lens on transportation therefore requires not just women professionals at the decision-making table, but also women from a range of backgrounds who would not be considered 'experts' in anything but their experiences. Community consultation can contribute this perspective, but too often it is co-opted as a token exercise, necessary for ticking the box, but not always having any significant ameliorating effect. Co-design is a more effective means to draw such people into the process of generating gender-sensitive design.

4.7 Transport and Co-design

Co-design involves placing user and participant engagement and testing continuously throughout a process that stretches from planning to implementation. It is not a matter of seeking 'input' at defined stages, but a methodology for the inclusion of users and multiple parties in the full design process. It brings together unconventional mixes of people with wide diversity in background, employment, ages, ethnicities, and genders, to take seriously the different needs of users of space and to prevent one set of needs dominating. Co-design is an equalising mechanism that disrupts traditional design hierarchies.

An example of the power of co-design comes from a follow up to the *Free to Be* project mentioned earlier in the chapter. Crowdsourcing is a form of co-design because it effectively harvests information from people whose voices are seldom heard. Learning of the project, the Melbourne office of the international engineering firm Arup Lighting were interested in whether there might be differences in the lighting between the perceived safe and unsafe places identified by young women and girls in Melbourne *Free to Be*. Increasing lighting is a standard response to places that are identified as 'unsafe' [28]. Yet brighter lighting may not be the answer. Some previous research has shown that women can feel highly exposed under floodlit conditions [6, p. 243].

Arup analysed the spaces on a number of key parameters, and concluded that there were distinct lighting differences between the safe and unsafe places of Melbourne [36]. Importantly, they confirmed that more light (high illuminance) did not necessarily make participants feel safer, but the *quality* of the light most definitely did [37, p. 32]. High levels of colour rendering, lower contrast ratios, and the layering of light relative to the materials it falls upon, strongly affect quality. An intriguing finding was that the safe areas all had a relatively narrow range of colour temperature in their lighting and it was within the warm range [p. 20].

Arup concluded that the quantitative evidence provided by the *Free to Be* data set challenges current assumptions and previous research about the relationship between high levels of light, cool colour temperatures and uniformity, with the perception of "safety". It is these assumptions that currently control the Australian lighting standards for pedestrians (essentially only specifying Lux/illuminance levels, similar to standards across the world). Such standards appear to be woefully insufficient both for safety and in the context of more sophisticated lighting options available in the twenty-first century. The knee-jerk reaction of simply adding more light to a place to 'improve' safety is wasteful of energy, contributes to light pollution, and, more importantly, is effectively counter-productive for safety. Once again, paying attention to women's experiences generates a more sustainable outcome.

Because the results counter some previous research, Arup are currently undertaking further investigation with Monash University researchers. The expansion by Arup of the *Free to Be* project was premised on the significance of the data set in collecting the voices and experiences of people who are seldom heard. The results strongly demonstrate the advantage and importance of going beyond the usual suspects when determining policy, standards and design.

4.8 Conclusion

This chapter has delved into the considerable research that has for decades detailed women's differential and often circumscribed access to and use of public transport. This affects their ability to be involved in public and social life, not to mention partake in economic and education opportunities. There are multiple barriers to women's mobility that range from cultural through to economic, but one of the most significant ones is unwanted sexual behaviour which causes women to feel unsafe. The spaces of public transport through their operation, design, and the behaviours of other people they can encourage, render them places where women fear harassment behaviour. This fear is a constant for women across the world.

The chapter then examines how it is possible to positively impact women's access to public transport through the consistent implementation of a gender lens. Every step of the process from evaluations prior to planning, through planning and the setting of policy needs a gender lens to keep the focus on the multiple and complex users of public transport, rather than simply those commuting for work. These then set the parameters for the gender-sensitive design of public transport infrastructure, which can influence perceptions and the experience of safety in public transport spaces.

Public transport design is often approached from a utilitarian and pragmatic viewpoint, but the chapter argues that widening that viewpoint to implement a gender lens on all aspects of public transport planning, policy and design has multiple beneficial outcomes beyond improvements for women's mobility. The chapter details some case studies that demonstrate how changes at a policy level can have both economic benefits for society and substantially move positively towards important sustainability goals. The more nuanced responses to safety for women in public transport provided by a gender lens and gender-sensitive design therefore becomes essential for the future viability and sustainability of public transport.

References

1. Currie, G., Stanley, J., Stanley, J. (eds.): No Way to Go—Transport and Social Disadvantage in Australian Communities. Monash University ePress, Clayton, VIC (2007)
2. Whitzman, C.: Women's safety and everyday mobility. In: Whitzman, C., Legacy, C., Andrew, C., Klodawsky, F., Viswanath, K. (eds.) Building Inclusive Cities: Women's Safety and the Right to the City, pp. 35–52. Earthscan, London (2013)
3. United Nations: Transforming Our World: The 2030 Agenda for Sustainable Development, Goal 5 (2015). https://www.un.org/sustainabledevelopment/gender-equality/
4. Hjorthol, R.: Daily mobility of men and women—a barometer of gender equality. In: Uteng, T.P., Cresswell, T. (eds.) Gendered Mobilities. Ashgate, Aldershot (2008)

5. Valentine, G.: The geography of women's fear. Area **21**(4), 385–390 (1989)
6. Loukaitou-Sideris, A.: Fear and safety in transit environments from the women's perspective. Secur. J. **27**(2), 242–256 (2014)
7. Dunckle-Graglia, A.: Women-only transportation: how "Pink" public transportation changes public perception of women's mobility. J. Public Transp. **16**(2), 85–105 (2013)
8. Uteng, T.P., Cresswell, T. (eds.): Gendered Mobilities. Ashgate, Aldershot (2008)
9. Guiliano, G.: Public transportation and the travel needs of women. Traffic Q. **33**(4), 607–616 (1979)
10. Greed, C.: Are we still not there yet? Moving further along the gender highway. In: Scholten, C.L., Joelsson, T. (eds.) Integrating Gender Into Transport Planning: From One to Many Tracks, pp. 25–42. Palgrave MacMillan, London (2019)
11. Loutaikou-Sideris, A.: A gendered view of mobility and transport: next steps and future directions. Town Plann. Rev. **87**(5), 548–552 (2016)
12. Bartomski, S., Smith, P.: Gender, fear, and public places: how negative encounters with strangers harm women. Sex Roles **76**, 73–88 (2017)
13. Levy, C.: Travel choice reframed: "deep distribution" and gender in urban transport. In: Scholten, C.L., Joelsson, T. (eds.) Integrating Gender Into Transport Planning: From One to Many Tracks, pp. 43–65. Palgrave MacMillan, London (2019)
14. Sheller, M.: Gendered mobilities: epilogue. In: Uteng, T.P., Cresswell, T. (eds.) Gendered Mobilities. Ashgate, Aldershot (2008)
15. Loukaitou-Sideris, A., Fink, C.: Addressing women's fear of victimization in transportation settings: a survey of US transit agencies. Urban Aff. Rev. **44**(4), 554–587 (2009)
16. Shah, S.B.: Open season: street harassment as true threats. Univ. PA J. Law Soc. Change **18**(5), 377–401 (2016)
17. Pain, R.: Whither women's fear? Perceptions of sexual violence in public and private space. Int. Rev. Victimology **4**(4), 300 (1997)
18. Stanko, E.: Everyday Violence: Women's and Men's Experience of Personal Danger. Pandora, London (1990)
19. Morrell, H.: Women's safety. In: Booth, C., Darke, J., Yeandle, S. (eds.) Changing Places: Women's Lives in the City. Sage, London (1996)
20. Ceccato, V.: Women's transit safety: making connections and defining future directions in research and practice. Crime Prev. Commun. Safety **19**(3–4), 276–287 (2017)
21. Davidson, M.M., Butchko, M.S., Robbins, K., Sherd, L.W., Gervais, S.J.: The mediating role of perceived safety on street harassment and anxiety. Psychol. Violence **6**(4), 553–561 (2016)
22. Plan International: Unsafe in the City: The Everyday Experiences of Girls and Young Women. Plan International, Woking (2018). https://plan-international.org/unsafe-city
23. Plan International: Free to Be: Mapping Girls' Safety in Cities. Plan International, Woking (2018). https://plan-international.org/publications/free-to-be-country-reports#download-options
24. Gekoski, A., Gray, J.M., Horvath, M.A.H., Adler, J.R., Edwards, S.: The prevalence and nature of sexual harassment and assault on public transport: an international review. J. Criminological Res. Policy Pract. **3**(1), 3–16 (2017)
25. Levin, L., Faith-Ell, C.: How to apply gender equality goals in transport and infrastructure. In: Scholten, C.L., Joelsson, T. (eds.) Integrating Gender Into Transport Planning: From One to Many Tracks, pp. 89–118. Palgrave MacMillan, London (2019)
26. Yavuz, N., Welch, E.W.: Addressing fear of crime in public space: gender differences in reaction to safety measures in train transit. Urban Stud. **47**(12), 2491–2515 (2010)
27. Cozens, P., Love, T.: A review and current status of Crime Prevention Through Environmental Design (CPTED). J. Plann. Lit. **30**(4), 393–412 (2015)
28. Lorenc, T., Petticrew, M., Whitehead, M., Neary, D., Clayton, S., Wright, K., Thomson, H., Cummins, S., Sowden, A., Renton, A.: Environmental interventions to reduce fear of crime: systematic review of effectiveness. Syst. Rev. **2**(30) (2013). http://www.systematicreviewsjournal.com/content/2/1/30

29. Walklate, S.: Reflections on community safety: the ongoing precarity of women's lives. Crime Prev. Commun. Saf. **20**(4), 284–295 (2018)
30. Ely, R.J., Meyerson, D.E.: Theories of gender in organizations: a new approach to organizational analysis and change. Res. Organ. Behav. **22**, 103–151 (2000)
31. Scholten, C.L., Joelsson, T. (eds.): Integrating Gender Into Transport Planning: From One to Many Tracks. Palgrave MacMillan, London (2019)
32. de Madariaga, S.: Public transportation: rethinking concepts and theories. Gendered Innovations Sci. Health Med. Eng. Environ (2012). http://genderedinnovations.stanford.edu/case-studies/transportation.html#tabs-2
33. Smidfelt Rosqvist, L.: Gendered perspectives on Swedish transport policy-making: an issue for gendered sustainability too. In: Scholten, C.L., Joelsson, T. (eds.) Integrating Gender Into Transport Planning: From One to Many Tracks, pp. 69–87. Palgrave MacMillan, London (2019)
34. Criado Perez, C.: Invisible Women: Exposing Data Bias in a World Designed for Men. Chatto & Windus, London (2019)
35. Greed, C.: Women in the construction professions: achieving critical mass. Gender Work Organ. **7**(3), 181–196 (2000)
36. Kalms, N.: More lighting alone does not create safer cities. Look at what research with young women tells us. The Conversation, May 28 (2019). https://theconversation.com/more-lighting-alone-does-not-create-safer-cities-look-at-what-research-with-young-women-tells-us-113359
37. Arup Lighting: Perceptions of Safety Lighting Study. Arup Melbourne (2019)

Chapter 5
Speeds, Deeds and Needs on Two Wheels (or Three)

Michael S. Wexler

Abstract The bicycle is back. For the last decade, cities across the globe have been turning to the bicycle to add to the array of transportation options for citizens. In an attempt at developing infrastructure that helps to legitimize the bicycle as a transport form in our urban centres, designers require a robust framework within which they can design appropriate facilities for bicycle users. Practitioners need to deeply understand the transport sphere that is occupied by the simple push bicycle in order to design effective solutions for its users. A design framework that has been adopted in a growing number of cities around the world today focuses on "AAA" infrastructure, or bicycle facilities for "All Ages and Abilities". While this aims at an inclusionary design approach for bicycle infrastructure, it is not often broken down into its component parts and investigated from a tangible and ground-level perspective. Who fits into this "All Ages and Abilities" model? How fast do they move? What kind of contraption are they riding? Where are they going and for what purpose? What do they need in order to feel comfortable enough to ride in the space? This deeper framework can help designers align form and function when developing new bicycle facilities for this AAA model. Whether the design user is riding in the peloton at 35 km/h or riding to the corner store at 15 km/h, this makes a significant impact on the design choices of the texture treatments, signal phasing and bends along the bicycle corridor. Whether the user is transporting themselves on a carbon fibre bullet or their two-year-old in a Danish Bullitt, this helps determine appropriate levels of protection of the facility from vehicle exposure. Whether the user has been riding since they were a toddler or just learned to ride in a straightline last month at the age of 30, this informs the design of a bicycle facility width for both to use the space efficiently and comfortably. It is through the examination of these speeds, deeds and needs that the urban designer can truly create spaces that are attractive, accessible and understandable to bicycle users of all ages and abilities.

Keywords Bicycle · Urban design · All ages and abilities · Speed · Observation · Behaviour

---

M. S. Wexler (✉)
Copenhagenize Design Co., 7236 Waverly Street #204, Montréal, Québec H2R0C2, Canada
e-mail: msw@copenhagenize.eu

S. Coxon and R. Napper (eds.), *Advancing a Design Approach to Enriching Public Mobility*, Intelligent Systems Reference Library 198,
https://doi.org/10.1007/978-3-030-64722-3_5

5.1 Introduction

Sitting at my local café in the Rosemont borough of Montréal on an afternoon like any other, can itself be a Master Class in urban design. Through two large double-paned window frames, passers-by send-off behavioural cues in plain sight that teach us important lessons about the built environment—whether they pass on foot or roll by on two, three or four wheels. We can observe the business man who chooses to stand and wait for a colleague under a tree, shaded from the bright sun, or notice the two students who whizz by on their bicycles, riding side by side and passing a pack of cigarettes from one to the other, while a delivery van behind them echoes a sharp honk in their direction. All of these urban citizens are simply going about their business, unaware they are offering us rich data with which we can better understand how to design the environments around ourselves. This practice of observation is key to the effective execution of design in our most public of spaces: our streets, parks, squares and outdoors that bind together our cities. As the foundation of a design tradition present in the urban planning world for a number of decades, observation has been used as a tool to better understand the intersection between our behaviour and our environments by some of the most prominent thinkers and writers, from the Danish Jan Gehl to the American William H. Whyte. As the mother of modern urban planning, Jane Jacobs wrote in a 1958 Fortune article, "there is no logic that can be superimposed on the city; people make it, and it is to them, not buildings, that we must fit our plans" [1] This idea of looking towards the people in our cities for the logic of how things should be planned and designed is integral to developing a public realm that works for the people that populate it and pass through it every day.

5.2 Observation

As urban designers who focus on the bicycle in city spaces, our firm has the privilege of working globally, and with that—observing public life and transportation in cities on a number of continents. Whether spotting an old man transporting plastic bottles on the handlebars of his bicycle from a sidewalk Shashlik grill in Shymkent, Kazakhstan or noticing the stream of suited commuters heading home with the bright blue bike share from a corner Taco bar in Long Beach, California—there is a wealth of information that can be absorbed by observing one's urban surroundings. An essential take-away from this amalgamation of global transport observations is in fact quite simple: there is an immense diversity in the people who ride bicycles in our cities. There is no archetype "cyclist". In fact, the word itself has become so skewed to describe one type of transport profile—most often the lycra-clad, sporty, confident, middle-aged man of a higher income who rides long distances and aims at reaching higher speeds. Everyone has had an experience with this particular type of cyclist, and the word has taken on almost a singular image that is often misaligned with the bicycle planning conversations we embark on in our work. In turn, this singularity

Fig. 5.1 Bicycle lanes only for the brave (*photo* Copenhagenize)

and misalignment leads to bicycle planning and urban design around the world that is ineffective at incentivizing a wide demographic of potential cyclists to use their bicycles at all. Due to this misalignment, we see painted bicycle lanes that are positioned beside and criss-crossing through three or four lanes of high speed vehicular traffic, or markings that ask the driver and the cyclist to just "share the road"—which only the brave few would venture to do on a bicycle. We have essentially, in many parts of the world, designed systems of bicycle facilities (using the word system loosely) that take to heart only the courageous, the bold and the overly confident cyclists. In turn, only a few brave cyclists venture onto many of our urban roads, and there remains the self-fulfilling prophecy and assumption that the only people who ride bicycles are fast, confident and do so for sport (Fig. 5.1).

5.3 Studying Real People

To remedy the singularity of this notion and broaden the way we think about "cyclists"—it might be helpful to describe a few typical bicycle users we might see on any given day through the window of the local café. First off, a common sight in many cities around the world is the student; let's call her Wendy. Wendy started riding her bicycle for transport one year ago when she commenced her bachelor's degree at university, as it is essentially free transport, though she isn't always comfortable on every street—especially those with heavy traffic. Her bicycle is a weighty, old mountain bike and has a big basket in the front, often overflowing with backpack and books. Secondly, we have Eve—a government employee by day, who has for many years now, been a long-distance sports cyclist in her free time. She rides her bike to work most days, and although is not decked out in all of her cycling gear on the way to the office, she keeps a fast and confident pace on her light, quick bike. Thirdly there is Xavier, who takes his daughter to school every morning in a wide trailer that clips into the rear of his bicycle. He rides much slower than Wendy or Eve as he has his daughter with him and tends to stick to streets where he feels it is safe enough for her. Lastly, we have Leonard, who is retired and rides his tricycle to the supermarket in the morning, fills up the rear basket with supplies for the day and continues to do his errands slowly through the neighbourhood. It is with this diversity of bicycle users in mind—the fast, the slow, the wide and the low—that we can begin to understand, as practitioners, how to design and plan a city that is friendly, safe and enjoyable in which to ride a bicycle.

5.4 All Ages and Abilities? Try Speeds, Deeds and Needs

In North America—a continent known for its car-centric infrastructure and decades of auto-oriented planning—there exists a new terminology for bicycle facility design that is finding root in a number of forward-thinking municipalities. This terminology or framework is often simply described as AAA—or more broadly, an All Ages and Abilities planning framework for bicycle facilities. The City of Vancouver in Canada has in the last number of years developed specific guidelines for all new bicycle facilities built in the city to follow this approach and in the process, mapped out for local practitioners what is and isn't considered worthy of a AAA standard design [2]. In the search to better understand how this approach translates into the lived experience on a bicycle—knowing that this is a relatively new planning and design framework for North America—we can look to world leading cities like Copenhagen for enlightenment. For context, Copenhagen is a city where decades of inclusive and observation-based bicycle facility design has created a system where 62% of Copenhageners use bicycles to get to work or school as their main mode of transportation [3]. The countless design lessons from Copenhagen have inspired new ways of thinking for urban cycling and planning in cities around the world. The systemic planning change in a context like Vancouver is aimed at encouraging people who might not cycle regularly or at all to consider riding a bicycle more often, by making the infrastructure inclusive to the diversity of existing and potential bicycle users—tapping into the latent demand for better cycling conditions. This chapter aims to look more deeply into the AAA framework in order to understand how urban designers can use the tools of observation to build better facilities that increase utility for as many bicycle users as possible. By thinking about the speeds, deeds and needs of Wendy, Eve, Xavier and Leonard and the countless other types of two-or-three wheeled riders when designing bicycle facilities, our cities can become places that truly encourage people of all ages and abilities to ride a bicycle.

5.5 Speeds

Let's begin this design assessment by looking at the concept of speed. If we were to only think about Eve—the quick, confident cyclist—we might assume that all bicycles in the city tend to whip by at a cool 28 km/h and design lanes that allow this type of average speed profile. In fact, when the City of Copenhagen studied speeds of cyclists on the busiest street Nørrebrogade, they found the average speed to fall under 16 km/h [4]—quite a bit slower than Eve's quick pace. We might say that perhaps Wendy and Xavier would both fall within one or two km/h of this 16 km/h average while Leonard might be taking his time comfortably closer to 10 km/h. This diversity of speeds should offer the design practitioner a cue to create bicycle lanes that account for not only a slower average, but also a wider spectrum of observable speeds. With that in mind, we can state with some confidence that: if the goal is to build facilities for all abilities and all speeds, then the bicycle lanes in question must be wide enough to accommodate both fast and slow riders and allow for peaceful cohabitation among cyclists. If designed too narrow, an eager Eve placed behind a leisurely Leonard might create frustration and friction between these users and potentially even a dangerous attempt of overtaking. It is for this reason that the standard width of a one-way bicycle lane is 2.0–2.2 m in both Copenhagen and through the AAA framework lens in a city like Vancouver. That is just enough space for the widest of push-powered vehicles like tricycles, cargo-bikes and trailers (usually measuring a metre from edge to edge of the wheelbase) to pass one another—showcased by the example of Xavier and his daughter overtaking Leonard safely in the bicycle lane. Taking this even further, the City of Copenhagen is trying since 2011 to make the network even more comfortable and encouraging to ride, by building out the PLUSnet—a portion of high-traffic bicycle lanes that get an even wider footprint, upward of 3.5 m in one direction [5] (Fig. 5.2).

If that wasn't enough, the City of Copenhagen has made riding a bicycle even easier for rush hour commuters by using speed as an important factor in their planning decisions. Along congested arterial corridors in the city, local traffic engineers tested a "green wave"—or synchronization of traffic lights, and adjusted them slightly higher than the average speed of the typical Copenhagener on a bicycle, 20 km/h. With this small technological change, bicycle traffic became more fluid than it had been previously on that corridor and the traveled route became over 2 min faster by bicycle, due to the lack of red lights encountered by bicycles at that speed [4]. A small improvement like this, rooted in the observable reality of the average bicycling speed, can translate into real gains for cycling, and more so as an incentive for commuters to get out of their cars and onto a bicycle.

Fig. 5.2 Wide one-way cycle tracks in Copenhagen allow for all types of users protected from motor vehicles by a simple concrete curb (*photo* Copenhagenize)

5.6 Deeds

If we dive a bit deeper and look beyond the observable speed of bicycles in our cities, but towards the deeds of the people who might ride bicycles, perhaps we can design better spaces for them to go about their business on two (or three) wheels. Funny enough, looking towards automobile design can offer some foundations upon which to design better bicycle facilities, and help shed some light on the transport preferences and assumptions we make as humans and designers. One thing is undeniable—cars are pervasive in modern twenty-first century society, and they have set a many number of assumptions and habits that we take to be true in our cities. One of these truths can be found when looking at the shape of your typical automobile—whether an electric, hybrid or gas-guzzler: cars have passenger seats. The modern sedan, and even the smaller SMART car, all have passenger seats. Mass produced vehicles today do not take inspiration from fighter jets, but rather are designed to promote social interaction. Some people take their friends to the mall, others drive around with a love interest aimlessly, just to be together, side-by-side.

What if we took a cue from this automobile design element and went so far to say that perhaps transport is enjoyed as a social activity, and then went one step further and said that perhaps your average citizen might enjoy biking a bit more in the city if they could be social while transporting themselves on two wheels. Well, this is exactly what the City of Copenhagen has been promoting in the development of their PLUSnet cycle lanes throughout the urban core. Copenhagen has codified the concept of social interaction while cycling into the now frequently used term of "Conversation Lanes"; new cycling facilities in the City must be wide enough for two bicycle riders to have a conversation on their way back from work or school, while a faster rider comfortably overtakes them. This design precedent is wholly rooted in the social cues and observable behaviour of urban citizens—a design inspired by the deeds and functions of social interaction in transport, that in turn, allow people to use infrastructure in a safe, enjoyable and social way.

Now if we wanted to afford bicycle users other amenities and comforts that allowed people to function better and achieve their daily goals while in transport, we might consider looking to the design of the bicycle itself. Look at a racing frame, for example: its function is for cyclists to move quickly, minimize weight and maximize aerodynamics. The angled body position, scooped racing handlebars and raised seat all promote speed. However, if the goal is to design a bicycle for urban transport, different daily functions are required that run counter to using a bicycle built for racing. First off, an upright body position with raised handlebars is much more conducive to observing the transport, shopping and social cues around you in a city on a typical commute. A sturdy and heavy bicycle frame that can hold panniers, baskets and other goods offers great utility for urban activities like shopping. Additionally, the design of the back-pedal or coaster brake is essential for allowing one hand to be free while riding in order to adjust one's scarf in the winter, rearrange a basket of groceries or wave at a friend, all while maintaining safe braking potential with two feet firmly on the pedals. It is no surprise why one of the world's most famous bike frames is colloquially called the "Dutchie", as this iconic bicycle design (seen by the tens of thousands in Dutch cities) has a straight-back upright posture, sturdy hold, coaster brakes, and a step through frame—that does not require you to throw your leg up over the frame—and is essentially the perfect bicycle design for urban contexts.

Having examined design thinking for bicycle facilities based on the observed speeds and deeds of bicycle users, we are left with images of comfortably wide lanes, populated by upright bicycles with coaster brakes and synchronized traffic lights. However, when surveyed, many urban citizens would often still choose not to ride a bicycle with those conditions alone. This is largely due to the perception of safety that we experience in different road contexts when we are free of the protective shell of an automobile. In 2006 the Portland Bureau of Transportation in the US conducted a large survey of local residents, of both cyclists and non-cyclists, and aimed at understanding who might consider riding a bicycle under certain circumstances. This study was repeated in a national US survey in each of the 50 biggest metro areas and similar results came forward: more than half of the urban population was deemed "Interested but Concerned"—whose concerns stemmed primarily from the perceived safety of roadways [6, 7]. When we look at many of the painted bicycle lanes in our cities today, most of the potentially interested cyclists would find their perception of safety to be quite poor due to 2 major risk factors: vehicle speed and vehicle volume. The more cars we cycle next to and the faster they are going, the more likely we are to negatively perceive the safety of that roadway.

5.7 Needs

As urban designers, it is incumbent to look at a roadway and determine what a bicycle user might need to feel safe and be safe enough to cycle along that corridor. Enter the protected bicycle lane, or cycle track. All roadways in the Danish bicycle planning methodology require designers to examine what vehicle volumes and speeds exist along that corridor, and from that, they can choose an appropriate bicycle lane design for the needs and perceptions of the user. The higher the speed and the volume of automobiles, the higher the necessary level of protection for the vulnerable bicycle user. This methodology works along a scale, so that along the motorways that connect the suburbs of Copenhagen to the urban core, fully segregated cycle tracks run parallel, nestled safely through a wooded area, some 50 m off to the side of the heavy traffic. It is this need-sensitive design approach that allows Copenhageners of all ages and abilities to ride from one corner of their city to the other, feeling at all times a high level of security on their bicycles, with opportunities for social interaction and equal transport opportunities for a diversity of bicycle speeds, widths and shapes.

5.8 Conclusion

It is only through looking at the speeds, deeds, and needs of bicycle users that design practitioners can understand how to truly create facilities that work for the large majority of potential bicycle users. This analysis shows us that the human experience and a user-based approach can shed light on the fact that in order for most average people to even consider riding bicycles in our cities, we need to aim for facilities that offer protected and separated lanes away from speeding vehicles, a lane width that offers a positive perception of safety within the facility, and design that aims to welcome a diversity of bicycle speeds to coexist and transport themselves socially through their city.

References

1. Jacobs, J.: Downtown is for People. (Reissue) viewed 12 May 2020. https://fortune.com/2011/09/18/downtown-is-for-people-fortune-classic-1958/ (1958)
2. City of Vancouver: Transportation Design Guide: All Ages and Abilities Cycling Routes. Viewed 12 May 2020. https://vancouver.ca/files/cov/design-guidelines-for-all-ages-and-abilities-cycling-routes.pdf (2017)
3. Cycling Embassy of Denmark: New bicycle Account: Bicycles Outnumber Cars in Central Copenhagen. Viewed 12 May 2020. https://www.cycling-embassy.dk/2017/12/05/new-bicycle-account-bicycles-outnumber-cars-central-copenhagen/ (2017)
4. European Cyclists' Federation: Green Wave for Cyclists. Viewed 12 May 2020. https://ecf.com/news-and-events/news/green-wave-cyclists (2018)

5. City of Copenhagen: Good, Better Best: The City of Copenhagen's Bicycle Strategy 2011–2025. Viewed 12 May 2020. https://www.eltis.org/sites/default/files/case-studies/documents/copenhagens_cycling_strategy.pdf (2011)
6. Geller, R.: Four Types of Cyclists. Viewed 12 May 2020. https://www.portlandoregon.gov/transportation/article/264746 (2006)
7. Dill, J., McNeil, N.: Revisiting the four types of cyclists: findings from a national survey. Transp. Res. Rec. **2587**(1), 90–99 (2016)

Chapter 6
Safety, Design and Law: A New Interdisciplinary Approach to Bicycle Rider Road Safety

Robbie Napper, Marilyn Johnson, and Vanessa Johnston

Abstract This chapter demonstrates a novel problem-solving approach to study the interaction at an intersection between a driver (turning left) and a cyclist (continuing straight ahead). According to the road rules in the Australian state of Victoria [RR141(2)], the bicycle rider must not ride on the left of a vehicle that is indicating and turning left at an intersection. However, road rules and the built environment were not designed in harmony and there is a high level of confusion about this road rule in the community. Further to these community concerns are professional ones. Previously, road safety and engineering pointed to the law (road rules), as being the source of collisions and injuries from this negotiation. From the position of the law, problems arose concerning how the road rules were applied in a road safety and engineering context. The road space and the road rules are two manifestations of controls and guidance. Safety science has identified that legal considerations are made externally to road engineering, if at all (Nicholls et al. in Cyclists and left turning drivers: a study of infrastructure and behaviour at intersections. Auckland, New Zealand, 2017 [1]). In order to develop solutions to the problem of left turn confusion and conflict, the research team hypothesised that a design approach (Lawson in How designers think—the design process demystified. Architectural Press, Oxford, 2006 [2]) could bring these fields together. The overarching method is the implementation of convivial design techniques [for example see Illich (Illich in 197 Tools for conviviality. Harper and Row, New York, (1973) [3]), Sanders and Stappers (Sanders and Stappers in Convivial design toolbox: generative research for the front end of design. BIS Publishers, Amsterdam, 2016 [4])] to bring together

R. Napper (✉)
Faculty of Art Design and Architecture, Monash University, Caulfield East, Victoria 3145, Australia
e-mail: robbie.napper@monash.edu

M. Johnson
Faculty of Engineering, Monash University, Clayton, VIC 3800, Australia

Amy Gillett Foundation, 80a Turner St, Port Melbourne, VIC 3207, Australia

V. Johnston
Graduate School of Business and Law, RMIT University, 124 La Trobe St, Melbourne, VIC 3004, Australia

S. Coxon and R. Napper (eds.), *Advancing a Design Approach to Enriching Public Mobility*, Intelligent Systems Reference Library 198,
https://doi.org/10.1007/978-3-030-64722-3_6

safety science, legal investigation of road rules, and design. Three specific design techniques were used. First, two-dimensional drawing was tested by the research team. Next, two-dimensional drawings and photographs of key intersection types were combined with three-dimensional scale modelling using LEGO® elements in facilitated discussions with road safety practitioners and experts. The third technique brought road rule considerations to the foreground by encouraging road rule annotation on two-dimensional drawings.

Keywords Design · Safety science · Law · Road rules · Drawing · Modelling · Interdisciplinary

6.1 Introduction—Bicycle Rider Safety at Left Turns

This work investigates the safety of people riding bicycles in a road space shared with motor vehicles. In particular, in an Australian setting at intersections where drivers intend to turn left and the cyclist is continuing to ride straight ahead as illustrated in Fig. 6.1. The nature of the problem has been defined in safety terms, which themselves relate to the road space design, the legal framework governing road user actions, and the way road users interact. This chapter is written in the context of the Australian state of Victoria. We note that this problem is not confined to Australia or left-hand driving contexts.

Throughout the road network, unsafe interactions occur simultaneously at the micro and macro levels. At the micro level, the specific, localised road space (e.g. road design, speed, roadsides etc.) makes it is easy to focus on the behaviour of individual road users to understand crash events. However, it is important to also consider the

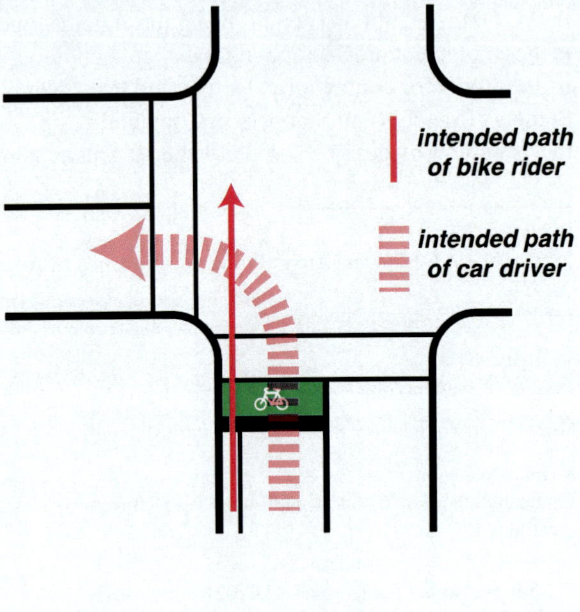

Fig. 6.1 Illustration of traffic situation whereby the bicycle rider wishes to travel straight ahead, and the car driver wishes to turn left

macro level in which an individual crash site exists. The broader context of social attitudes, cultures, technical complexity and local or time-based factor (e.g. weather, travel demand) also need to be taken into account. This chapter presents how design methods integrate the subject matter of three domains into a single research and development project. Namely, that methods from design were used to bring together safety, law and road design in a manner analogous to how the concerns of these three domains co-exist in application, out on the street.

The next section will describe the problem from each of these viewpoints. Section 6.3 describes the precedents for a design approach to the problem. Sections 6.4, 6.5 and 6.6 set out the three methods of drawing, 3D modelling and legal annotation respectively. Section 6.7 discusses the implications of this approach and offers conclusions.

6.2 A Problem in Three Parts—Safety, Law and Design

The problem of bicycle rider safety at left turns could be approached in isolation as a problem of safety, law or design. This section sets out the nature of the problem from these three perspectives.

6.2.1 *Safety—Problem Definition Through Detailed Observation*

Fundamental to road safety in Australia is the Safe System approach. This conceptual framework underpins road safety at all levels of government and elevates our understanding of road safety issues from an isolated event that occurs due to unsafe road user behaviour, to a failing of a system that requires consideration of system-wide factors. The key guiding principles of the Safe System are:

1. People make mistakes that can lead to road crashes
2. The human body has a limited physical capacity to tolerate crash forces before harm occurs
3. A shared responsibility exists amongst those who design, build, manage and use roads and vehicles, and those who provide post-crash care to prevent crashes resulting in serious injury or death
4. All parts of the system must be strengthened to multiply their effects, and if one part fails, roads users are still protected [5].

In this investigation of the interaction between cyclists (continuing straight ahead) and drivers (turning left), the primary focus from a behavioural perspective was to understand how cyclists and drivers navigated the interaction at a range of different intersection types. To gain an in-depth understanding of the interaction, particularly

Fig. 6.2 Site setup showing mounted video camera (left, circle) and field of view (right, cropped). *Photo source* Sarah Chancellor-Goddard

when people negotiated the interaction safely, we conducted a series of observational studies using covertly positioned video cameras at several intersections across Melbourne. A small GoPro camera with custom designed, 3D printed housing was mounted at intersections using a standard 190 mm circumference roadside pole. The video camera faced the intersection and filmed all road users on approach and through the target intersection (Fig. 6.2). Observations of two hours on non-sequential weekdays were conducted in the typical peak commuter travel times (7:00–9:00 am or 4:30–6:30 pm) (±15 min).

The strength of covert observations is that the video camera captures the behaviours of all road users, making it an excellent method of investigating behaviours as the video footage can be repeatedly reviewed and re-analysed [6]. This is preferable when investigating cyclist-driver intersections over other methods (e.g. surveys, interviews) which require road users to remember details of their behaviours, which can often be difficult when actions are habituated (e.g. driving) or nuanced behaviour that are key details to understanding the event are not considered important enough to report. In the case of cyclist-driver interactions, there is a risk that drivers in countries with low cycling participation, like Australia, are less likely to look for cyclists on the road and this may limit their ability to report on the interaction. Covert video observation also removes the behavioural bias introduced by the Hawthorne effect that is when people change their behaviour because they know they are being watched [7].

Outcomes from observational studies conducted in metropolitan Melbourne reported a wide range of factors that affected the cyclist-driver interaction including: presence/absence of continuous bike lane, volume of traffic (motor vehicle or cyclist), dedicated left turn signal, pedestrians crossing the adjacent road, other road markings (e.g. hook turn wait lines) and other road infrastructure (e.g. tramlines). The most consistent behaviour by cyclists and drivers was repeatedly observed at locations with a dedicated bike lane on approach to the intersection (e.g. Fig. 6.2). This type of road design maximised intuitiveness and was likely to assist both cyclists and drivers to accurately predict each other's behaviours and reduce the likelihood of conflict [8–10].

It is important to note that there was considerable variation in which road users gave way across the different designs on approach to the intersection. It is likely that this is related to confusion about the road rules, particularly that cyclists are not permitted to travel on the left of a left turning vehicle. However, this type of infrastructure positions the cyclist to the left of a left turning motor vehicle. This contradiction between the road design and the road rule highlights the important of an interdisciplinary approach that considers the road rules as an integrated part of the investigation.

6.2.2 Law—Consistency of the Framework and Relation to Built Environment

The law examines this problem in terms of the legal requirements of the left turn manoeuvre, namely any rules that apply to drivers or riders completing this manoeuvre in different road environments.

Unsurprisingly, the rules that govern the actions of drivers and riders on roads and road-related areas are complex. In Victoria, road rules are prescribed by the *Road Rules Road Safety Road Rules 2017* (Vic) ('Victorian Rules'). The Victorian Rules are enacted under the *Road Safety Act 1986* (Vic) ('RSA'), and form part of a national scheme to harmonise road rules in all Australian states and territories under the model Australian Road Rules developed by the National Transport Commission in 1999. Without legal effect themselves, the model Australian Road Rules must be enacted by law in each Australian state and territory to have legal effect, and while equivalent laws enacted in most states and territories follow the model rules, some variations exist.

Importantly, from a legal perspective, understanding the specific requirements placed on a left turn manoeuvre by the Victorian Rules is key to setting out the parameters within which road design and safety considerations operate. At a high level, the Victorian Rules reflect the social, political, and economic conditions within which they are enacted, which continue to change over time. More specifically, the Victorian Rules comprise 408 individual rules and explanatory materials over more than 550 pages. While established principles of statutory interpretation can be called upon to interpret the Victorian Rules and resolve any ambiguities, the complexity of the Victorian Rules and the circumstances to which they apply, mean that it isn't possible to govern left turn manoeuvres with a single rule. Importantly, a rule relating to specific interaction between road users cannot be understood in isolation but may refer to terms defined in the Dictionary to the Victorian Rules, or in a diagram, which form part of the law. Alternatively, concepts relevant to a rule may be explained in a notation or in the 'Readers' Guide' which can influence road user behaviour, although are not legally binding. In many cases, rules must be interpreted and applied within the context of other rules for specific actions required to complete the manoeuvre (e.g. turning left); the specific road environment (e.g. turning left at a slip lane; intersection; or T-intersection), or to the particular road user (e.g. driver of a motor vehicle, rider of a motorbike, or bicycle). Lastly, the Victorian Rules should be applied in ways that are consistent with the objects of the RSA, as the enabling Act, to set out the general obligations of road users for responsible road use, and provide for the safe, efficient, and equitable use of Victorian roads (RSA s 1).

From a legal perspective, the left turn manoeuvre is a key example of how difficult it is to articulate how various rules interact with each other to affect the contemporaneous behaviour of different road users in this situation. The key rule affecting the left turn manoeuvre is Victorian Rule r 141(2), operating as an exception to the general rule about overtaking to the left of a vehicle, stated in r 141(1).

Victorian Rules r 141(2) states:

(2) The rider of a bicycle must not ride past, or overtake, to the left of a vehicle that is turning left and is giving a left change of direction signal.
Penalty: 3 penalty units.

Note: ***Left change of direction signal*** is defined in the dictionary.

While Victorian Rules r 141(2) appears both clear and simple, it is deceptively so. Interpretation of Victorian Rules r 141(2) relies on (i) the interpretation of defined terms including 'rider', 'bicycle', and 'vehicle'; (ii) the explanation of actions in other rules including 'overtake', 'turning left', and 'giving a left change of direction signal'; and (iii) the satisfaction of cumulative conditions to trigger the obligation to 'give way' in Victorian Rules r 141(1), 'turning left **and** giving a left change of direction signal' (Fig. 6.3).

In the process of identifying how Victorian Rules r 141(2) operates in the context of these further rules, it became apparent that while Victorian Rules r 141(2) could be articulated *as a matter of law*, it was very much affected by road design, and road safety factors, which were regulated by other disciplines. Moreover, some of the terms defined by law (e.g. turning left) were perceived or defined differently as a matter road design or road safety. While these interpretations weren't necessarily incompatible, they created inconsistencies in how the stated law applied in practice. Accordingly, it was possible that desired behaviours according to road safety or road design principles would be 'illegal' according to the Victorian Rules; or conversely 'legal' behaviours to be 'unsafe' according to road design and safety principles. From a legal perspective, the latter interpretation would be rejected as a direct contradiction to the objectives of the RSA. These risks highlighted the importance of treating this issue as an interdisciplinary problem. Working together, there was not only the potential to identify where inconsistencies arose, but to find new ways to address the problems that road users experienced during left turn manoeuvres.

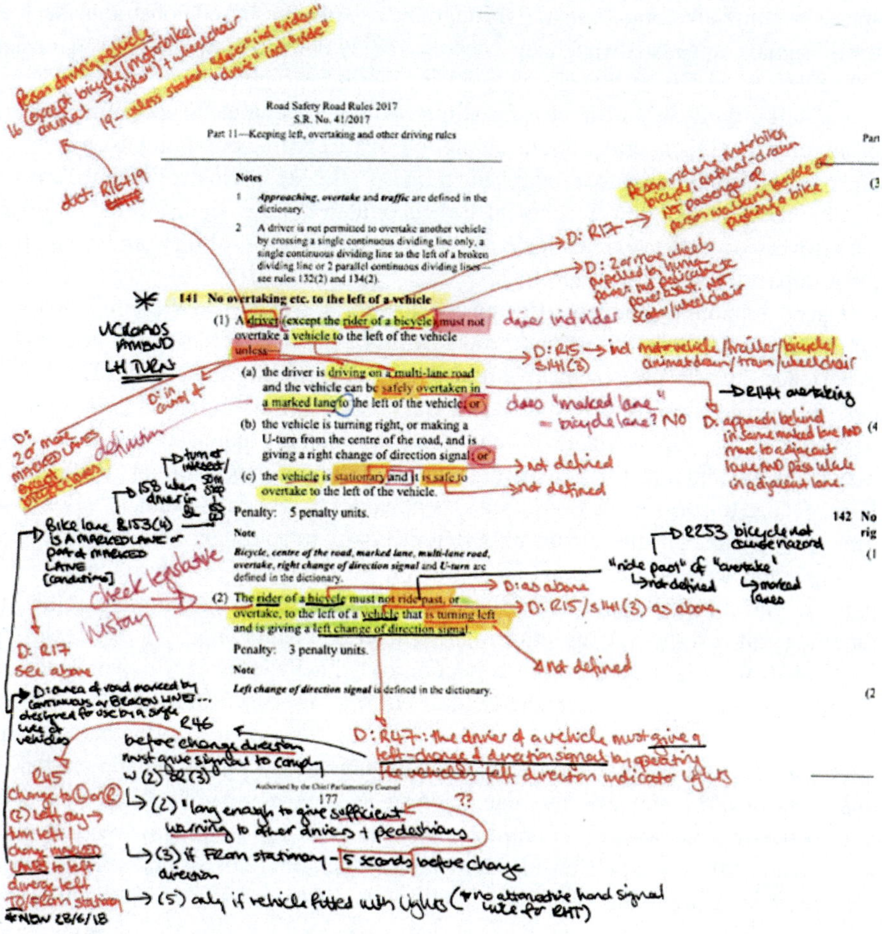

Road Safety Road Rules 2017
S.R. No. 41/2017
Part 11—Keeping left, overtaking and other driving rules

Notes

1 *Approaching*, *overtake* and *traffic* are defined in the dictionary.

2 A driver is not permitted to overtake another vehicle by crossing a single continuous dividing line only, a single continuous dividing line to the left of a broken dividing line or 2 parallel continuous dividing lines—see rules 132(2) and 134(2).

141 No overtaking etc. to the left of a vehicle

(1) A driver (except the rider of a bicycle) must not overtake a vehicle to the left of the vehicle unless—

(a) the driver is driving on a multi-lane road and the vehicle can be safely overtaken in a marked lane to the left of the vehicle; or

(b) the vehicle is turning right, or making a U-turn from the centre of the road, and is giving a right change of direction signal; or

(c) the vehicle is stationary and it is safe to overtake to the left of the vehicle.

Penalty: 5 penalty units.

Note

Bicycle, *centre of the road*, *marked lane*, *multi-lane road*, *overtake*, *right change of direction signal* and *U-turn* are defined in the dictionary.

(2) The rider of a bicycle must not ride past, or overtake, to the left of a vehicle that is turning left and is giving a left change of direction signal.

Penalty: 3 penalty units.

Note

Left change of direction signal is defined in the dictionary.

Authorised by the Chief Parliamentary Counsel

177

Fig. 6.3 Initial analysis of road rules 141(1) and 141(2)

6.2.3 Design—Technology, Behaviour and Culture

A design approach to this problem seeks to combine the technical requirements of the left turn manoeuvre, the designed and built environment in which it takes place, and the human-related factors of the people carrying out actions within it. As a broad discipline, design serves the ‘two cultures’ of art and science by being adept with creative processes and human perception at one end, and with the technical at the other. It is a mediating discipline to some extent—not claiming to offer more depth in either of these fields than, say, engineering or psychology, but rather a means of combining and creating meaning in the space between.

From a technical perspective, the left turn problem contains geometric issues much like any other design problem, requiring certain things to fit into finite space. The space—an intersection—is shared through the use of time-based constraints such as traffic signals, and also through an access hierarchy defined by the law. As identified above, this hierarchy at present requires bicycle riders to give way to vehicles being turned left provided they are also signalling. Another technical aspect of the design problem is that it concerns vehicles in motion. If we confine our thoughts to bicycles and cars (putting aside trucks, buses and trams as other actors in the problem) we can see that these vehicles have technical aspects of their own which affect the problem. The drivers of these vehicles have differences in vision, visibility, and protection from impact to name just three.

User behaviour is the second design consideration. It is foolish to attempt or assert an understanding of all users and how they do or may behave. As many designers experience, human users will thwart attempts at corralling behaviour. The best a designer can do is make the desired behaviours easier, and attempt to constrain out the undesirable ones [11]. Variation in behaviour is difficult to predict, but generalisations may be useful and may provide a broad understanding of most—of not all—actions. We also need to consider that variation in behaviour is beyond individuals; at different times of day and in the pursuit of different goals, people behave differently. For example, the actions of an adult bicycle rider can vary a good deal based on whether they are carrying children on their bicycle or not. Further considerations might be the impact of rush-hour, congestion, and weather on behaviours.

Thirdly, we might consider the collection of behaviours that form over time and establish norms to be a culture. The culture of road users will impact this problem because it sets out some of our community expectations. Our individual actions contribute to, and are affected by, culture. We must take care in the application of culture however; it may give us some guidance as to expected individual behaviours but it offers no guarantees. At an intersection level, a culture may be evident only at particular times of day. Consider for example the culture of an intersection when school students are using it—it may have a very different look, feel and collective behaviour than from 7 to 8 am when it may have been a high-speed thoroughfare. In order for a design proposal to be generated, the culture of the place needs to be considered if we are to propose something acceptable.

The design of an intersection which might be safer for users will consider the combined space of technology, behaviour and culture.

6.2.4 Combining Approaches: Why Design?

Road design has historically been the domain of engineering, and with good reason. The discipline of engineering has provided a foundation—literally and theoretically—to making roads. The necessary technology to build durable and dependable infrastructure such as roads and bridges cannot be ignored and these fields of development continue, despite the current state of reliability achieved in many developed countries. The theoretical foundation provided by engineering is important too. Having a working theory to describe the movements of goods, people, and vehicles on our roads or railways—to use but two examples—allows us to do more with less. For example, to move more goods using less energy. These advances have secured engineering's place in matters of road design. The dominance of engineering approaches is not conceptualised here as a failing of this particular discipline, but rather as a characteristic of many mature domains that have a significant depth of expertise.

In the period since the invention of the automobile, and the development of this mode of transport, statistics suggest that Australia has adopted the car wholeheartedly, as a technically and culturally dominant mode, with 75% of journeys to work in the 2016 census being by passive private modes like the car and a corresponding 5% by active transport [12]. More recently, especially in the 21st Century, there have been reasons to re-examine engrained practices in engineering as the culture this discipline serves has changed. In particular, the dominance of the car, and of roads dominating our urban landscape has been called into question because of concerns about urban space, energy intensity, carbon and other tailpipe emissions locally, and their contribution to global environmental issues such as climate change. Some political change is in the air, especially as evidenced by election results in cosmopolitan places like cities. Car mode share is still very high in Australia, which provides a contested cultural backdrop to the changes being demanded by citizens in some areas—demands for more and safer cycling to use the present example, set in a culture that prioritises motor vehicles both on the road and through associated means such as Australia's taxation system which provides financial incentives to motor vehicle use. So, as well as being a technical issue, there is a cultural element. A further example from the driver licensing process in Australia is that cyclists have been constituted as 'hazards' which reinforces the notion that roads are for cars [13]. Engineering is a largely technical domain. It is not ignorant of culture; however, the collective behaviour of people is not a principal concern. The best engineering education, and engineers, will already be sensitive to needs like this.

Both design and engineering deal with the built environment and are charged with the broad task of dealing with ideas, to develop "that which does not yet exist" [2]. In the creation of material interventions both domains carry a responsibility for aesthetics, however this is a point of departure for their similarities. In the same way that engineering grasps more of the technical and physical concerns of a problem—for example applying the laws of physics; design tends to grasp more of the aesthetic and cultural concerns—for example the acceptability of an intervention's appearance or manner of carrying out a process. In issues such as road safety we are dealing with minute, personal variations in behaviour. In striving for a behavioural change in culture, design offers new methods for achieving some of these goals.

The engineering and design domains have many common elements. There is an elegant tension between engineering and design, captured neatly by pioneering industrial designer Henry Dreyfuss [14]:

> The engineer is the best friend and fiercest critic of the industrial designer.

This tension is useful and symbiotic when at its best. Designers need engineers and engineers need designers. Both need many other domains of expertise as well, if the general goals of these two domains are going to be met. When we look at the goals of a road safety investigation, the contribution of design in addition to the established approaches of engineering become clearer.

Firstly, design can contribute new methods to investigate a situation by bringing fundamental techniques—such as drawing—to the novel situation. Especially relevant to design disciplines which deal with three dimensional objects and space, such as industrial, spatial and urban design, there is a tendency in design training and practice to emphasise the human users in investigative drawings. When we consider that the cars and bicycles people are using in the road space are the outcomes of industrial design, we have all the more to learn from a design approach as these very vehicles are subject to close scrutiny by designers in order to achieve usability. Two-dimensional drawings often progress to three-dimensional models. These models enable a richer understanding of the road safety situation by more accurately showing sightlines, understanding scale and space. Drawing and three-dimensional modelling are two of the methods identified and explained later in this chapter.

Secondly, these drawing and modelling techniques extend to the generation of new ideas. While design contributes to the investigative work in a road safety project, it makes a more profound contribution in the generation of new ideas. In design, the generation of new ideas can take a technical guise in a similar way to engineering. Alternatively, it can be a path to an open and creative mindset; engaging with some core concepts of creativity through the use of defined methods. These methods, such as scenario-based responses, are used as examples in this chapter of how the methods were applied.

Thirdly, and perhaps most importantly, these design methods are examples of how design, as a method, has a predisposition for bringing together relevant factors into a single work package. Design does not have a sole claim to this ability to induct. Any applied domain will have the same necessity to do so. Where design differs is that there are specific methods to bring relevant factors to bear in the creation of new things. A designer is required to generate creative responses to a client's brief, while at the same time acting within constraints, such as relevant standards. These standards may be specific and technical; often these are tackled at the same time as cultural and ethical standards. In this research we applied road rules and road design standards at the same time using a design approach. The three specific convivial methods used are detailed in the next sections, showing how the accessibility of these methods to users, as well as their capacity for trading-off demands from different disciplines, offers a path forward in this complex problem.

6.3 Using Two-Dimensional Drawing

6.3.1 Characteristics

Insights from safety, law and design contributed to understanding safe outcomes at intersections, particularly for a bicycle rider when a driver is turning left. While there is a history of a siloed approach to road safety in Australia [5], it is pleasing to acknowledge some exceptions when the considerations of these three fields—and many more besides—all come together in the space, and moments, where people interact with each other on the road. The research team sought a method to bring these ideas together such that they might be considered in relation to one another rather than separately. In essence, seeking a research approach that was closer to the reality on the street.

Two dimensional drawing is the first method used for this purpose. An example of this drawing is shown below in Fig. 6.4. These drawings are characterised as follows:

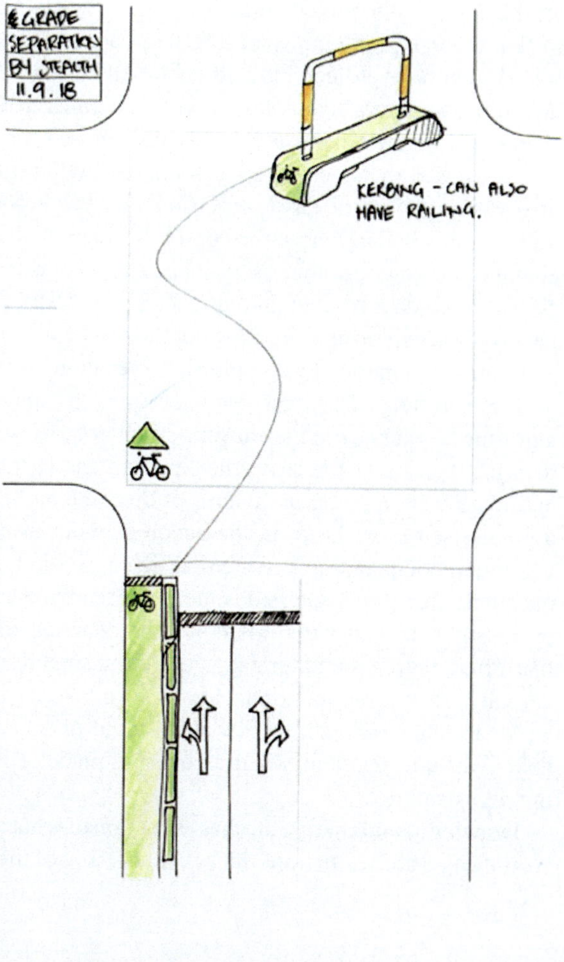

Fig. 6.4 An example of the two-dimensional drawing used in this research

1. Plan view. Most of the drawings are of the road space and are made in the perspective which places the viewer above the road looking straight down.
2. Of no particular scale. The drawings do not attempt to capture or adhere to any particular size representation.
3. Showing users. Humans without vehicles and humans on/in vehicles such as bicycles, cars and buses are shown in some drawings.
4. Capturing motion. Arrows and gestural devices communicate where users and their vehicles are intending to travel.
5. Rough. The method dispenses with formality, correctness and strict accuracy in favour of capturing the essence of ideas in the moment. Tools such as rulers are seldom used.
6. Exploratory and developmental. The drawings were not created for formal presentation, but were made during meetings in the context of discussion.

6.3.2 As an Exploration Method

At the outset of the project, drawing helped the team come to a shared understanding of the research problem. Road safety research has been conducted leading up to the interdisciplinary project and had documented the nature and extent of the problem according to several variables, such as the infrastructure provided at a given intersection [1]. Drawing was used to explain these scenarios at a broad level. Drawing was also used to sketch detailed scenarios within the broader types. For example, at particular times of day, pedestrians are the highest mode share users at many intersections in the Melbourne central business district (CBD) and this significantly changes the flows of other users such as bicycle riders and vehicle drivers. Drawing provided a means to explain these scenarios. Explaining the research problem thus gave way to exploring it in great depth and detail.

In the explanation of a complex topic, drawing was also used by the research team to pose questions. These questions became the topic of discussion to a large extent, and thus we acknowledge the important role of verbal discussion accompanying drawing. Questions to clarify the situation and to pose "what if…?" type responses and ideas were a common feature of discussions. In these cases, drawing provides a greater sense of clarity to the accompanying discussion. It can help to ensure a common understanding between participants—in a field such as road safety where questions are set in a detailed context with moving actors, this shared understanding of questions and answers was important. The experience of using drawing showed that often, before something was drawn, a shared understanding was not achieved verbally. The most common incidence of this was when the research team discussed a scenario and although there was a shared understanding at some broad level, the detail was not shared until a drawing was made. The drawings exposed differences in understanding.

Drawings of this nature are an abstraction of what is going on in the physical world. As indicated above, an important characteristic of these drawings is their informality.

They are not scaled, neat, or strictly conforming to a drawing protocol. The informality of the drawings and their nature as an abstraction of the road space worked, despite these characteristics which may be interpreted as failings. One possible reason for this success is that the drawings were of something that the research team had a lot of experience with—the road space. We might assume that any adult member of an urban community carries such experience too, if slightly less detailed and analytical. Drawing exploited the common grounding shared by the research team and used it as a means to delve into more detailed understanding.

Each intersection in our road system is somewhat unique. Although a typology of intersections exists in theory, the world provides the variations which lead to this uniqueness. For this reason, the use of drawing as a tool to achieve shared understanding and explore all these variations presents no real ending to such a task. While the exploration is somewhat ongoing, the research turns to a development phase, in which two-dimensional drawing has some uses.

6.3.3 As a Development Tool

This research follows a considerable effort in problem definition and is positioned at the development of solutions. As discussions continued in the project, the subject matter evolved into the development of possible solutions to the left turn problem.

By this stage the method of drawing was firmly engrained in our approach. We achieved a level of comfort and fluency with drawing that comes with experience, in this case quantified as regular, fortnightly meetings of two to three hours duration, over the course of nine months. A level of familiarity and collegiality is important to illuminate at this point because of the shift from discussing a problem to developing new ideas. Proposing new ideas in an unfamiliar group setting requires courage, training, and experience—the sort of skills that are directly taught and practiced in the relative safety of a design school. The above use of drawing as an exploration tool assisted in the non-design researchers acquiring these skills as well as establishing the working norms of a team who needed to be comfortable sharing drawings and ideas.

Developing new ideas was observed to become a natural extension of the exploration drawing and discussion process. In a drawing of an existing real-world scenario, the same method of drawing can be used to propose something "which does not yet exist" [2]. These are broadly characterised as 'interventions' into the problem, as the physical ideas took on many forms, including road markings, signs, road geometry changes, and new signals. Non-physical ideas were also drawn, such as signal phasing programs and road rule changes.

An unexpected outcome of the drawing process was an intervention that the research team did not initially consider to be part of the research scope. In one discussion meeting, a new knowledge gap was reported from a member of the public with the simple-sounding question of "what is turning left?" This question had profound consequences for the team, as although there is a basic shared understanding of the road and vehicle space, it proved much more difficult to provide a detailed, legally accurate and defensible answer to this question. The question was asked in a legal context following a case where road rule 141(2) was applied. Discussing, and drawing the question, and a possible answer, led to the idea of a public communication campaign, the engagement of a graphic design team, and the creation of the infographic poster titled 'This is turning left.'

Communication Tool

At certain stages in the research process it was necessary to report progress externally. In these instances, drawing was used as a communication tool in ways very similar to existing protocols and conventions of engineering and design. More formal drawings were created using the rough drawings as a base. The formal drawings were made by the team, and informal processes still ensured a shared understanding of what we were communicating. These processes are not new; rather they are within the normal realm of how drawings are produced in a professional design setting.

Once initial barriers such as fluency and comfort are overcome, the research demonstrates that drawing can become part of a research team's skill set. We especially note that two out of three researchers in this project are not designers by training but have developed their existing drawing ability to use drawing as a research tool. The medium is exploited as a discussion and development method to great effect in this project, while the role of drawing as a presentation and communication tool remains strong.

6.4 Using Three-Dimensional Modelling

6.4.1 Characteristics

We used drawing to overcome a siloed approach to road safety. Three-dimensional modelling was used for similar reasons. Additionally, three-dimensional modelling gave us access to an enriched visualisation and experimentation tool. In a design setting, and in particular industrial design, three-dimensional (3D) modelling is a common approach sitting across development, iteration and documentation. There is no single type of 3D modelling, but rather a family of methods. An exhaustive list is not provided here, however readers may consider that rough paper-and-cardboard mock-ups, sophisticated simulations such as appearance models, and working prototypes are all types of physical 3D modelling. Computer based 3D modelling is widely used in the development and documentation of designs, for example to provide instructions for manufacturers. The type of 3D modelling used in this project is of a prototyping kind, and in particular is based on the use of LEGO™ elements and models.

An example of the LEGO™ modelling is shown below in Fig. 6.5. The characteristics of these models are as follows, the models:

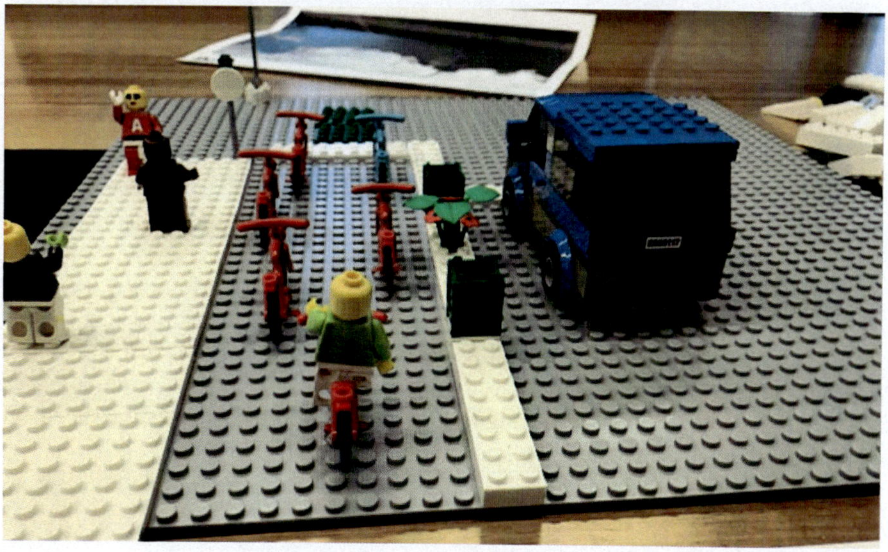

Fig. 6.5 Typical LEGO model used in this research process

1. Use LEGO™ elements such as bricks, plates and human figures to represent infrastructure and people. Model vehicles were used to represent cars and buses.
2. Are partly two-dimensional, for example showing the road and footpath space in plan view.
3. Identify particular functions of the intersection in 3D, for example vehicles, hard separation such as kerbs, grade changes, and traffic signals.
4. Allow the depiction of grade separation.
5. Identify users and vehicles suggestive of their size difference, rather than specific scale.
6. Are rough, showing only enough information as needed. They are not formal or presentation models.
7. Do not inherently capture motion. Rather, they allow the researcher or participant to demonstrate motion and stages of road user interaction by moving the model's parts.
8. Are often used in mixed-media approaches to 'animate' a two-dimensional drawing.
9. Engage with play-based learning to the extent that the LEGO™ format acts as an ice-breaker because of its familiarity and market position as a toy.

6.4.2 *For Exploration*

3D modelling brings into focus one of the key factors of the left-turns problem; the size difference between motor vehicles and people riding bicycles. While not intending to be exactly scale-accurate, the models show the relative size of each road user and their vehicle. Early exploration of the problem was enhanced by this rough scale. For example, in examining how a driver of a car of some 5 m long negotiated a turning manoeuvre in the same space as a bicycle and rider the research team observed that there was no single point in time or space that could be defined as a 'turn'. Further to this was the observation of how both vehicles possessed different dynamic characteristics which affected how they interacted in this situation, for example the difference in swept paths.

The 3D models were used in repeated demonstrations of the left turn by the driver. This afforded the research team the ability to look at the nuances of turning in great detail. For example, exploring the difference made by having a bicycle stop ahead of a car, as is the case in an advanced stop box; or in the case of negotiating the intersection during a green traffic signal, where we explored the relative speed and position of the vehicles as they approached the intersection.

6.4.3 For Development

As with the drawing method, the research progressed from problem definition to the development of possible solutions. Developing ideas in the 3D modelling method was a similar exercise to drawing, in that a 'sketch' mentality was adopted. Ideas were suggested, tried out and modified during discussions using the 3D materials.

3D modelling enhanced understanding the nature of the left-turns problem by demonstrating—closer to reality than drawings—the finer details of an idea. One example of these finer details is the use of 3D models to capture sightlines when designing interventions. Visibility between road users is an important part of crash avoidance.

The models provided a means to animate an intersection treatment with the movement of people and vehicles, and importantly to freeze this animation at any point in time. The nuance provided by 3D modelling in exploration activities continued into development. The LEGO™ materials allowed the demonstration and assessment of potential intersection treatments that could account for vehicle and human movements at any stage in the intersection traffic cycle. The models also allowed demonstration of the variations in user behaviour, for example a cautious rider or driver who may demonstrate behaviour that is safer than assuming another road user will give way; or when a bicycle rider was animated to suggest they would not stop at an amber light.

6.4.4 As an Engagement Tool

The 3D modelling technique used in this research is highly accessible, based on two main observations. First, the method of building a road space model out of

LEGO™ elements is quite simple as most cases are based on linear road alignments, and this geometry is part of the fundamental mechanics of the LEGO™ system. Roundabouts and non-perpendicular intersections are more difficult to capture. The second observation is that using LEGO™ as a building material was familiar; it is a common toy in our culture.

The method was used in meetings with the project funding body to ensure their ongoing familiarity with the problem and developments in the research. Rather than presenting project information in an abstracted graphical format such as drawings on a screen; the 3D method showed itself to be a more tangible and approachable type of abstraction. For example, once a road design is represented in LEGO™ it can be tested out with representations of users, and the road rules brought to bear on their simulated actions. An important side note is that LEGO™ materials are, in our observations, irresistible. In all meetings, participants across all professional background and levels of seniority picked up the LEGO™ elements to build their concepts. The 3D modelling tools provide a successful means for the project themes to cross disciplinary boundaries. This is further demonstrated in the third method.

6.5 Using Road Rule Annotation

6.5.1 Characteristics

As described above, maximising safe interactions between cyclists and driver at intersections is not confined to a single discipline. Safety science, law and design are being utilised to extend our understanding. Since the road transport system concerns these and many more disciplines the research team set out to find ways of bridging these domains. One of the methods used was to annotate the project work with road rules.

The essential characteristics of this method are that the annotations:

1. Are numerical. They use the shorthand number reference of a rule and or sub-rule. These are looked up during discussions.
2. Accompany spaces. Arrows are used to locate the rules in the spaces where they apply.
3. Are connected to particular road features, for example paint used in an intersection to invoke rules about a special purpose lane.

An important characteristic of the road rules is that drawings are used alongside text, to describe the rules. Unusually in a legal context, these drawings form part of the rules themselves, a fitting approach when we consider the rules apply to the spatial and temporal arrangement of people and vehicles in the road space. The research team used these drawings as a lead to introduce the road rules into the more general drawing activity described above in Sect. 6.4. Drawings of existing and proposed road designs were annotated with road rules relating to particular situations. See Fig. 6.6 for example.

This annotation allowed us to clearly identify and visualise the specific places where the road rules apply. This gave a partial understanding of how they are applied. To gain a full—or at least fuller—understanding of road rule application the drawings then illustrated particular road interactions between users. These interactions invoke the application of situation-specific rules such as Victorian Rules r 141(2) which describes the requirement for bicycle riders to give way to vehicles turning left.

Annotating drawings with road rules was a learned skill, cultivated by the research team especially as a method of communicating across traditional disciplinary boundaries. It was easy to learn, as the technique consisted of some very simple drawing methods such as arrows and road rule numbers featuring in drawings. The places where road rules apply were identified with vague circling, or detailed carefully

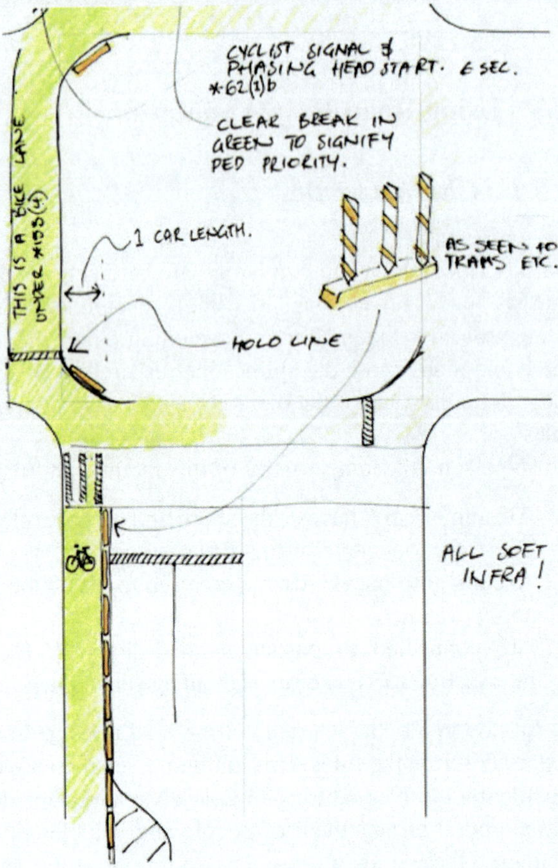

Fig. 6.6 Intersection design sketch using road rule annotation

drawn boundaries, where appropriate. See Fig. 6.6 for example. Once the road space and the road rules were identified in a drawing, it was often the case that users and their vehicles were represented in the drawings too, in order to apply the rules to traffic interactions.

6.5.2 For Exploration

There are 408 road rules in the state of Victoria, Australia. It would be very challenging for anyone to recall subsets of these rules as they apply to any given situation. Undertaking a legal analysis, the team were able to create a subset of rules to apply to the left turn manoeuvre. This subset could then be mapped on to a drawing to give it additional context and to facilitate discussion. As with above creative techniques, the discussion was integral to the development of possible solutions.

When a subset of rules were used to annotate a drawing, we can see how those rules apply in a space. Spatial definitions are also used to see how a user passes through the smaller areas within an intersection in which particular rules apply. The same goes for the spaces which lead up to, and out of, intersections. This exploration led to two important insights. First, the understanding of how users flowed through various legal obligations on their way into, through and out of an intersection. Second, it enabled a visual understanding of the ways road rules interact with respect to space—in particular how some rules set out required behaviour in a blanket sense (e.g. Victorian Rules r 125: Unreasonably obstructing drivers or pedestrians) and others only apply in specific places (e.g. Victorian Rules r 73: Giving way at a T-intersection).

Looking more deeply at the first of these insights, and with particular reference to the act of turning left at an intersection, the drawing method gave a better understanding of what turning left involves than the road rules alone could do. Three stages in the process of turning left for a driver of a car and a rider of a bicycle were identified as prepare, approach, turn; each with a suite of road rules and obligations. Comparing the three stages for cyclists and drivers identified road features and road rules at points where previous research had identified that crashes were more likely to occur [1].

6.5.3 For Development

In addition to the advantages of drawing already described above, the annotation of drawings with road rules allowed the development of more complete 'interventions' rather than road designs, or rule changes alone. Since the transport system consists of physical and legal elements, it stands to reason that if we seek to develop improvements to this system we ought to use all the means possible. Road rules can change, as can road designs; the annotation of descriptive and analytical drawings set the scene for developmental drawings which considered new road rules and road designs.

6.5.4 For Engagement

Drawing the road rules makes them more overt. Although they are freely published and form a part of the licencing requirements for drivers, the road rules are poorly recalled by the public [15]. The pervasive practice of generally observing the road rules is characterised as a benign observance by most road users, rather than a detailed understanding as is the nature of this research.

6.6 Discussion and Conclusion

This chapter proposes design methods as a convivial means of investigation. In particular, using methods which are descriptive and have a low barrier to entry for non-designers. Drawing, 3D modelling using LEGO™ elements, and annotation of drawings with road rules are proposed as three possible convivial means, and have been tested out as a research method through the analysis, development and engagement activities as we sought to develop novel solutions to the conflict between bicycle riders and vehicle drivers at left turn intersections.

The three techniques were introduced with some aids for users to gain confidence. Techniques of tracing and use of templates were important in overcoming any initial reluctance to draw, as the performative aspects of drawing can be a deterrent to some people. We also note that because one of our team is a designer, the other team members were initially reluctant to draw because of the intimidation when drawing in a team with a designer. After some comfort is achieved with drawing, the technique became part of the 'language' of the research. The technique is practical, cheap and suits the road safety problem space well. One limitation is that drawing does not capture motion well.

The above observation regarding comfort in drawing relates across the interdisciplinary frame of the project. Our team's experiences show that the reluctance to draw is manifested in much the same way that it takes non-legal experts some time to grow comfortable discussing the law; and non-safety experts time to become conversant with safety science. This insight could be a contributing factor to the problem introduced in this chapter—that siloed approaches are the norm in situations like these.

Three-dimensional modelling assisted in demonstrating spatial–temporal problems, such as sightlines between drivers of different vehicles in queues and turning. This three-dimensional modelling extended into role-playing to provide an effective means of communication for movements which sit at the core of road safety research. An interesting aid for users in road rule annotation was that diagrams of the road rules already exist as part of the rules. Where our work extended beyond this was in the study of real, multi-rule scenarios.

As noted in the introduction, there are many domains with important insights to improving road safety outcomes. Three out of the many are safety science, law and design, which have depth and boundaries created by the skills necessary to practice them. This research has identified methods to bridge boundaries between these fields. Another way to conceptualise this development is that the fields have been brought closer together and the boundaries removed. Such development is evidenced by the ability of each researcher in the team to partially work in the domain of another, such as when the designer and safety scientist use road rule annotations and propose new road rules. In simplified hindsight, three stages of change from the new methods are observable. An initial state of separation is followed by the domains getting closer as our team became more conversant with each other's fields, and an emerging shared space of interdisciplinary practice. The state we reached in this research is that the domains of expertise are maintained but that the boundaries are significantly diminished. A significant body of expertise remains in each of the domains, which is useful for the detailed investigation and development of new interventions. These three stages are communicated in Fig. 6.7. While we acknowledge that it is a simplification, it aims to illustrate the progression of the three domains as we applied the methods above.

The road transport system presents a highly diverse array of interactions to users and relies on the design of the road space and road rules as principle methods of control and guidance to create safe and effective transport. For such a system to operate with harmony, this research proposes that investigative and developmental endeavours should be conducted with similar harmony.

Combining three fields is observed to provide a setting in which safety science, design, and the law can co-exist and thrive. This ensures that one field does not dominate the discussion, nor was the offer from any of the fields diminished. The complexity of the road environment, variety of users and vehicles means that a broader range of disciplines must be utilised to find new and innovative solutions. This research shows that design techniques can be used as a means to carefully and effectively unite disciplines which to date have worked largely in isolation.

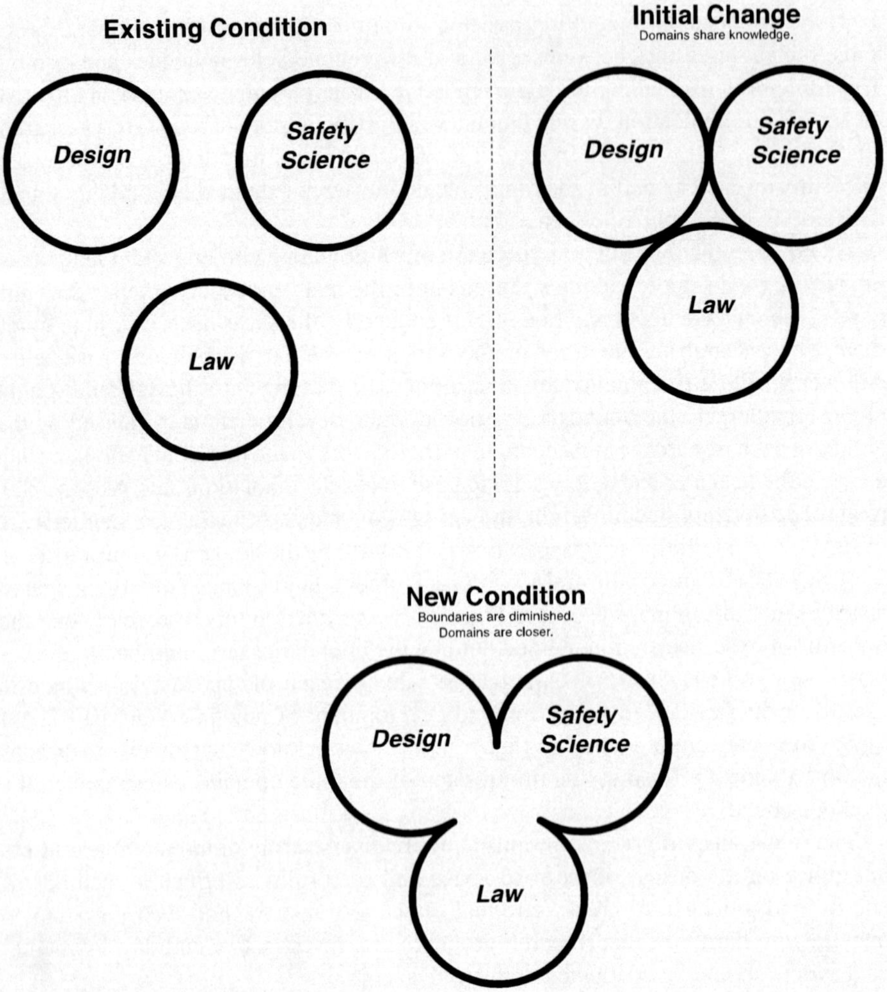

Fig. 6.7 Representation of the three fields combining

References

1. Nicholls, H., Rose, G., Johnson, M. and Carlisle, R.: Cyclists and left turning drivers: a study of infrastructure and behaviour at intersections. In: 39th Australasian Transport Research Forum (ATRF). Auckland, New Zealand (2017)
2. Lawson, B.: How Designers Think—The Design Process Demystified, 4th edn. Architectural Press, Oxford (2006)
3. Illich, I.: Tools for conviviality. In: Anshen, R.N. (eds.) World Perspectives, vol. 47. Harper and Row, New York (1973)
4. Sanders, E.B.N., Stappers, P.J.: Convivial Design Toolbox: Generative Research for the Front End of Design. BIS Publishers, Amsterdam (2016)

5. Woolley, J., Crozier, J.: Inquiry into the National Road Safety Strategy 2011–2020. https://www.roadsafety.gov.au/sites/default/files/201911/nrss_inquiry_final_report_september_2018_v2.pdf (2018)
6. Johnson, M.: Cyclist safety: An Investigation of How Cyclists and Drivers Interact on the Roads. Monash University. Thesis (2011).https://doi.org/10.4225/03/5890198046314
7. McCambridge, J., Witton, J., Elbourne, D.: Systematic review of the Hawthorne effect: new concepts are needed to study research participation effects. J. Clin. Epidemiol. **67**, 267–277 (2014)
8. Johnson, M., Johnston, V., Nicholls, H., Stephan, K., Napper, R., Taylor, W.: Drivers turning left, cyclists going straight: an interdisciplinary study combining behaviour, design and the law to improve safety. In: International Cycling Safety Conference, Brisbane, 18–20 Nov 2019
9. Chancellor-Goddard, S., Johnson, M.: Cyclists and left turning drivers: the impact of road infrastructure on safety and efficiency. In: 41st Australasian Transport Research Forum (ATRF), Canberra, 30 Sept–2 Oct 2019
10. McMenamin, G., Johnson, M.: Left turn negotiations between cyclists and drivers: an investigation of safe and legal. In: 41st Australasian Transport Research Forum (ATRF), Canberra, 30 Sept–2 Oct 2019
11. Norman, D.: The Design of Everyday Things: Revised and Expanded Edition. Basic Books (2013)
12. Bureau of Infrastructure, Transport and Regional Economics. Yearbook 2019: Progress in Australian Regions, p. 262. Canberra (2019)
13. Bonham, J., Johnson, M., Haworth, N.: On constituting cyclists as 'hazards.' J. Transp. Geogr. **84**, 102675 (2020)
14. Dreyfuss, H.: Designing for people. Skyhorse Publishing Inc. (2003)
15. Tierney, P.: Review of Victorian Cycling Related Road Rules and Legislation. https://www.vicroads.vic.gov.au/-/media/files/documents/safety-and-road-rules/reviewofvictoriancyclingrelatedroadruleslegislationmarch2015.ashx?la=en&hash=E46A6BEE94671370F802D02D71630D62 (2015)

Chapter 7
Innovation Opportunities of Compact Vehicles for Urban Transportation

Yang Yan and Robbie Napper

Abstract The rapid motorization of China has led to widespread traffic issues. While solutions exist, transport providers can often create new problems while solving another. Currently two choices dominate the public debate; firstly the automobile as a comfortable transport mode that offers the idea of good mobility for the individual, an offering degraded by traffic congestion; and secondly the alternative of high-efficiency public transport which offers the consumer less door to door comfort but moves more people. This chapter investigates opportunities in the third choice of smaller vehicles. Such vehicles take into account efficiencies, comfort, and door to door access, but require both design innovation and regulatory innovation to be a more attractive option for urban commuters in China. We conceive of a traffic system which can balance the high efficiency of public transport with the mobility and comfort of personal vehicles.

Keywords Low speed vehicle · Individual vehicle · Electric bicycle · Electric motorcycle · Electric moped · Quadricycle · Urban transport · Public transport · China transport · Vehicle innovation

7.1 Introduction

Chinese cities have entered the age of the automobile in the twenty-first century. This change has occurred with unprecedented speed. To the general alarm of the people, the automobile has brought the corollaries of traffic congestion, traffic accidents, energy waste and air pollution. Subsequently, Chinese cities have quickly turned to public transport and achieved remarkable results in a decade. Public transport and urban agglomeration of super-large metropolitan clusters have changed the way

Y. Yan (✉)
Academy of Arts and Design, Tsinghua University, QinghuaYuan 1, Haidian District, Beijing 100084, People's Republic of China
e-mail: yanyang@tsinghua.edu.cn

R. Napper
Faculty of Art, Design and Architecture, Monash University, 900 Dandenong Road, Caulfield East, VIC 3145, Australia

S. Coxon and R. Napper (eds.), *Advancing a Design Approach to Enriching Public Mobility*, Intelligent Systems Reference Library 198,
https://doi.org/10.1007/978-3-030-64722-3_7

people commute, namely that the availability of high-speed services have increased distances rather than reduced commuting time. There exists a need for a third type of transport offering which combines some efficiencies of public transport with some of the conveniences of the private car.

7.1.1 The Past and Present of Urban Transportation in China

The last 40 years of social change in China have been profound in scale and speed. The change has been led by cities. Western travellers to Beijing in the 1960s described few tall buildings in the capital city of China, with the only exceptions being palaces and constructions from Ming and Qing dynasties; those situated in the centre of the city with golden glazed roof tiles and scarlet walls. The remainder were low, grey brick buildings which when submerged in the mist of winter and spring made Beijing look like a "big village".

The impression of Beijing's road network from this time is of a North–South grid. Roads are flat and straight, with trees on both sides to ward off cold winds and dust from the North in winter and spring, and to be thickly shaded in summer. Not until the 1990s were cars ordinary for a Chinese family. Urban residents travelled by bus, bicycle and foot. 1970s Beijing had only a single subway line with the remainder of public transport being served by buses. Bus services were systematically developed, and affordable for ordinary residents, however the travel experience was unpleasant. During peak hours stations and stops were crowded with dozens of passengers, anxiously awaiting delayed services and swarming on board when services arrived. In what are familiar scenes to many public transport systems, crowding led to delays as passengers struggled to board and attempted to force their way on to the bus. In addition, early buses were not equipped with any air conditioning and as a result the crowded experience was made more painful, especially in winter and summer (Fig. 7.1).

In contrast, cycling was easier and more comfortable than bus travel in this period. Beginning in the 1950s, Chinese cities built unique roads for bicycles, called "non-motorised lanes". Unlike narrow facilities built in some European cities as part of sidewalks, these non-motorised lanes are formalized, three to six metres wide, isolated from the left motorised lanes and are 15 cm lower than the right-side walking path. While these facilities are notable, so too are the conditions. In the long cold Northern winter, cycling on icy roads in the cold wind was a dangerous undertaking. Despite the challenges, bicycles represented the only feasible private vehicles for Chinese families before cars came into reach. The bicycle was the mode of daily life and facilitated shopping, outings with children and families, and even carrying patients to hospital. From the second half of the twentieth century a three-level urban road system was formed in China, consisting of motorised, non-motorised, and walking lanes [1] (Fig. 7.2).

China's first domestically manufactured car was made in 1958, but until the 1980s these locally produced cars were a special supply to government officials under

Fig. 7.1 Bus in Shenzhen in 2008. Photo by Yang Yan

extremely limited production. This changed in 1984 when AnDing Li, a Xinhua News Agency reporter suggested to the government that cars be sold to the public as usual goods. Furthermore, it was suggested that foreign automobile enterprises be introduced to the Chinese market. The result was the market reform of China's automobile industry, and the subsequent "arrival" of the family car in China. From 2 million vehicles sold in 2000 this market has grown to 28 million in 2018, an increase of 14 times in 18 years (Fig. 7.3).

Further to this market reform and expansion, the production of cars and the infrastructure on which to drive them has promoted economic growth; for example, through the construction of roads, bridges and parking lots. As a result, the car industry has come to be regarded as an essential economic growth mechanism by local governments. Especially since the 1990s, motorisation has brought a focus on urban construction in China, with a particularly frenzied pace of construction in the period between 1990 and 2010. Old narrow urban roads have repeatedly been broadened and overpasses have become a new feature of China's urban landscape. Car ownership has led some cities to convert non-motorised lanes into motorised lanes. At the same time, parking spaces have become scarce in proportion to demand. Broadly speaking, China's urban road structure is changing from "bicycle cities" to "automobile cities" (Fig. 7.4).

The enjoyment of automobile life has been short lived for city residents. Building new expressways has only induced new travel demand, leading to the same congestion the construction was aiming to diminish. This process has been rapid in the more modern Chinese cities of Beijing, Shanghai, Guangzhou and Shenzhen with new expressways congested within six months of opening. A UBS study in 2014 found that globally, the slowest driving speeds are to be found in Chinese cities. Beijing averages 12.1 km/h [2]. This slow speed has an impact on the amount of time commuting trips

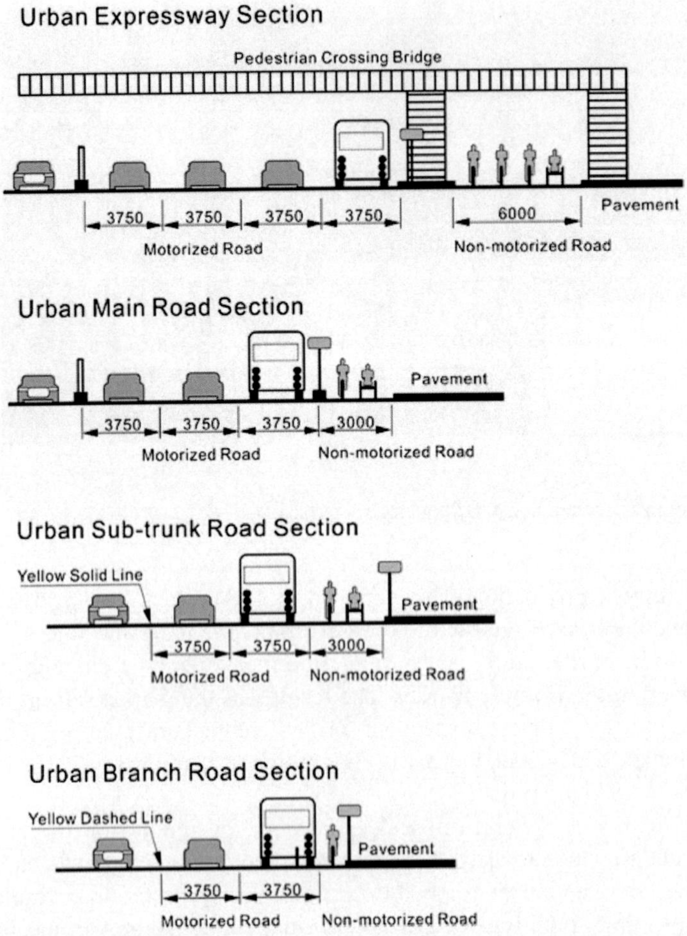

Fig. 7.2 Urban road sections of China cities. Illustrated by Yang Yan

take in a car, with Beijing's morning peak commute taking 47.2 min; the evening, 48.2 min [2] (Table 7.1).

The problems caused by larger roads affect more people than those driving on them. Larger roads have caused severance in old urban areas, with elderly and low-income residents disproportionately affected; it is harder for them to cross the road. Existing roads in the old city centres are effectively narrowed when cars are parked along them. Parking spaces have become a scarce commodity, reflected in price growth. An 18 square metre parking space in Beijing and Shanghai can now cost between 500,000 RMB to 1,200,000 RMB and these spaces are regularly the subject of disputes (Fig. 7.5).

Following the rise of motorisation and its attendant problems, many Chinese cities are now turning to de-motorisation. In 2003 the Beijing Municipal Commission

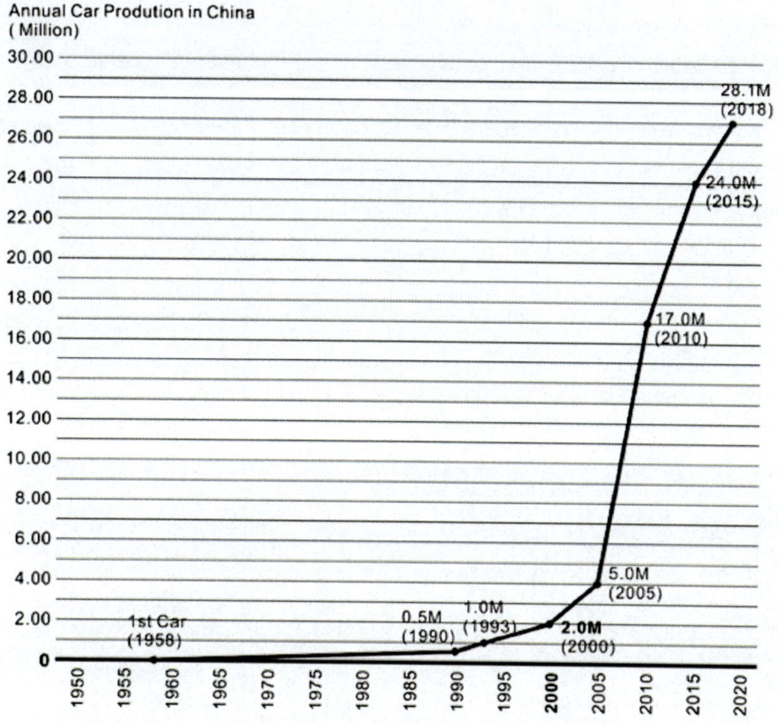

Fig. 7.3 Annual automobile production in China. Illustrated by Yang Yan

Fig. 7.4 Car congestion in Beijing. Photo by Yang Yan

of Transportation put forward the *Development Outline of Beijing Transportation 2004to 2020*. This outline proposes to:

> ...build a comprehensive transportation system dominated by public transport.

Through this, the commission aimed to achieve a mode share of 50–60% for public transport through the development of subway, BRT and super bus. In Beijing, as in

Table 7.1 List of time and distance of different travel modes

Mode	Average distance (km)	Average time (min)		Average speed (km/h)	
	Whole day	Morning peak	Evening peak	Morning peak	Evening peak
Bus	11.0	66.5	67.6	9.7	9.2
Subway	17.9	74.9	75.6	15.3	14.4
Taxi	9.4	43.2	45.2	11.6	10.3
Car	14.4	47.2	48.2	16.1	15.2
Bicycle	3.0	22.9	24.3	9.1	8.9
Walking	1.0	11.6	11.9	5.4	5.2

Source Beijing Traffic Development Annual Report 2019

Fig. 7.5 The narrow roads inside the old city become narrower. Photo by Yang Yan

other medium and large cities, this movement has led to the construction of subways. According to the Report on Statistics and Analysis of Urban Rail Transit in 2019 issued by China Association of Metros [3] 40 cities in Chinese Mainland had built subways by the end of 2019, totalling 5180.6 km. For comparison, prior to 2000, only three Chinese cities had subways and the total track length was only 253.7 km. Shanghai and Beijing have reached 669 km and 617 km respectively, their subways are now the largest in the world. According to the plan, the total length of subways in three super-large cities, Beijing, Shanghai and Guangzhou, will eventually exceed 1000 km, while Shenzhen, Wuhan and Chengdu will exceed 800 km.

Fig. 7.6 Subway in Beijing. Photo by Yang Yan

The de-motorisation movement has also brought improvements to bus services and passenger experiences. Air conditioning, wireless internet and low floor vehicles are now common. At the same time, Government policy has kept public transport cheap, and passengers over 60 years old ride for free, and cashless fare payment systems are widespread. These measures have combined to make the bus more attractive with bus Travel in Beijing now having 73% mode share [3] (Fig. 7.6).

7.1.2 *The Last Mile Problem and the Rise of Illegal Low-Speed Powered Vehicles*

In 2017 there were 30 Chinese cities with a population greater than 8 million, and a further 13 cities with a population of more than 10 million people. Three additional cities, Chongqing, Shanghai and Beijing have populations greater than 20 million people. For these super-large cities, public transport cannot spread widely enough to reach every inhabitant. Personal mobility modes are necessary for people to reach local destinations, especially since these facilities such as schools, shops, restaurants and hospitals are not concentrated into clusters in new urban areas. In 1995 the electric assist bicycle was welcomed as a viable mode of transport, especially by low-income users. These electric bicycles were speed limited to 20 km/h by traffic rules, and were required to combine human power with electric assistance. These rules did not prevent modifications that exceeded them, and gradually these electric bicycles evolved into bigger, heavier, faster machines without pedals at all, and became a new machine, the electric motorcycle.

Capable of speeds up to 60 km/h and weighing around 100 kg, these electric motorcycles are quite distinct from the vehicles they evolved from. They are larger than a bicycle, but are still a convenient and practical short distance vehicle suited to the urban environment. To some users, they function as a family car that carries three or four people; to others the vehicle is a pickup truck, carrying proportionately large quantities of goods through streets and alleys. Chinese traffic laws require a class-F driver's licence for electric motorcycles but in reality few users possess one. Drivers range in age from teens to the elderly demonstrating the appeal these vehicles have, for a wider age range of users than the bicycle (Fig. 7.7).

Despite this appeal and widespread use, the electric motorcycle does bring some problems. Initial dissatisfaction from the motorcycle industry led to the development of a general specification which brought vehicles of this type with a curb weight exceeding 40 kg and speed capability over 20 km/h under the aegis of the motorcycle industry. This means that the vehicles would have to pass a unique safety standard and be operated on motorised roads. Licences would be required to drive electric motorcycles. Development of this specification aroused corresponding dissatisfaction from the electric bicycle manufacturers because the new rules would place some 1800 manufacturers at risk by altering their production requirements and standards. Users were not pleased about the licencing requirement either, with the time and money spent on obtaining a class-F licence being perceived as onerous. Safety concerns were also raised by the requirement to ride the vehicles in space that was shared with

Fig. 7.7 The electric bicycle evolved into electric motorcycle. Photo by Yang Yan

cars. This view was reinforced by the police who were already busy with motorised vehicle crashes without introducing a new, vulnerable road user type into that space. Although the law was planned for introduction in 2010 it was postponed until 2019.

A concurrent vehicle development to the electric motorcycle was low speed, four wheeled, electric scooters which have been of particular benefit to the elderly. Before motorisation and cars became widespread, the elderly depended on public and human powered two and three-wheeled transport. This meant that their travel range was subjected to physical constraints such as accessing public transport, and weather constraints around comfortable bicycle use. Appearing on the market around 2000, the four wheeled electric scooter was readily accepted by the elderly. These vehicles are characterised by having a comfortable seat, battery electric drive, four wheels, and being imported. In a similar evolution to the electric bicycle, these vehicles have gradually changed to be capable of speeds around 30 km/h or higher, and their occupancy has increased from one to two or three seats. In part, this change was motivated by early vehicles having low power and speed, and thus causing difficulties in traffic. Manufacturers were encouraged to increase power output and attempts were made to revise the relevant laws. The vehicle now approaches the standard of EU category L6e Light Quadricycle [4] and weighs around 300 kg (Fig. 7.8).

These vehicles were supported by some local governments. Weakness of existing laws and regulations have essentially allowed the consistent breaking of limits on performance. Some instances of elderly scooters can now reach speeds of 60 or 80 km/h which for comparison is the top of EU category L7e for heavy quadricycles [4]. Despite these power and speed increases, the vehicles are still not subject to traffic rules and their drivers do not require licences. These are important factors in crashes.

The electric motorcycle and the elderly scooter have been welcomed by urban residents since the 1990s as convenient, low cost vehicles. As part of the rapid motorisation of China they have caused some traffic problems, most notably through their incompatibility with existing laws and regulations. The resulting state of traffic was somewhat chaotic for some time; although one benefit of the chaos is that city administrators can learn from how it exposes contradictions and helps in the pursuit of

Fig. 7.8 The evolutionary history of the elderly scooter. Illustrated by Yang Yan

appropriate solutions. In this case, and after years of discussion, three new national standards were introduced in 2019. "Technical specifications for safety of power-driven vehicles, GB7258-2017" [5], "General Specifications for Electric Motorcycles and Electric Mopeds GB/T 24158/2018 [6]" and "Safety Technical Specification for Electric Bicycle, GB17761-2018 [7]". The three standards are based on European Motorised Vehicle Categories [4] and clearly define the specifications of different vehicle types. A further standard, Low-Speed Four-wheeled Battery Electric Vehicle-Specifications (Draft) [8] is currently in discussion with implementation expected in 2021.

7.2 Future Trends of Urban Traffic in China

Development in China is the subject of local government "stage plans". These plans underwrite the development rate of China and as recent history shows once they are ratified the plans are rigorously implemented. We can take a perspective that the current plans therefore provide a reasonable picture of what the near term future will be like so far as development is concerned. This section describes transport and demographic trends of interest to urban traffic.

7.2.1 *The Path of Urbanisation Based on Urban Agglomeration*

One technique used determine the level of development in a country is to examine the proportion of its population that live in urban areas. The US has the world's highest level of urbanisation with 80% of the population living in urban areas. The average for developed countries is around 75%. The urbanisation rate in China has increased from 30.4% in 1998 to 59.6% in 2018, with linear growth of 1.46% per year. The Chinese government is aiming for 75% urbanisation by 2035, meaning that in that time 200 million people will move from rural to urban areas (Fig. 7.9).

In the past 40 years China's cities have developed separately, and the benefits of living in these cities is not shared with those remaining in rural areas. This example of the Matthew Effect [9] has created an economic gap between cities and rural areas. The current approach to improve urbanisation China has adopted an urban agglomeration approach. The 13th Five-Year Plan of National Economic and Social Development of People's Republic of China (2016–2020) [10] promulgated by the State Council of China in 2016 shows that China will build 20 urban clusters. Among these, five super-large urban clusters with populations of more than 100 million are planned. The plan aims to solve the problems of individual city development and achieve sustained development. The plan's aims are characterised by the following changes.

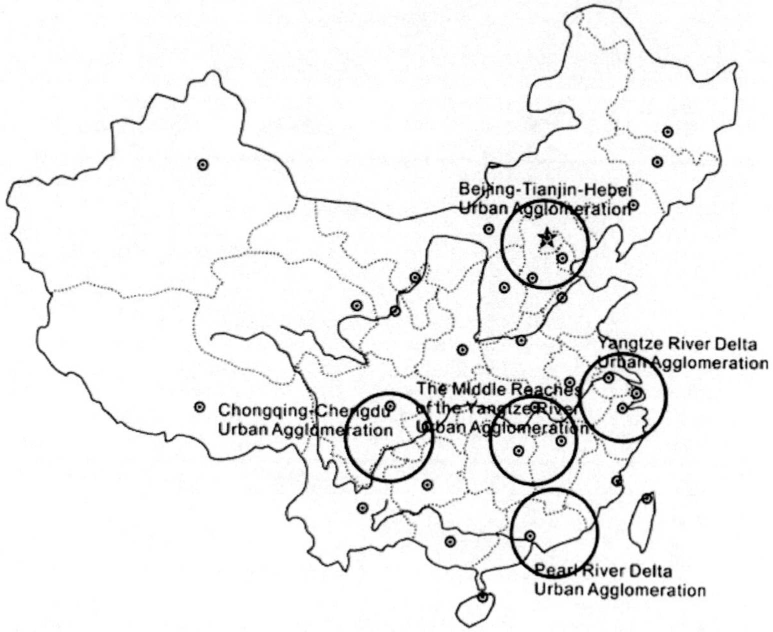

Fig. 7.9 Five super urban clusters according to the 13th five-year plan of national economic and social development of People's Republic of China (2016–2020). Amended by Yang Yan

A change from single-core to multi-core cities aims to distribute opportunities beyond a single city core. Beijing, Shanghai, Guangzhou and Shenzhen are examples of single core large cities in China. They all concentrate employment in the centre and disperse residential areas in surrounding, and increasingly remote areas. The traffic flows are "morning in and evening out" with a typical distance being 50 km one way. As a dominant travel pattern in a very large city this leads to capacity challenges that even very large subway systems struggle to cope with.

In contrast, the distance between the cities in the multi-core urban clusters is generally between 50 and 200 km, a distance at which a metro or subway type system is inappropriate. There are plans to introduce high speed railways, maglev systems and regional aeroplane services for parts of these multi-core agglomerations.

7.3 Intercity Daily Commuting

The availability of new transport services has led to new travel patterns. One pattern is the "urban pendulum clan" referring to those who commute using high speed intercity railways. This group are broadly characterised as living in Tianjin, but working in Beijing, and travelling the 120 km between the two cities using high speed railway capable of 300 km/h. This allows access to jobs in Beijing and lower living expenses in

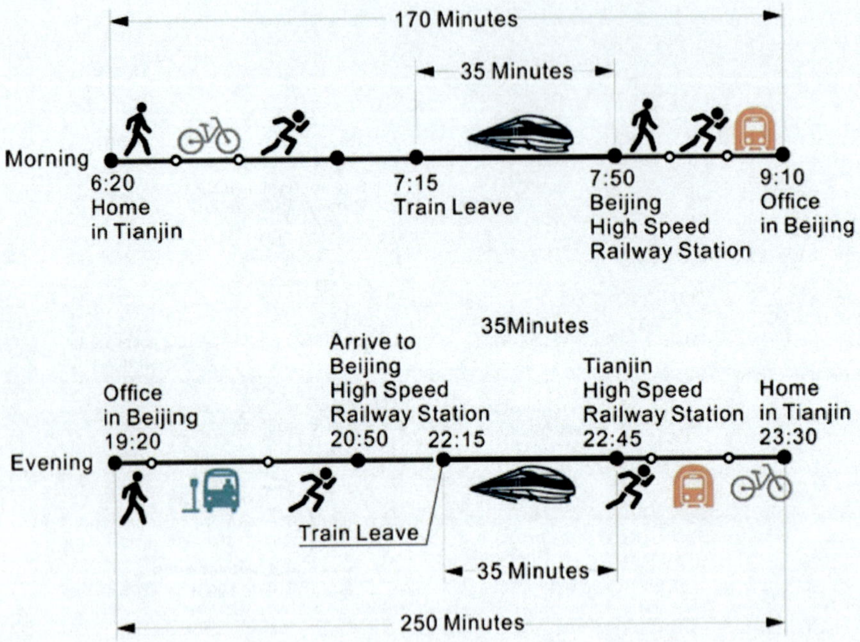

Fig. 7.10 Li Hao's Commuting Schedule. The original picture comes from the Paper.cn, 2018.08.09. https://www.thepaper.cn/newsDetail_forward_2323941/. Amended by Yang Yan

Tianjin. Despite this high speed mode being part of their journey, the urban pendulum clan commute for up to seven hours per day. Figure 7.10 shows the commuting schedule of Mr Li Hao, to illustrate the journey [11]. With the development of urban clusters it is possible we will see more patterns like this, and it is plausible that air services could become part of daily commuting.

7.3.1 *Multi-Mode Shifting*

From the commuting schedule of Mr Li, we can see walking, bicycle riding, subway, bus and high speed railway modes all used in one day. Although a great distance is covered, the high speed railway mode takes only 70 min per day or only 16.7% of commuting time. This travel pattern reveals two bottlenecks; urban traffic and the time taken to transfer between services, both of which may be fertile ground for innovations.

7.3.2 *China's Ageing Society*

China implemented a one-child policy in 1978, which has served as designed to reduce the rate of population growth, but has also served to accelerate the ageing rate beyond what would otherwise have naturally occurred. Modelling indicates that China will enter a period of severe ageing from 2030 that will peak by 2050 by which time it is estimated that there will be 430 million people in China over the age of 60 [12]. The number of elderly would exceed that of the United States, Europe and Japan combined, which will lead to a variety of challenges including mobility (Fig. 7.11).

The modelling goes on to predict that a typical Chinese family will be structured in an inverted pyramid of 8/4/2 with the total maintenance ratio (minors and elderly) of 66% by 2030 and 88% by 2050 [12]. Under the pressure of medical expenses and insufficient pensions, a foreseeable forecast for mobility is that unlike today's recently retired elderly who travel for leisure, people of the same age in 2030 and 2050 will travel for work (Fig. 7.12).

In this future context we need to consider what traffic modes might be suitable for an ageing society. Of particular concern are the need for physical strength and condition, and cognitive power and their role in creating achievable mobility. If we assume, for example, that a 75-year-old works at a location 10 km away two or three days per week, what reasonable mobility solution is available, considering the

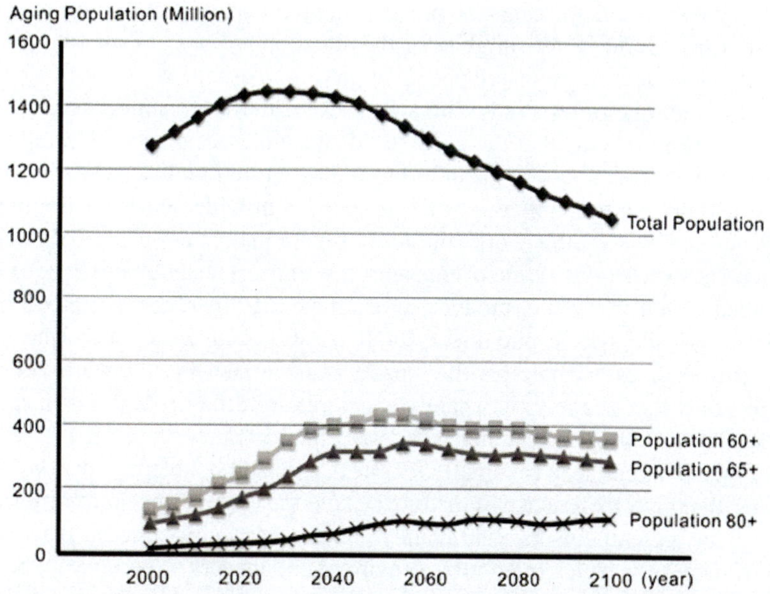

Fig. 7.11 The aging trend of China's Population in 21 Century by Peng Du, Renmin University. Amended by Yang Yan

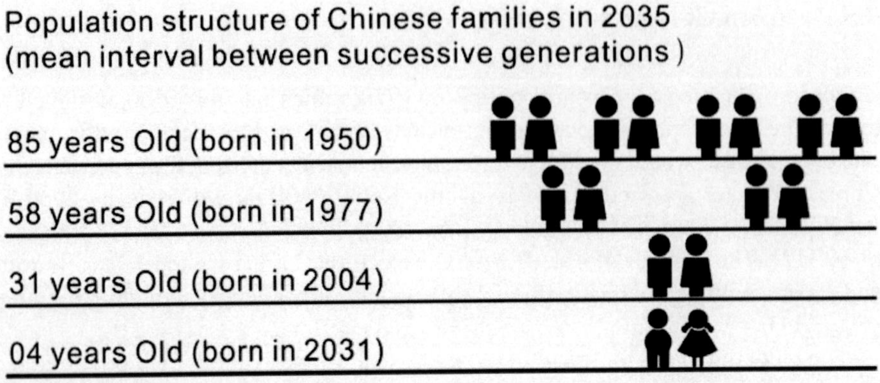

Fig. 7.12 Population structure of Chinese families in 2035. Illustrated by Yang Yan

limitations discussed above such as access to public transport nodes? For China's emerging cities of more than 20 million people this is a difficult goal to achieve.

7.4 Individual Vehicles

China has arrived at its own version of the urban traffic problem. The most pressing aspect of this is spatial, the car is too big. A vehicle conceived for multiple occupancy in a pre-urban world is not suitable for the situation we see in China's, as well as many of the world's cities, where people are trying to use the car for a single person to travel in multiple times a day. China's car use patterns are similar to places with similar traffic problems, that the car is used by single occupants on a daily basis, by two occupants on a less frequent basis, and by entire families on weekends. The cars people drive are nominally 4 m long and 1.8 m wide, and with Beijing's car occupancy rate of 1.09 this works out to 24 square metres per person of road area. This represents a serious waste of transport resources. The automotive industry has responded with new, smaller models, for example the Smart car with 2.695 m long and 1.559 m wide, but despite this, a small car still occupies a whole lane and the Smart car occupies a whole parking space. Cars smaller than this are difficult to conceive of if they are to satisfy safety standards so further reductions in scale are not likely to have much effect, nor be likely (Fig. 7.13).

Despite an urgent need for small, flexible, single and double occupant passenger vehicles, the travel experience of the bicycle and motorcycle appear to be unacceptable to many potential users, who mainly cite climate and weather as barriers. The array of modern vehicles in an urban environment appears to be diverse, but from a vehicle design perspective this is not the case, as this array of vehicles fits into a narrow category of cars, all of which share very similar intrinsic characteristics despite the efforts of car manufacturers to differentiate the vehicles based on surface design and market positioning. What is under-explored is the possibility for vehicles

Fig. 7.13 One of the smallest car: smart. Image by Michael Rivera—Own work, CC BY-SA 4.0, https://commons.wikimedia.org/w/index.php?curid=38800114

outside of the automobile, motorcycle and bicycle categories. New vehicle segments should be conceived to avoid disadvantages of the three dominant segments. New segments could, for example, more smoothly connect with public transport, and have higher passenger occupancy for their size.

Three new niche areas for vehicle innovation have been identified among established national standards;

- "Technical Specifications for Safety of Power-driven Vehicles, GB7258-2017" [5],
- "General Specifications for Electric Motorcycles and Electric Mopeds, GB/T 24158-2018" [6],
- "The Safety Technical Specification for Electric Bicycle, GB17761-2018" [7] and,
- "Low-speed Four-wheeled Battery Electric Vehicle-Specifications (Draft)" [8].

These emerging standards acknowledge the elderly scooter, the "low speed four wheeled battery electric vehicle", that has been in use in Chinese cities for more than a decade. As a replacement for the car, this vehicle has the ability to meet the travel needs of urban transport if it can appeal to a wider range of users.

7.4.1 To Meet Urban Travel Demand

Commuting accounts for 44% of Beijing's transport [3]. The nature of commuting travel is firm in time and place, because people have less choice in commuting than other travel over their destination or arrival and departure times. This is a large determinant of peak travel times, and as such it is interesting to note that people will choose less comfortable modes of transport than the car, such as bus and bicycle

Fig. 7.14 Sending children to school is one of the main purposes to buy cars for urban families. Photo by Yang Yan

to overcome the time-based problems of traffic jams, and to avoid deferring travel times. Commuting vehicles should prioritise parking and transfer to other modes.

Another part of urban transport which a lot of Chinese families see as a reason to purchase a car is to take passengers, for example driving children to school. Witness the large number of cars waiting at school gates, and similarly, around hospitals. A small non-car vehicle could be a good solution to this problem, for example the low-speed four-wheeled electric vehicle since they have room for one or two passengers and have easier access and egress than very small cars, which tend to have very cramped conditions in the back seat (Fig. 7.14).

Shopping capacity needs to be considered in the development of suitable urban vehicles too. Very compact car bodies may not have enough carrying volume for practical trips. Again it is conceivable that a non-car vehicle design could have a better ratio of carrying capacity to vehicle size because of the relatively smaller space requirements a low-speed and lower-range electric vehicle would have over a car.

Weather, and the attendant issue of comfort is a final consideration for a new urban vehicle. Cars offer an enclosed space that is climate controlled, and as such set a high standard for alternative vehicles if they are to appeal. As a large country, the challenge in China, as for many places, is that seasonal and geographical variation in climate is vast, with freezing conditions in the North contrasting against tropical conditions in the South.

7.4.2 Road and Parking Space

A smaller vehicle has the potential to increase the capacity of the roads, so long as the occupancy rate per unit of road space increases. The road traffic law does not permit vehicles to travel side by side in the same lane, so as discussed above there is little use in reducing the width of a car to increase capacity. There may be an ability to save area in parking spaces however. In 2000 Beijing established park and ride facilities on the outskirts of the urban area, in a bid to encourage transfers to public transport. These parking lots were underutilised because they were too small. A small vehicle, such as a low speed electric quadricycle, could use smaller spaces. Three small vehicles could be parked in the space normally set aside for one car.

If vehicle sizes were known and more closely controlled, for example through the establishment of the new standards, it is also possible to conceive of parking which has smaller gaps between vehicles. This could diminish the wasted space between vehicles today, which is essentially there to accommodate the differences between different car styles. While a closer control system is not without problems of its own such as increased complexity, it may be possible to save even more space.

7.4.3 Allowing More People to Drive

In the past decade, driving a low speed four wheeled battery electric vehicle has not required a licence. This is a matter of great convenience to elderly users who have never been trained in driving. It is also a matter of great safety concern, since untrained drivers have caused crashes. Driver licencing has been debated for some time. We do not object to the idea that drivers need to be trained, and would like to focus on what an appropriate level of training is for this vehicle type. In Europe, similar vehicles are divided into two classes; Heavy quadricycles (EU category L7e) with a maximum speed of 80 km/h, and light quadricycle (EU category L6e) with a maximum speed of 45 km/h. The corresponding Chinese classification is Low speed four-wheeled battery electric vehicle specifications (Draft) [8], which stipulates a maximum kerb mass of 750 kg, a speed range in the order of 40–70 km/h, and that occupants are provided with seat belts and airbags (Fig. 7.15).

Several alternatives are available to deal with the issue of licencing. One approach is to apply the same licence requirement for these vehicles as for the car (Class C licence) to those people wishing to drive the Low Speed Electric Quadricycle, however a counter argument to this is that it would exclude many teenagers and elderly people with access to independent mobility. An alternative is a less difficult driving licence test for this new type of vehicle; a test suitable for the vehicle's limitations and also the target market of teenagers over 14 and those people older than 70 [12]. This could be part of an enabling strategy for these smaller vehicles to be part of an urban transport system.

Fig. 7.15 Renault Twizy, EU category L6eL7e quadricycles. Image by Arnaud 25—Own work, CC BY-SA 4.0, https://commons.wikimedia.org/w/index.php?curid=40965339

7.5 Innovation Opportunities

7.5.1 Low-Speed Four-Wheel Electric Vehicle

Research in the Transport Design Cluster of the Academy of Arts & Design, Tsinghua University (TH Trans D) has studied the relationship between the specifications in this vehicle category and the vehicle package outcomes that they precipitate. The study finds that if a Low-speed Four-Wheeled Battery Electric Vehicle is designed around too many occupants, the resulting vehicle could end up being larger than a car from the A00 segment. This would diminish the ability of the smaller vehicle to offer advantages such as high-density parking in urban environments. It would, in effect, be another car. Reversing this approach, and putting the vehicle package first in the design process has determined that a small vehicle could still be designed if the occupants were limited to two adults and one child under 12 years old. By controlling the number of occupants, the vehicle body can be kept within a tighter package dimension of 2800 × 1400 mm. In turn, this enables us to realise the goal of increasing storage density by 3 times when compared with the car (Fig. 7.16; Table 7.2).

7.5.2 All Weather Electric Motorcycle

Bicycles and motorcycles are ideal urban personal vehicles mainly because of their size, however the drawback of being exposed to the elements discourages some people from using them. Some of these weather issues can be extreme, for example in the North of China a user can reasonably expect between five and seven months of icy conditions, and rainy, hot summers are common in Southern China. Adding

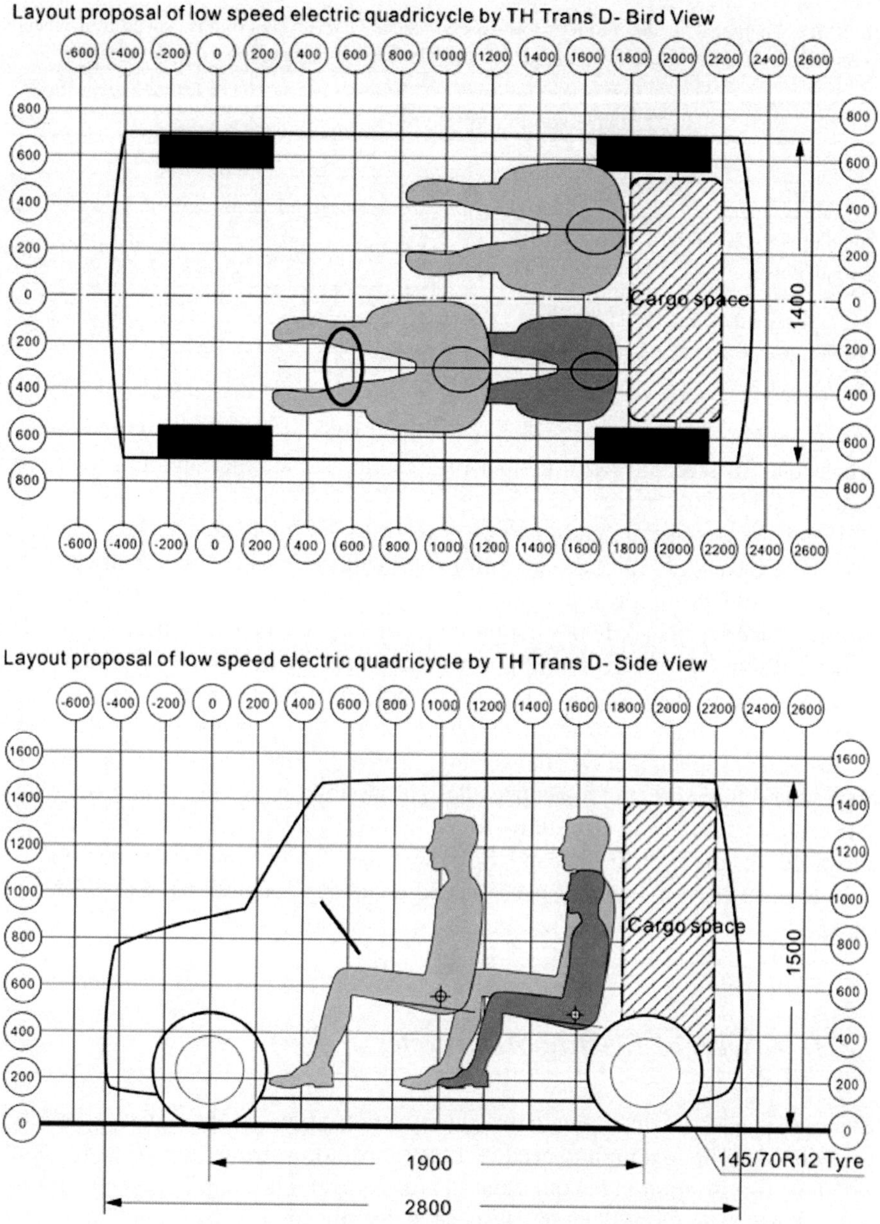

Fig. 7.16 Layout proposal for low speed quadricycles by TH Trans D. Illustrated by Yang Yan

Table 7.2 Technical specification of low-speed electric quadricycles by low-speed four-wheeled battery electric vehicle-specifications (draft) and TH Trans D suggestion

Technical specification	Low-speed four-wheeled battery electric vehicle-specifications (draft)	TH Trans D suggestion
Speed	40–70 km/h	40–70 km/h
Curb weight	<750 kg	
Number of passengers	<4	2+
Overall dimensions	3500 L * 1500 W * 1700	2800 L * 1400 W * 1600
Power	Lithium-ion battery	

Source Low-speed Four-wheeled Battery Electric vehicle-Specifications (Draft)

a roof and bodywork to a two-wheeled vehicle such as the motorcycle or bicycle is possible, for example resulting in a velomobile; and another option is to create a motortricycle.

Motortricycles are divided into two categories: Two-wheeled motorcycles with side-car (EU category L4e) and motortricycles with three wheels symmetrically (EU category L5e). The motorcycle with sidecar is about the same width as the Low Speed Electric Quadricycle, so it does not have an obvious advantage in space saving. The symmetrical motortricycles have different characteristics depending on whether they have two wheels at the front or back. Two wheels at the front results in a vehicle more suited for high speed and two wheels at the back for low speed. Both offer drivers and occupants the car-like advantage of static stability, which for the driver makes parking easier, and also opens design possibilities for protective bodywork. The vehicle package and size study shows that with a fully enclosed body the length and width of a motortricycle can still be less than 2100 × 800 mm enabling four such vehicles to fit in one car space, or under "no gap" conditions, six vehicles per car space.

7.5.3 *Drivers of Electric Motorcycles*

Electric tricycles are divided into two classifications; Electric Motorcycles and Electric Mopeds. The maximum speed of Electric Motorcycles is limited to 120 km/h and the driver is required to hold a class-D (motorcycle) licence. The electric Moped by contrast has more limitations on speed and application being limited to 50 km/h and being ridden on sub-trunk roads. In turn this leads to the relatively less limited requirement for class-F (light motorcycle) licence. In other words, the level of restriction placed on the vehicle type is inversely proportional to the level of restriction in the licencing to drive it.

Electric motorcycles require more skilled drivers because of the high speed performance when compared to Mopeds. The respective licencing requirements both span from 18 to 70 years of age. If we consider the travel needs of potential Moped riders who fall outside of this age range, and also their abilities, it may be possible to shift some of the local urban transport task from the car to these smaller vehicles. First, the case of the middle-school student, aged around 15. Although generalisations will always have exceptions, the typical Chinese middle school student has achieved proficiency with bicycle riding some years prior and this experience is often gained in crowded cities. The longer trip to middle school may not be feasible by bicycle, and so this trip is taken as a chauffeured trip in the family car. If this trip were made by electric moped it would alleviate the car trip to school, as well as potentially eliminating the empty chauffeur trip home by that same car, as the student could travel to school by themselves. Second, at the other end of the age range, it is evident that many older Chinese people aged between 70 and 79 have better physical and mental capabilities than those people of similar age a generation ago. As discussed above, China's demographics also mean that more people in older age brackets will be working. The health characteristics of China's development have meant an extension to healthy lives and thus we might also be able to extend the car-replacing capability of mopeds to this older age bracket as well.

7.5.4 Proposed Layout of Motorcycle

Putting aside their higher cost, the F2R1 (two front wheels and one rear wheel) motortricycle could provide an ideal platform for all-weather designs. The vehicle can be ridden by two occupants and the front wheels, if provided with a tilting steering system, can provide similar dynamic characteristics to two-wheeled motorcycles. Current examples in the market show this outcome to some extent, for example the Piaggio Mp3. In contrast to this F2R1 layout, the F1R2 layout does not need a complex tilting steering system and may be more suitable for low speed applications (Figs. 7.17 and 7.18).

7.5.5 Electric-Tricycle with Increased Shelter

It would be difficult for a road planner, or anyone, to conceive of the range of future transport modes and vehicles that could one day use a road which is being planned in the present day. The roads we currently have are similarly a victim of their time, and it is possible to conceive that the potential of China's "non-motorised roads" extend far beyond the bicycles for which they were originally designed. One such example is the e-bike, which has rapidly become a fixture of sustainable urban transport in developed countries and are ridden on infrastructure that has not been developed with these higher speed, higher mass bicycles in mind. Unfortunately, the e-bike specification "Safety Technical Specification for Electric Bicycle, GB17761-2018"

Fig. 7.17 Piaggio Mp3. By FaceMePLS from The Hague, The Netherlands—Piaggio MP3, CC BY 2.0, https://commons.wikimedia.org/w/index.php?curid=51520089

stipulates that the benefits of the electric assist feature cannot be added to tricycles. Adding electric assist to a tricycle may seem to be a minor issue, however when we consider the user needs of those living above 40° North and how they may choose non-car modes during the five to seven winter months, there is an opportunity. The weather protection offered by the car is a huge comfort advantage for this mode. Adding some minor weather protection to a tricycle would take it some way to being more protected, but would add mass to an already large human powered vehicle, a disadvantage which would be overcome to some extent with the addition of electric assist. The static stability of a tricycle may also offer advantages in icy conditions. Figures 7.19 and 7.20 illustrate potential layouts of such vehicles. Figure 7.21 sets out the motorised vehicle classifications in China against motorised and non-motorised roads.

7.5.6 Transport Systems for Very Large Urban Agglomerations

There is little doubt as to the role public transport plays in China's super-large urban agglomerations, since some commute distances will exceed 100 km and therefore be dependent on higher-speed trunk routes to achieve some form of feasibility to fit into the working day. The role of smaller vehicles in such agglomerations is still

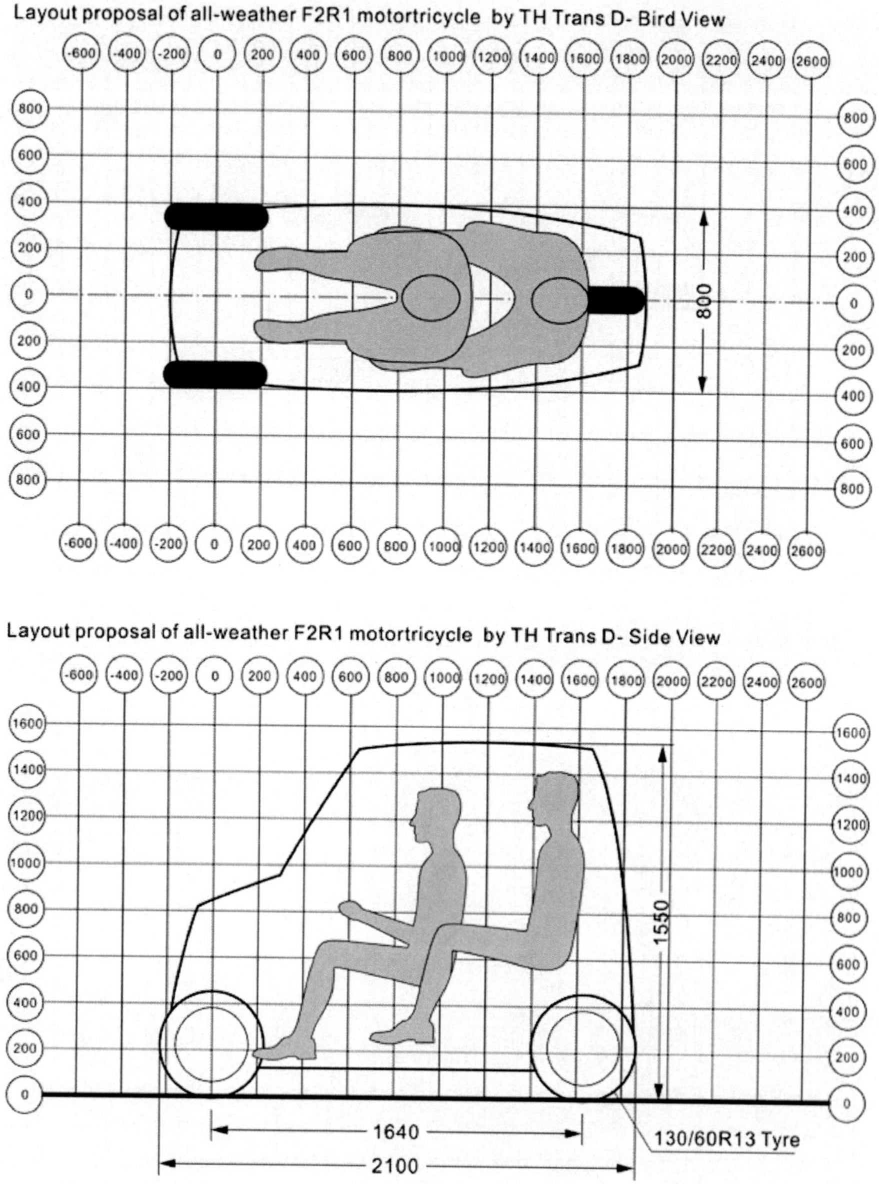

Fig. 7.18 Layout proposal for F2R1 motorcycle by TH Trans D. Illustrated by Yang Yan

important though—and here we discuss their role in relation to the public transport system in two regards. First the role of smaller vehicles as a feeder mode for public transport's "last mile" problem, and second the provision of short local trips which are not served by public transport.

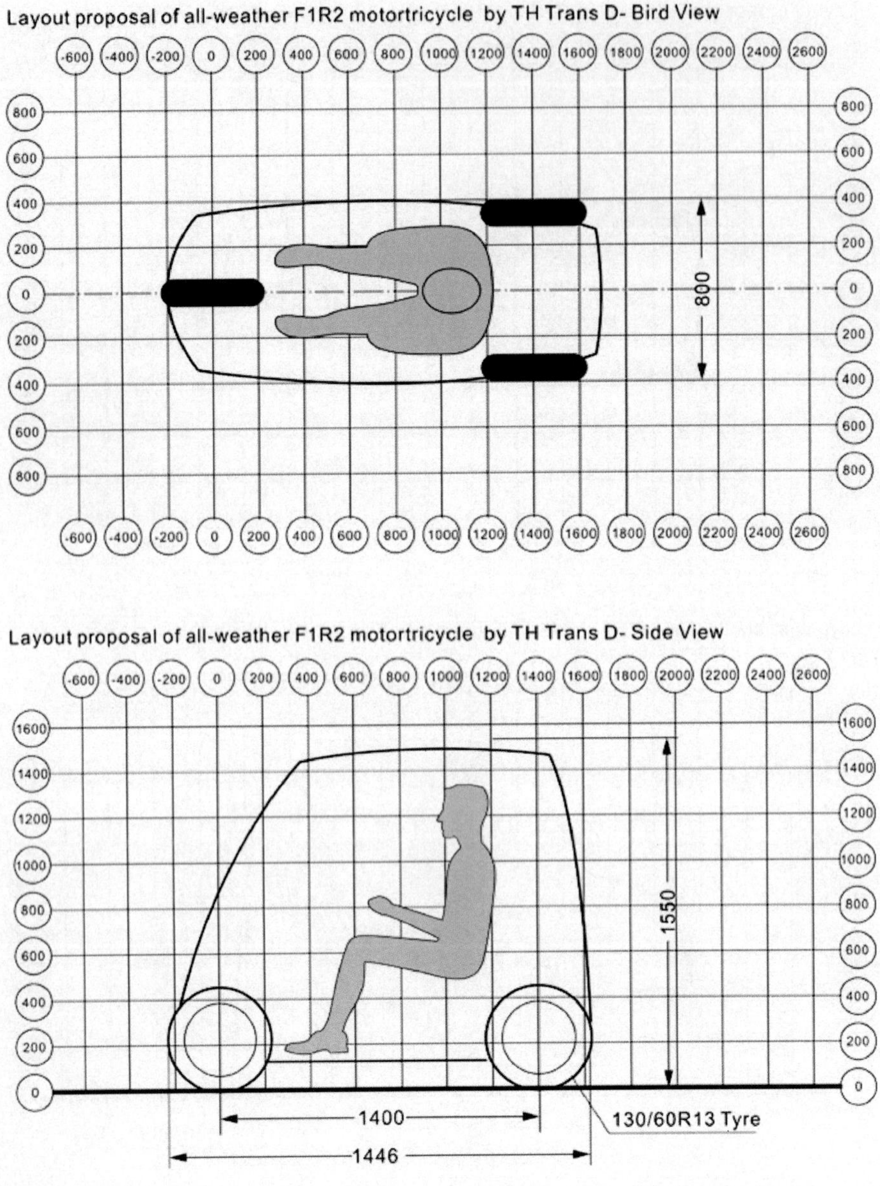

Fig. 7.19 Layout proposal for F1R2 motorcycle by TH Trans D. Illustrated by Yang Yan

7.5.6.1 High Density Storage at Public Transport Nodes

The trunk capacity of public transport is underwritten by geometry—by using a single large vehicle many people can be moved with spatial efficiency. This is in

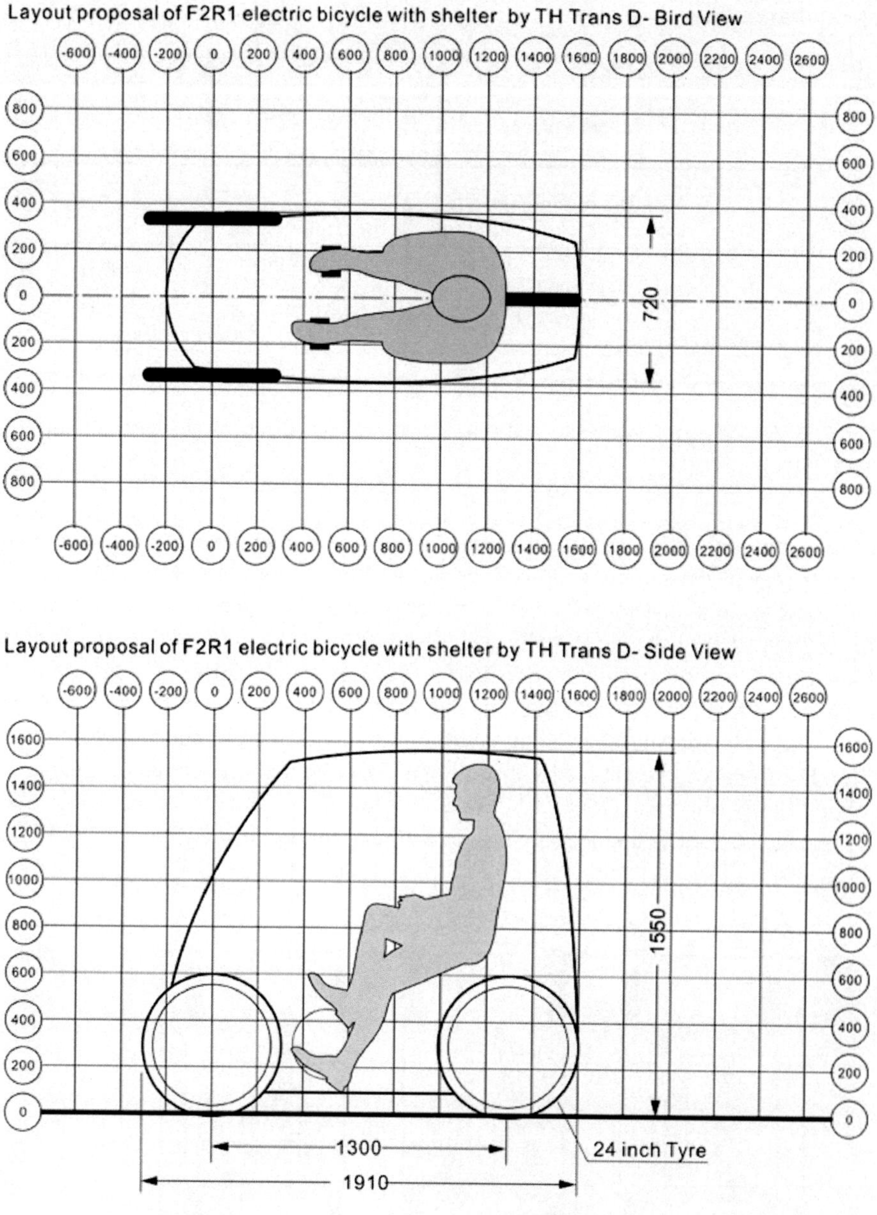

Fig. 7.20 Layout proposal for F2R1 electric tricycle. Illustrated by Yang Yan

stark contrast to the spatial efficiency of single-occupant cars. And yet, in many cases the car is relied upon as a feeder mode for public transport, creating a spatial problem at public transport nodes, where many cars are parked all day, and only serve two journeys; those to and from the public transport node. One advantage of a smaller vehicle may be to increase the spatial efficiency of node parking. If we

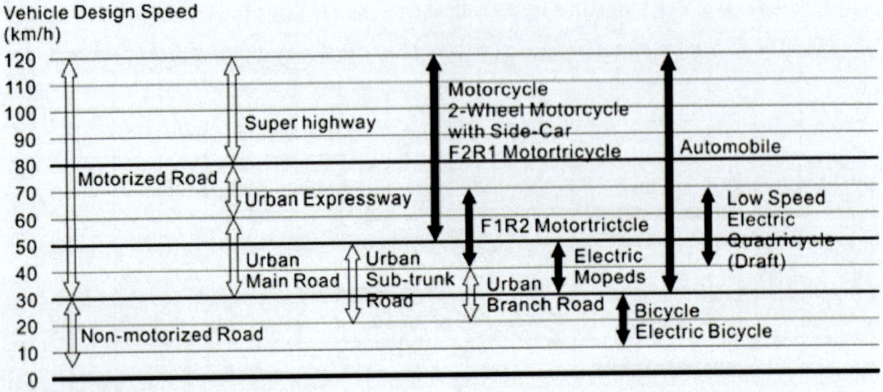

Fig. 7.21 Motorised and non-motorised vehicle and road classifications of China. Illustrated by Yang Yan

change our consideration of vehicle package from vehicle development to how the vehicle may be densely parked, we see that the storage density of low speed electric quadricycles and all-weather electric motortricycles can respectively be from four to six times that of the car. Smaller vehicles can also be conceived, with a concept for a Slim-Escooter designed specifically for high-density parking being up to 18 times more spatially efficient than the car. Simple layouts reveal that 40 low-speed four wheel electric vehicles, 60 all-weather three wheeled electric vehicles, or 180 Slim Escooters can be parked by installing a two-metre automatic parking garage at a height of two metres above or below the ordinary subway station building. Such an arrangement takes up the same space as 10 car parks (Figs. 7.22, 7.23, 7.24, 7.25 and 7.26).

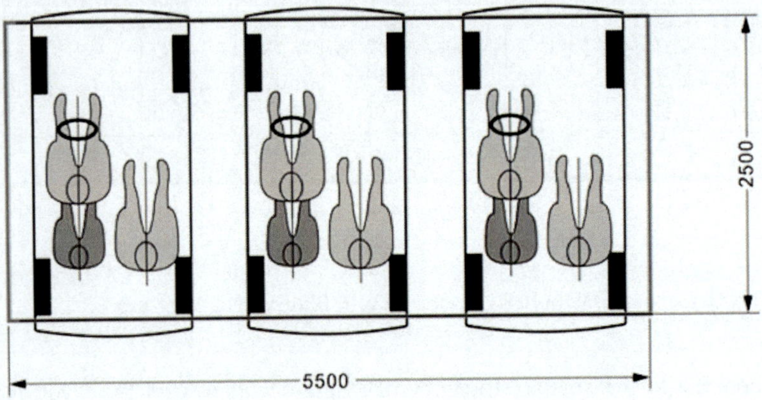

Fig. 7.22 High condensed parking for quadricycles. Illustrated by Yang Yan

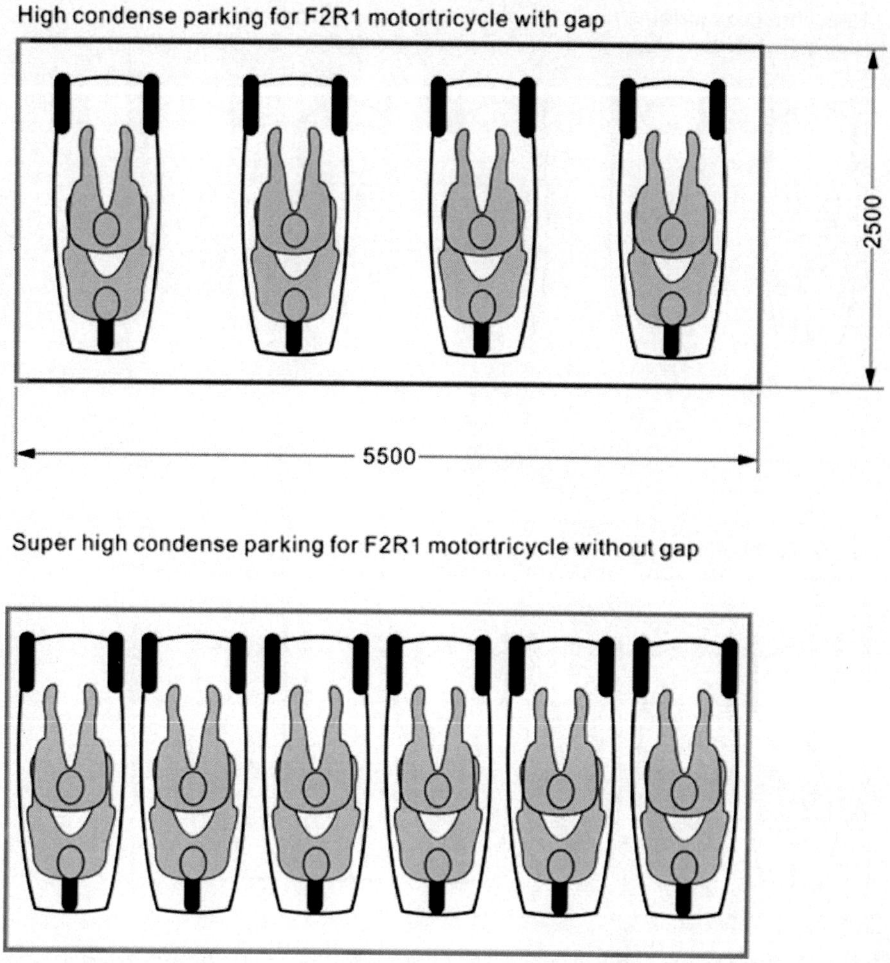

Fig. 7.23 High condensed parking for F2R1 motortricycle. Illustrated by Yang Yan

7.5.7 System Design

The objectives of the above research have been to consider, and exploit, novel vehicle package and configuration designs that answer the needs of the transport system in a way that is more considered than meeting individual needs. An area for future research is to consider the role of data in how vehicles are used. For example, if recent developments in AI were applied into the problem of transport, we could conceive of a future where the supply and demand of share vehicles is determined and executed in a way that offers vehicles—both in type and quantity—in line with user needs. Similarly, advances in autonomous vehicles could develop into transport system designs whereby garaged vehicles are shared, and yet become available when the owner returns to their park-and-ride location for the journey home. Similar systems

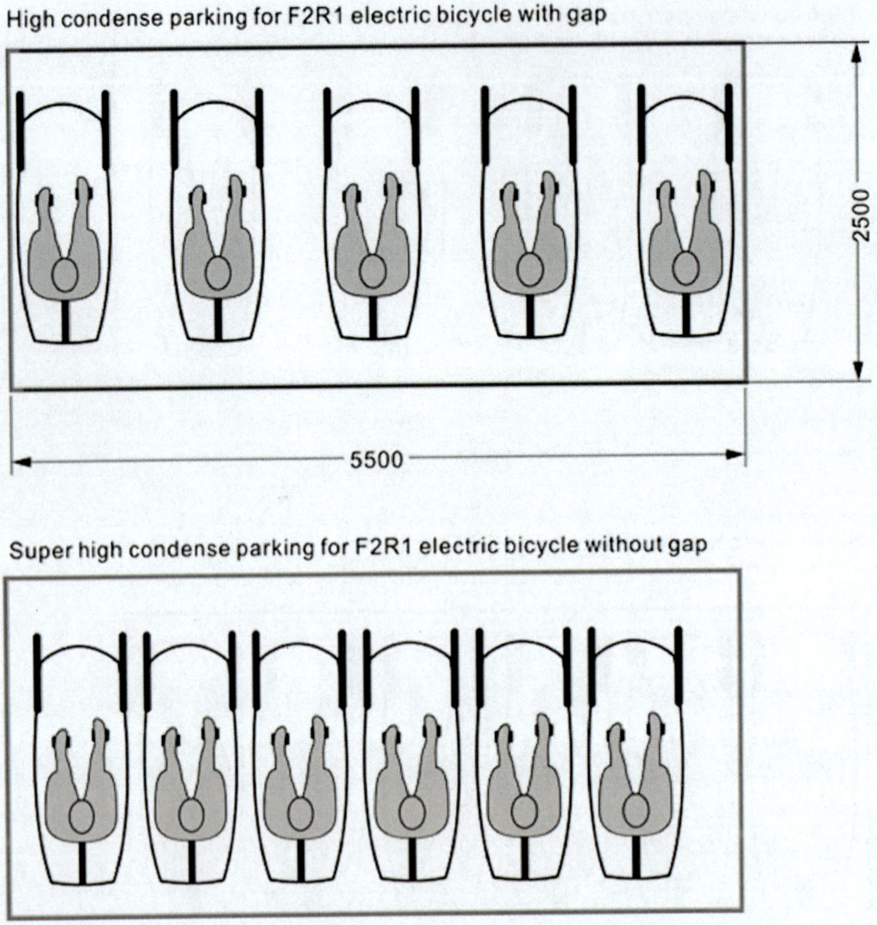

Fig. 7.24 High condensed parking for F2R1 electric bicycle. Illustrated by Yang Yan

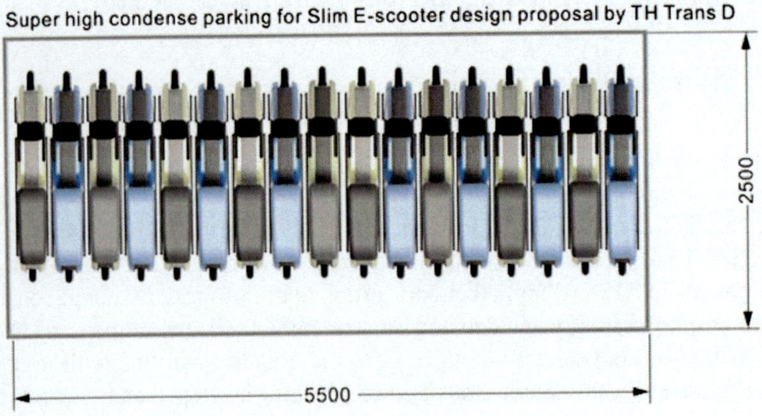

Fig. 7.25 High condensed parking for Slim Escootor. Illustrated by Yang Yan

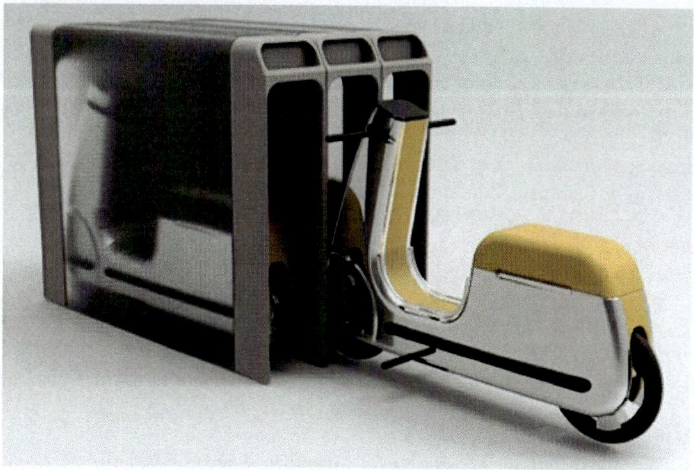

Fig. 7.26 Slim Escootor. Design and rendering by Yang Yan

are already deployed for bicycles in advanced cities as part of Mobility as a Service (MaaS). There is an opportunity for vehicle design to integrate with MaaS to form more intelligent system responses to the needs identified in this chapter.

7.6 Conclusion

China's development in recent decades has led to the rise of automobility, and the rise in the attendant problems of congestion, pollution and impacts on the availability of urban space for transport and parking. At the same time, vehicle regulation and licencing requirements have not kept up to the pace of vehicle development. In this chapter we've presented novel licencing, regulation and vehicle designs as potential solutions to some of China's problems, especially those of transport in urban super agglomerations. We conclude that solutions such as those presented may provide a way forward; however the broader conclusion offered is that the design of vehicles can no longer afford to be carried out in isolation of regulatory and infrastructure considerations, and that the development of a holistic transport system can encourage the design, and use, of more efficient vehicles.

References

1. Guideline on the Design of Traffic Operation Of Urban Roads GB/T 36670-2018. The Ministry of Public Security of the People's Republic of China. https://std.samr.gov.cn/gb/search/gbDetailed?id=77D9F23EFB9D58E2E05397BE0A0A2879
2. https://news.163.com/14/0117/07/9IPB2U0L00014MTN.html

3. Urban Rail Transit Statistics and Analysis Report in 2019. China Association of Metros. https://www.camet.org.cn/tjxx/5133
4. Directive 2002/24/EC of the European Parliament and of the Council of 18 March 2002 Relating to the Type-Approval of Two or Three-Wheel Motor Vehicles and Repealing Council Directive 92/61/EEC, 2002L0024-EN-17.05.2005
5. Technical Specifications for Safety of Power-Driven Vehicles, GB7258-2017. National Standardization Technical Committees
6. General Specifications for Electric Motorcycles and Electric Mopeds GB/T 24158-2018. National Standardization Technical Committees. https://std.samr.gov.cn/gb/search/gbDetailed?id=7643B2F25185267CE05397BE0A0AAF6A
7. Safety Technical Specification for Electric Bicycle, GB17761-2018. National Standardization Technical Committees. https://std.samr.gov.cn/gb/search/gbDetailed?id=71F772D827C1D3A7E05397BE0A0AB82A
8. Low-Speed Four-Wheeled Battery Electric Vehicle-Specifications (Draft), National Standardization Technical Committees. https://std.samr.gov.cn/gb/search/gbDetailed?id=5DDA8B9D9F5D18DEE05397BE0A0A95A7
9. Rigney, D.: The Matthew Effect: How Advantage Begets Further Advantage. Columbia University Press
10. Outline of the 13th Five-Year Plan of National Economic and Social Development of People's Republic of China (2016–2020), National Development and Reform Commission. https://www.ndrc.gov.cn/fzgggz/fzgh/ghwb/gjjh/201603/P020160318564052484043.pdf
11. A Tale of Two Cities: The Pendulum in Beijing and Tianjin, The Paper.cn, 2018.08.09. https://www.thepaper.cn/newsDetail_forward_2323941
12. Peng, D., Zhenwu, Z., Wei, C.: The aging trend of China's population in 21 Century. Population Res. **29**(6) (2005)

Chapter 8
ReinCARnation: The Emergence of Time Over Ownership as the True Brand Value and Driving Force for Design

Mark Johnson and Selby Coxon

Abstract Is the one-hundred-year love affair with the privately-owned motor car drawing to an end? Are we entering a period where unfettered enthusiasm for the peak of personal prestige and perceived freedom is tempered by a world of climate anxiety, economic challenges, densely populated cities and shifting cultural priorities? Within the design industry, automotive design has long been the popular and jealously guarded elite of industrial design practice. From the genesis of mass production on the assembly lines of Henry Ford to the elite robot populated super factories of Volkswagen. The push to appeal to consumers through the exceptionalism of developing form and surface has been the backbone of the automotive industry's business model. However, as we move into the third decade of the twenty-first century the supremacy of the privately-owned car is under attack. Developing nations are seeing a decline in young people taking up licenses and while former developing countries such as China and India are embracing the motor car, sales in other parts of the world are in decline. More of the world's population live in cities than at any other time in human history and corresponding challenges for mobility have not been kind to the car. The motor car's expansive footprint relative to occupant capacity has congested streets and choked arterial roads making the former tactile pleasure of 'motoring' seem like a distant fantasy. All the while, an accompanying shift away from fossil fuels with their incumbent problems of pollution, expense and finite life. If design has been such an influence on the past successes of automotive industry then how will it address new business models predicated upon shared use, a decline private ownership, new power trains and oversee a smooth transition to a new engagement with mobility in general.

Keywords Automotive design · Brand · Time related ownership · Access economy · Car share · Experience design

M. Johnson (✉) · S. Coxon
Mobility Design Lab, Monash University, Caulfield East, Melbourne, VIC 3145, Australia
e-mail: mark.n.johnson@monash.edu

S. Coxon and R. Napper (eds.), *Advancing a Design Approach to Enriching Public Mobility*, Intelligent Systems Reference Library 198,
https://doi.org/10.1007/978-3-030-64722-3_8

8.1 Introduction

> The car is still something that shows who you are. It is important, because it enables you to move around, and it makes you more free. It's like fashion: different types of people are attracted by different brands and models, and your car is a status symbol to show who you are, what you think and what you would like to be. Humans don't change: a man will always want to show himself, he wants to spend his money, and show what he prefers by his choice of car. A car can do that.

Car design master Giorgetto Giugiaro

Few inventions have had as profound an effect upon society as the motor car. The central historical narrative of human transport has been the reduction of travel time with a corresponding increase in attainable distance. The motor car did this dramatically and with these advances provided more opportunity for life experiences such as leisure and work and all to a wider variety of places. The motor car spawned new manufacturing techniques beyond the car industry alone and cemented a place in our society and culture.

The profound effect of the step by step assembly line approach not only made the motor car affordable but many other consumer products as well. Fordism is in the lexicon of the English language and the cornerstone of a consumer society. Nearly four million people work in the car industry directly [1] and many more in industries that service it such as oil and gas. The expansion of our cities and suburbs that surround them, are all made possible through the availability of the motor car.

In examining the human trait for the accumulation of material goods, the ownership of a motor car is second only to owning land and property. The purchase of a motor car hints at the admiration we stand to receive from our neighbours and friends. The automotive industry has long understood that to survive it needs to continually appeal to the human frailty towards the desirability of the 'new'. How does one maintain a market after it has become saturated? Forms of planned and dynamic obsolescence fed by teams of designers working on facelifts and styling affections before full new model roll outs kept the aspirational public fully engaged in keeping up with the latest offerings.

There are drawbacks to this position. Styles that fail to capture the imagination of the public backfire and require hasty retreats and re-workings while losing market share. Styles that garnered success have articulated the dynamics of movement; streamlining, tail fins and a 'slippery' form that evoked movement but also some sort of imagined future. Styling is an essential part of an effective marketing tool creating spin off segments from family wagons to sports cars and so on. Motor car form language, certainly in the early decades reflected national brand sensibilities. Italian cars seen as flamboyant and exciting, with brands reflecting the culture of motor sport and a carefree youthful optimism. While German brands developed a Teutonic style to complement their impressively engineered cars suited to fast autobahns. Motor cars occupy the same political space as a national airline, Citroen, General Motors, Ford, Rover etc., all embodied in their motor cars the technical and artistic brands of the nations from which they emerged. Until the end of the twentieth century this seemed set in stone.

8.2 The Flipside

In late 2019 Washington city councillors produced a bill to encourage employees to trade in their subsidized car parking spaces in return for additional monthly salary. In a country so imbued with the culture of the car to see the introduction of incentives to encourage other forms of mobility and reduce congestion is a colossal change of 'cultural' heart [2].

There are 7.7 billion people in the world and over half of them live in cities. And while they don't all own or use cars there is a sufficient number in our largest metropolises to clearly indicate that the freedom of the open road is rapidly passing into mythology. A single occupant in a car takes up a comparatively inefficient amount of space compared to a group of pedestrians or a well occupied bus [3].

Building regulations in many countries insist that every house built needs a car space and quite a number need two [4]. What is effectively the second largest room in the house and one that's not insulated or furnished, is kept over for a motor vehicle. Housing affordability is affecting cities and population movements. One way to make homes more affordable is to strip back 'unnecessary' elements such as parking spaces. cars take up an enormous amount of space while parked relative to humans going about their business. Planners have been devising cities for cars not really for people. The conflict lies in that many people like the independence, as they see it, that comes with car ownership. Motor cars create redundant space since they can only be parked in one place at a time while often being 'counted' as needing space in two locations e.g. parking at home and parking at the workplace. While one is occupied the other most likely redundant.

Car ownership is declining on average in the state of Victoria, Australia 1 in 15 households does not own a car. Garages cost about $20,000 AUD to a build and sit on land that could be worth around $30,000 AUD. Unless it's used as a table tennis room or a place to store other possessions, then the absence of a car makes this space an expensive affection. One should also include the stamp duty bill of $25,000 AUD and interest payments on any mortgage of perhaps around $38,000 AUD over the lifetime of a loan [4]. Despite the potential productivity offered by the motor car's presence, the cost of having car spaces is a drag on the economy. While the removal of a garage might not change the situation overnight for low income earners seeking access to the housing market, blanket provision for adding garages certainly does not help the situation [4].

There is a scene in the iconic film 'American Graffiti' in which the character Toad claims that he'll *".love and protect this car until death do us part,"* his life transformed by a 1958 Chevy Impala cruising the main drag, racing the back streets in machines that embody speed, prosperity, freedom and adulthood. This film was set in an age when owning a motor car was a norm deeply desired and relatively achievable for the young. During the 1970's 86% of American 18 year olds held a driving licence. In 2010 only 61% of 18 year olds Americans' held a driving license. What was once a rite of passage appearing to no longer hold the same appeal. This is undoubtedly a concern for the automotive industry who will be wondering where their future customers will be coming from. Causal factors are only now being explored. The average rate of decline in license uptake globally is 0.6% per annum with the highest rate in Australia. It appears that there are multiple causes; changes in life stage aspirations and living arrangements with parents, the expanding role of e-communications and motorcar affordability amongst them [5]. This speaks to a fragility with the future model of motor car ownership.

All motion at speed renders the person experiencing it at risk of injury in the event of a sudden and unexpected stop. While the development of the motor car has seen extraordinary improvements in safety especially around the mitigation of the effects of kinetic energy upon the human body, there remains a very large global road toll of about 1.25 million per year [6]. Lives ended by their own misadventure or unlucky chance at the hands of road traffic. And still many more injured. The overwhelming cause is speed. And the highest risk factor amongst younger inexperienced drivers.

The predominant fuel source, petroleum, produces a raft of harmful by-products in both its sourcing and especially its combustion. 20–30% of greenhouse gas emissions are produced by the global vehicle fleet [7]. There are slow steps to turn this around with the increase in EV's (Electric Vehicles) and experiments in hydrogen powered vehicles. However, their market penetration remains slow and the infrastructure to get up to speed will be grinding with remaining issues over the environmental credentials of the source of the electric power, while coal and gas dominate.

Urban planners are increasingly re-examining city centres to expand areas with restricted car access. With a longer-term view to encouraging active mobility options. This has pushed commentators to observe that we may be reaching 'peak car' [8] an economic notion that the motor car private ownership market will begin to decline in preference to new and emerging modalities and ways of accessing and engaging with motor cars. The success of new mobility services, while small now, is steadily chipping away at the traditional private ownership market.

8.3 Growth in New Ways to Engage with the Motor-Car

Progressive elements of the automotive industry are examining how the future might unfold. How the benefits of a motor car, largely carrying capacity, privacy protection from the elements and each other, in comfort, might remain while managing out the manifold issues crowding in on the current status quo. The growing numbers of non-car households have not all taken to the pushbike but an increasing examination of 'shared' options.

Car rental services have existed for decades. The utilization of cars for renting for very specific purposes and over short amounts of time, has been attempted in major cities since the end of the second world war but only the advent of the smartphone and a digitally connected network has the concept gained traction. The App culture enabled controls for reservations, access return and the location of stations around the city. Apps have also enabled a blurring of the edges between commercial enterprise and private ownership [9, 10].

Rising levels of congestion faced by city dwellers is a contributing factor in the growth of car sharing. This is shifting generational mindsets about private ownership. A number of phenomena are converging. Successful car sharing requires population density. Lower density areas are more difficult to serve where distances between stations are greater. Car sharing services can be incorporated into new buildings as an alternative to multiple car ports, parking and garages. Replacing private motor cars with shared ones directly reduces demand for parking spaces. A key issue of motor car ownership is that a vehicle typically remains parked for 95% of the time [11]. Sharing a motor car may reduce traffic congestion with the monitoring of costs is an incentive to reduce driving. Car sharing differs from the older rental model not least because it is not subject to office hours. Reservations and pick up as well as return are all self-service. Big data informing decision making is now possible with computer processing power handling a growing array of back office functions not previously available.

Fleet owners can brand their vehicles and associated experiences to increase revenue or price surge in responses to demand, as the Uber taxi service does now. Car sharing is more likely to invest in electric cars with their stations also being charge points. In July 2018, Volkswagen announced its intention to launch an all-electric car-sharing service by 2019 [12]. An important technical innovation may well lie in driverless vehicles which are very unlikely to be owned by individuals but shared.

Car share programs fall into one of four models:

- Round trip beginning and ending at the same place.
- One way—Beginning and ending a trip at different designated stations or places.
- Peer to peer—similar to a round trip but the vehicles are privately owned and potentially leased to a third party.
- Fractional—this is co-ownership with another private individual. Useful in urban environments where public transport walking and cycling are available, but a car is needed for the carrying of goods and or longer trips out of the city.

What are the benefits? A reduction in the overall consumption of resources. Lowering consumer costs. Increasing access to goods for those who cannot afford ownership. Reduction in monetary entry barriers. The monetizing of unused assets such as renting out of one's own car or becoming an Uber driver and turning one's own car into a taxi.

> In almost every case, what compels people to open up their homes and cars to complete strangers is money, not trust. … Tools that help people trust in the kindness of strangers might be pushing hesitant sharing-economy participants over the threshold to adoption. But

what's getting them to the threshold in the first place is a damaged economy and harmful public policy that has forced millions of people to look to odd jobs for sustenance. [13]

8.4 Re-designing to a New Kind of Motor-Car Procurement

What is the impact upon design of collaborative consumption rather than outright ownership of future world fleet? Designing not for individual procurement but for a shared economy. There are two significant design challenges that the automotive industry will need to embrace.

1. The characterisation of the motor vehicle as a utility rather than as an object of desire.
2. Projecting a brand identity based more upon experience of use than styling cues.

A significant challenge that many users will need to overcome in relation to a non-owner shared economy is 'stranger danger'. A collaborative consumption share economy relies upon people unknown to each other to make an exchange. But they also have to trust the digital platform upon itself as well as the goods [14].

As a consequence of the internet, a generation of people who had been consumers, planning their lives around the premise of acquisition and ownership now look to find meaning and satisfaction in having access to 'things' and experiences. The shift from the material to the experiential. Motor cars have been good at this since the pleasure derived from driving has been central to their advertising and marketing. Evidence shows that life experiences make people happier than material goods of the same value [15, 16]. This post-materialistic society has profoundly impacted the core of the global business offerings, placing experiences as a distinct form of economic output [17].

Automotive designers considering the challenges of the shared economy need to re-evaluate where the value is for their customer. The key challenge is that so much of experience is intangible. Therefore, experience is quite challenging to express compared to traditional physical approaches.

There are a number of ways we try to articulate experience design. There is the moment by moment tactility of haptics, an aesthetic sensitivity and imbued meaning. There are the behaviours that accompany the use of the motor car. The contribution this makes to an individual's personal narrative. Then there is the consumption of the experience, how one feels about accomplishment, especially important in the sense of skill and expertise that accompanies driving a motor car. Design orchestrates these design elements to implant positive memories.

Designers articulate experiences by leveraging the techniques of the storyboard. Looking at the 'why' before the how? Brand is re-enforced beyond an easy usability. The designer is seeking something that will have impact. The consequences of use and how they can be influenced in ways different to how it has been addressed so far. Designers cannot fully control all that might happen to a driver and their passenger, but they can try and choreograph the complex interactions that solicit emotional and physical responses. There isn't a clear road map (Figs. 8.1, 8.2 and 8.3).

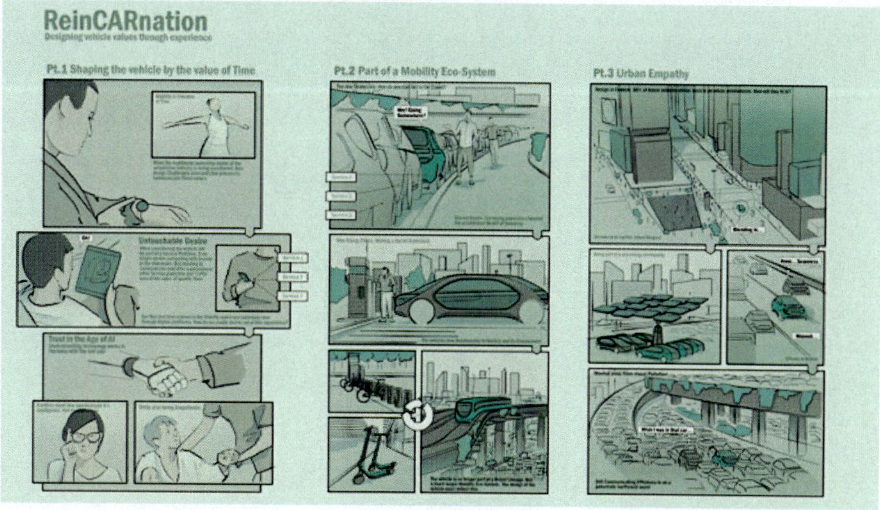

Fig. 8.1 Developing a new narrative around our relationship with cars as articulated by technique of the storyboard. Image M. Johnson

Fig. 8.2 Changes in vehicle form language to reflects utility. Image M. Johnson

What all these methodologies have in common is that they are people-centric, non-linear and iterative, which means that they have a human-centred approach to design that uses rapid prototyping to test solutions with real users and iterate until reaching the desired outcome for people. As services and products are intertwined through a series of physical and digital interactions, striving to create coherent and meaningful stories for people. Experience Design is not a new design discipline per se

Fig. 8.3 Motor manufacturers are increasingly seeing the opportunity to address the first and last mile of a journey as it dovetails with other modalities.

but rather a new mindset resulting from the shift from technology-driven innovation to social and human-driven innovation. Experience designers are empathic product and service designers, who think about the consequences of their design before crafting experiences that will positively impact people's lives.

8.5 Conclusion and Discussion

Motor car ownership has been the backbone of our mobility choices in a consumerist economy for a hundred years. As we move through the first quarter of the twenty-first century a combination of technological and sociological changes have conspired to change the traditional model of the automotive industry. A decline in car ownership has led to innovative thinking about how we engage with this former pinnacle of personal independent mobility. Traditional design methodologies which focused on dynamic obsolescence and the demand driven *shock of the new* is being re-examined in terms of experience design where ownership is seen through the lens of accessibility not as property.

The experience design paradigm is challenging the way designers think about their work. How design is articulated and expressed. How automotive manufacturers re-engage with new customers with new cultures of behaviour their needs and wants. How to appeal to a variety of users none of whom are going to be the sole owner of the vehicle. It is clear that this development in the automotive industry is new and emerging and there will be pains along the way q.v. the demise of the Holden brand in 2020. History has shown us that as one door closes another opens, and it could be that a mobility revolution is beckoning us in.

References

1. Hoffman, K., Kaplinsky, R.: Driving Force: The Global Restructuring of Technology, Labor, and Investment in the Automobile and Components Industries. Routledge, New York, NY, USA (2018)
2. Tuck, A.: Car and Away, Monocle Magazine, Anders Braso, Accessed 9th Mar 2020
3. Coxon, S., Napper, R., Richardson, M.: Urban Mobility Design, p. 184. Springer, The Netherlands (2019)
4. Tivendale, K.: Housing Affordability and Housing Cars, Planning News, Australian Institute of Planning, vol. 45, No. 2 Mar 2019
5. Delabosc, A., Currie, G.: Causes of youth licensing decline. Synth. Evidence J. Transp. Rev. **33**(3), 271–290 (2013)
6. World Health Organization (WHO): Global Status Report on Road Safety 2018. Accessed 17th Apr 2020
7. Institutue of Concerned Scientists: Car Emissions and Global Warming, Accessed 14th Apr 2020 (2018)
8. Metz, D.: Peak car and beyond the fourth era of travel. J. Transp. Rev. **33**(3), 255–270 (2013)
9. Shaheen, S., Sperling, D., Wagner, C.: Carsharing in Europe and North America. Past Present Future Transp. Q. **52**(3), 35–52 (1998)
10. Navigant Research: Navigant forecasts global carsharing services to grow to $6.2B. In: Green Car Congress. Accessed 20 Mar 2020 (2013)
11. Bartsche, A.: How to Cut Costs Through Carsharing, CFO Insight. Archived from the original on 12 Oct 2012. Retrieved 14 Apr 2020 (2012)
12. Korosec, K.: VW Plans to Launch an All-Electric Car Sharing Service Next Year. Techcrunch.com. Retrieved 14th Apr 2020 (2018)
13. Roose, K.: The Sharing Economy Isn't About Trust, It's About Desperation. New York Magazine. Retrieved 15th Apr 2020 (2014)
14. Morris, D.Z.: Today's Cars Are Parked 95% of the Time, Fortune. Accessed 15th Apr 2020 (2016)
15. Boven, L.V., Gilovich, T.D.: To do or to have? That is the question. J. Pers. Soc. Psychol. **85**(6), 1193–1202 (2003)
16. Carter, T.J., Gilovich, T.D.: The relative relativity of material and experiential purchases. J. Pers. Soc. Psychol. **98**, 146–159 (2010)
17. Pine, J.B., Gilmore, J.H.: A leader's guide to innovation in the experience economy. Strat. Leader. **42**(1), 24–29 (2014). https://doi.org/10.1108/SL-09-2013-0073

Chapter 9
Implementing a User Experience Design Approach in Melbourne's Public Transport System: The Case of Wayfinding and Customer Information Design at Flinders Street Station

Carolina Gaitan

Abstract It is well known and documented by research that besides network and service design of public transport—like travel time and reliability, passenger information design and delivery is a key driver of use and satisfaction with the system. In other words, from the user perspective a usable public transport system is one that provide services that are easily understood and internalised. When networks start to grow from a few lines and services to a comprehensive and extended network, a user experience approach is key to design, development and delivery of public transport's information to drive use and experience, as demonstrated by the case study. The application of a colour wayfinding system significantly improves the usability and comprehension of the system for all user types reducing the amount of working memory resources required to find their way.

Keywords Wayfinding · User experience design · Customer information design · Colour · Usability test · Built environment prototyping

9.1 Introduction—The Problem and the Project Brief

Flinders Street is one of the busiest train stations in Australia, Melbourne's most iconic building and the heart of the city's train network providing metropolitan and regional services. Nearly 200,000 people pass through the gates of Flinders Street station each weekday, streaming in and out under the clocks to the concourse or direct onto one of 13 platforms via subways or ramps. As the first train station built in Australia—opened on September 12th, 1854 with a single platform for services to Sandridge (now Port Melbourne), its design and construction kept evolving to the needs of the time. The main building was completed and officially opened by 1910 platforms and services were added over time, and multiple failed plans of redevelopment can also be added to its history.

C. Gaitan (✉)
Victorian Department of Transport, Melbourne, VIC, Australia
e-mail: Carolina.gaitan@transport.vic.gov.au

S. Coxon and R. Napper (eds.), *Advancing a Design Approach to Enriching Public Mobility*, Intelligent Systems Reference Library 198,
https://doi.org/10.1007/978-3-030-64722-3_9

Why do we care about the station's history? Even though this is not the focus of this chapter, it is important to contextualise how the way it has evolved over time has created a lot of functional and operational challenges, adding complexity and constraints to any new upgrade projects. For example, it was already reported as the busiest passenger station in the world by 1926 with problems of crowding and increasing demand we still face today.

In 2016, as part of a $100 million project to upgrade the station ('the project'), Public Transport Victoria (PTV) created a multi-disciplinary team dedicated to review and improve the wayfinding and information. The core project team (the 'project team') was a mix of visual/graphic designers (the 'design team') and human centred designers (the 'user experience team') that had to work alongside operations, architects and engineers (the 'upgrade project team') to incorporate signage into the built form.

- The design team knew they had a broken and an outdated wayfinding system design with multiple and clearly identified problems (see Fig. 9.1):
- Crowding and passenger flow conflict as a consequence of patronage growth and the station being the centre of operation of the whole network with all services starting or finishing there.
- Competing messaging with advertising driven by contractual agreements with the operator.
- A 10+ years signage system that hasn't been updated with compliance and asset-driven standards inhibiting modernisation.
- Mixed messaging including old operational data on fixed signage and signage blocking displays or vice versa.
- Heritage constraints given the station is listed on the Victorian Heritage Register to preserve its important historical elements. This adds complexity and constraints

Fig. 9.1 Some of the problems identified at flinders street station

to the wayfinding functional needs of today, like signs that are part of the elements that refer to services that do not exist anymore but need to be preserved, like some of the clocks under the main entrance on Swanston Street and Flinders Street and signs on Degraves underpass pointing to services in Port Melbourne that do not exist anymore.

- Space Constraints like Elizabeth Street subway that was divided in two to serve different purposes: an unpaid area to allow people to go from Elizabeth street to Southbank and a paid area for passengers that need to access the station from this entrance or just want to interchange.
- A station with multiple functions, not only for public transport users—for train services and to connect to/from bus and tram services—but as a place of destination as a popular meeting point for locals ("I'll meet you under the clocks"), tourist attraction for visitors and the portal to many other places of interest in the immediate precinct like Federation Square, The Arts Centre, Southbank and many others.
- Multiple entry points with people moving in different directions: starting their journey, finishing their journey and interchanging from train to train or to bus and tram.
- Multiple types of users from daily commuters and frequent users to infrequent users, tourists and visitors and the accessibility community.

9.2 We Know the Problem and We Have the Solution

The design team also thought that they already had the solution, but the proposed new design had already been rejected for a couple of years before the project, due to some operational constraints that were unique to the way the rail network was currently operating in Melbourne. The design team wanted to apply a colour-based map and wayfinding system that London and many other jurisdictions around the world have already applied with success. However, direction changes along the day and last-minute platform arrival changes to services at Flinders Street Station made the application of the colour map almost impossible.

9.3 The Station Mock-Up

Even though the concept of a colour map was adapted to Melbourne's context following its success in other jurisdictions around the world, it had never been tested with users to define the performance of the system for users or to iterate, adapt and improve it. For this reason, the first idea presented by the user experience team to apply design research methods and understand the potential impact on the user experience and usability of the public transport network, was to test the concepts in a live environment at the station. However, rail safety regulations didn't allow the project

Fig. 9.2 Some images of the test

team to proceed, given the uncertainty of results in behaviour change that could put people at risk, especially around platforms. This led them to proceed with a second option of creating a station mock-up.

Applying the concept of rapid and low-res prototyping the concept of the mock-up station started with a small room where the critical environments of the station could be recreated: concourse, platforms and the interchange experience. However, the mock-up escalated quickly given the complexity of the environment we needed to recreate, and the project team ended up using a big warehouse at the Meat Market facilities in North Melbourne. The facilities not only had a similar heritage look and feel but open spaces with high ceilings that allowed the team to recreate the signs at real scale. Figure 9.2 includes some photos taken during the test.

As the project team needed to bring all types of users (regular and non-regular users of Flinders Street) to the test, they decided to invent a 'new' station within the current network map called Melbourne Station to create a common ground for all and avoid comparisons with the real stations across the network. The project team also determined that only the visual elements of all touchpoints in the wayfinding system were going to be tested. This meant that the design of the experiment excluded elements like face to face and on-line customer service interactions. See Fig. 9.3 on the elements incorporated in the research design. Even though these elements are used and combined in real life situations the team needed to stress test only the wayfinding and signage and how well that would support user needs. The hypothesis was that if the visual elements worked in isolation, then when combined with other touchpoints they would work better.

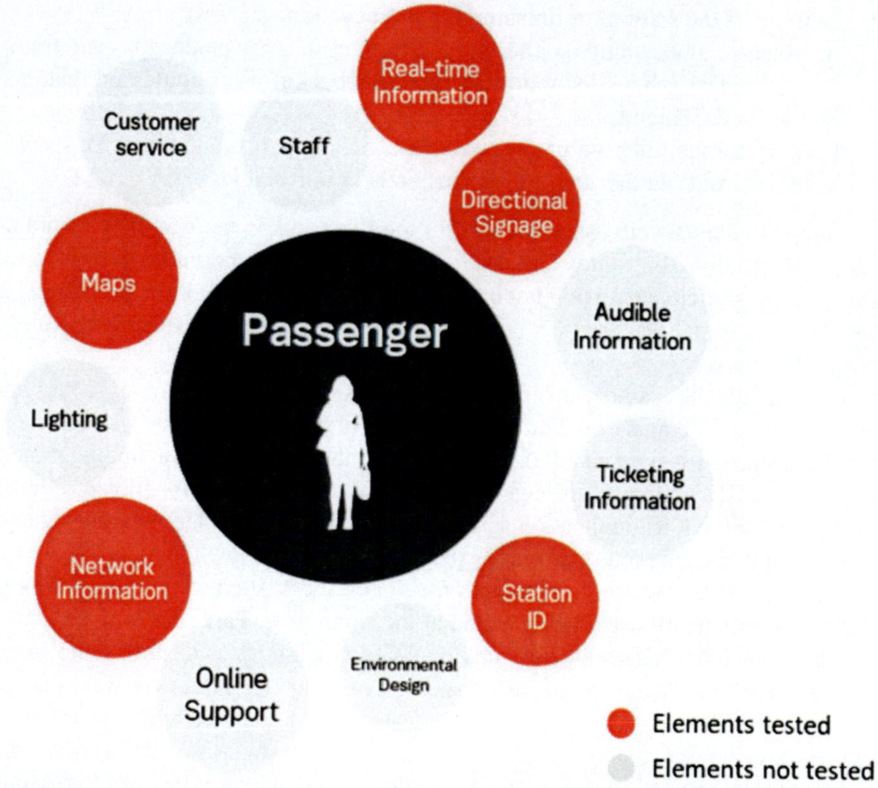

Fig. 9.3 Isolation of the elements to be tested

9.4 The Study Design Approach

Over 3 days of testing, 100 members of the public were invited to participate in a series of wayfinding tasks. Applying a completely new way for the design team to test designs with users, the study was designed to understand people's behaviour instead of preferences. Traditionally, the design team had relied on market research techniques that asked people's opinions about the designs in front of them with questions like 'do you like this sign?' or 'what do you think about this sign?', which usually prompted people to express their opinions about the look and feel of the design instead of its utility. For this reason, the level of insight of those techniques usually didn't support the design team in the iteration and improvement of their designs.

The user experience team designed this study around typical tasks to allow the project team to test all possible interactions of users with wayfinding and information to support their journey. After careful consideration of all possible needs, the project team created 5 tasks that combined all possible needs:

1. Starting the journey at the station

2. Getting off the station (or finishing the journey at the station)
3. Interchange at the station, either train to train or to other modes (bus and tram)
4. A specific task where disruption occurred to check if users could find that their journey had changed
5. Looking for facilities within the station like ticketing, lifts and toilets, or specific landmarks outside the stations like hospitals or universities.

The participants were asked to perform the five tasks in the way they would do naturally and think out-loud. The tasks were assigned in random order and varied from asking participants to find not only services and facilities within the station but also landmarks and places of interest outside the station like the MCG stadium or a hospital.

Even though the design team knew the problems with the current system a baseline was required for comparison. For this reason, the first day of the study was dedicated to test the current system and during days 2 and 3 the new colour wayfinding system was introduced and iterated throughout the day. The new system not only introduced the colour map but its application across directional signage, identification signage and real time information or digital displays as shown on Fig. 9.4.

Each participant was guided by one facilitator and participated on the test only once, with around 10 users being tested at the same time. Participants were filmed through a CCTV system and with individual GoPro cameras as well. In some cases, the GoPros™ were used by the participant and in others the facilitator was wearing it. This allowed the team to get rich video material that could be analysed and used to showcase insights to key stakeholders. The CCTV system also allowed the team to bring to the test environment key stakeholders and observe without interrupting, making participants feel as close as possible to a real situation, which means only one person observing them directly.

In terms of the types of participants the project team made sure all types of users were incorporated in the study and represented across all iterations given each iteration had new participants. It included not only train users—ranging from regular

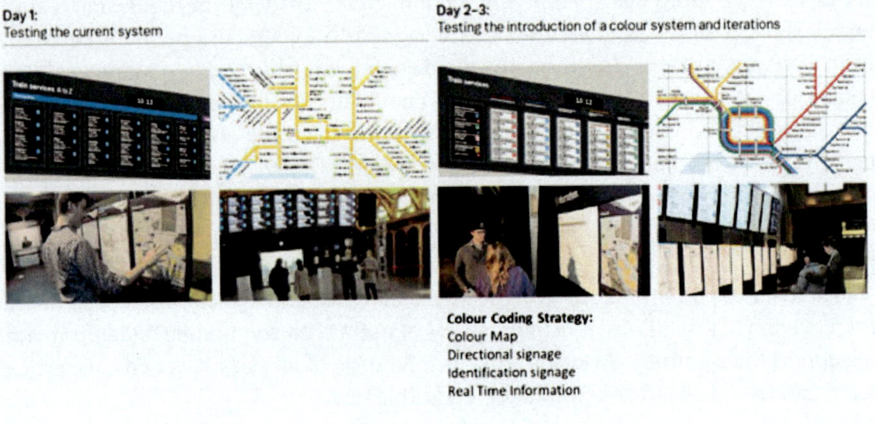

Fig. 9.4 The elements tested

commuters to infrequent users—but tourist and visitors and even Melbournians that never use public transport. It is also a good standard practice at PTV to always incorporate the accessible community, but in this case the recruitment emphasised people that required the use of visual signage. One of the misconceptions at the beginning of the process was that the only group required to participate was the user group, but throughout the process the design and upgrade project teams learned the advantages of testing with the extreme user that have no clue about the system, besides the first-time visitor/tourist (see Table 9.1). If a system is designed with the needs of non-users in mind, it will serve everybody.

Table 9.1 User cases

	User Type		
	Regular commuter	Tourist	Melbournian that never uses public transport
Observation	Even though this person travelled in the network to work during week day, when one of the tasks asked him to use a different line from his, he immediately behaved as a non-user and required a map and the help of the rest of the wayfinding system to finish the task	As expected, this user type needed the use of all tools at his disposal to perform the first task. But after the first task they started using information collected from previous tasks They will also use knowledge and assumptions of other transport systems they have used but won't know specifics about Melbourne that locals do. We observed tourist looking for a tram inside the station	Even though this person doesn't understand the train network, uses her knowledge of the city to facilitate the task e.g. the use of the map is facilitated by her previous knowledge about the geographic and spatial location of suburbs However, they need information to understand how the rail network works like a tourist
Debunked assumption	A regular commuter knows the whole network	Everybody knows the difference between a train and a tram	Residents will be able to navigate the network easier than a 'newcomer'
Reality	They only know the part of the network they use regularly	Some people won't know that in Melbourne trams only move on the road network while trains are segregated networks We can't assume anything with first time visitors	The level of knowledge and understanding of the network is specific to the individual, therefore designing for the 'average user' instead of extreme users is not useful

When it came to the tasks of recreating the space, keeping the balance between the elements that were required at a relative high resolution vs. the ones that could be low resolution was critical. In this case the project team only developed mid to hi-res artefacts for the elements to be tested including the use of digital displays and hard signage. To create the look and feel low to mid-res elements were incorporated. For example, only black curtains and standard chairs were used to recreate the train and the Myki gates (ticketing validation machines) were just plain boxes, as none of these elements were being tested, but it's location in the testing environment helped the participant to simulate the navigation of a real-life situation. The team also incorporated some sound recorded from a real station as background and other printout elements and accessories where placed throughout the station even though they were not going to be used for the tasks to support the look and feel of a station.

The final mock-up included a concourse divided with paid and unpaid area, two functional platforms with a subway to allow the interchange experience—but signage with multiple platforms, and multiple entrances/exits. The inclusion of these elements allowed the team to recreate the complexity of Flinders Street Station. See Fig. 9.5.

By design the group of facilitators for the tests was going to be internal: the project team itself instead of outsourcing as it is usually done. Even though the design team were not experts in the field of research, the objective was to give them firsthand experience with real users instead of being distant observers or recipients of a research report. The user experience design team, as the experts in the field, knew that one of the risks of this approach could be the potential mistakes the unexperienced facilitators could introduce that would bias participants and the results. For this reason, they got trained with a simple but effective technique: the facilitator was only allowed to explain the tasks specifically in the way it was written in the guide—to make sure they didn't incorporate new information without noticing—and when

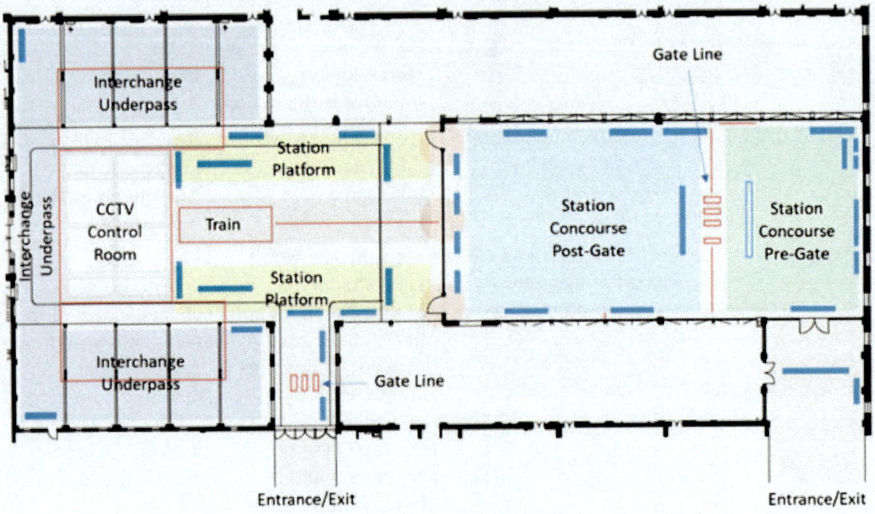

Fig. 9.5 Final mock-up station layout

participants were performing the task they could only remind them to think out-loud or respond to their question with these two questions: 'What do you think?' or 'What would you do?'. Participants were not only informed that the facilitator wouldn't be able to help them but explained that they were not being tested—only designs were being tested, so there were no wrong answers and if they couldn't perform the task it wasn't their fault but the design.

The project team also debriefed after each session to make sure all details were captured and reflected into following iterations of the design. During these sessions they could also capture what were the common themes across different user groups as well as the differences.

In terms of data collection, the research guide included not only observation of each one of the tasks throughout the whole experiment, but participants were asked few questions after each task to gather their perceptions about the level of difficulty of the task and general feelings. Video footage was also used to measure the time the tasks took.

9.5 The Initial Results

The use of a mix of qualitative and quantitative measures coming from the observations and questions asked to participants at the end of each task, helped the project team to enrich the insight generation developed from the data collected. Figure 9.6 shows a comparison of results in terms of how the new colour system outperformed the current system. While the quantitative data showed how the tasks were performed in a shorter period of time and the perception of participants improved, the qualitative data of behavioural observations showed the design team what the improvements were and why. The colour system reduced significantly the cognitive load of

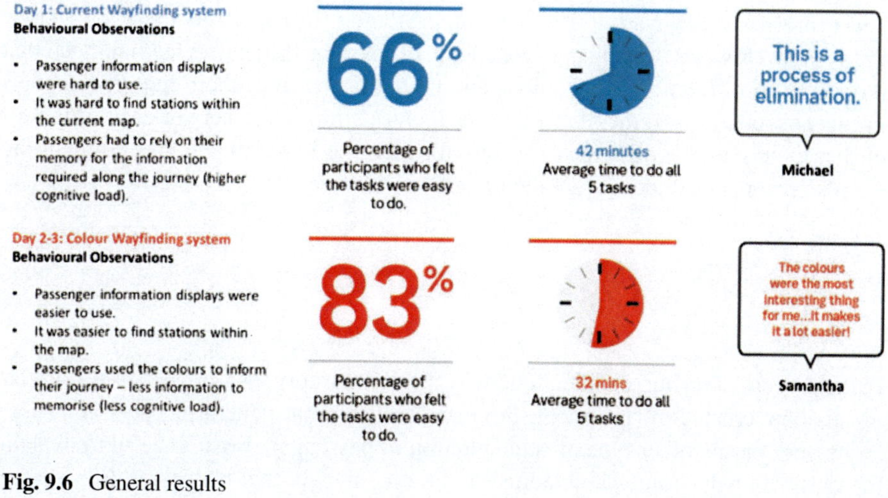

Fig. 9.6 General results

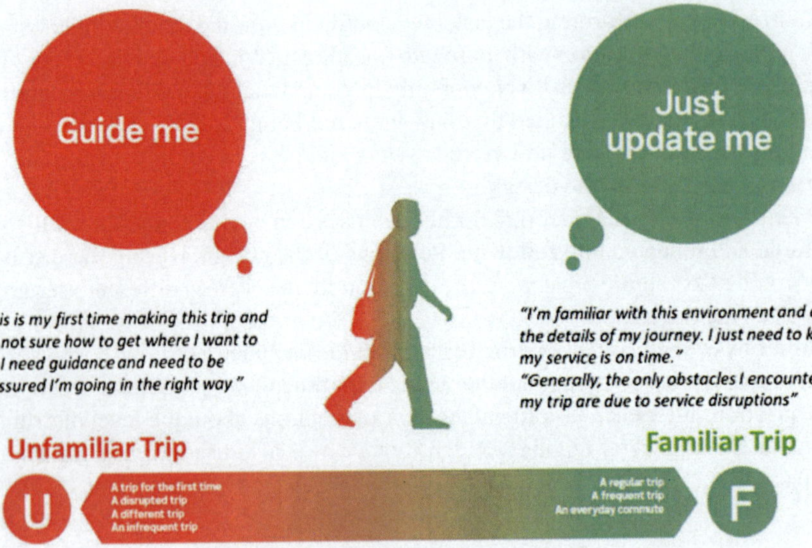

Fig. 9.7 The core insight: familiar/unfamiliar trip

participants while performing the tasks, specifically in terms of less reliance on their working memory. They didn't have to memorise various station names (destination station and the name of the line—which goes by the name of the termini station). Instead, when the colour system was introduced, they just had to remember the colour of the line. Definitely much easier! Overall, the new designs changed the feeling of anxiety and distrust in the system to feelings of confidence and general ease of use.

In terms of how different types of users behave, the design team had a hypothesis that each type of user would have different needs and would use the wayfinding system in different ways. The preconceived idea was that regular commuters will only need real time information as they already know the network and that tourists and visitors will always require more support and information as they don't know the system. However, during the first couple of sessions the project team noticed that there was no difference in behaviour among them and all of them required support in this new context—remember this was a new station in the network, but it wasn't clear why. Figure 9.7 describes 3 different user cases in detail and how some of the team's preconceived ideas were debunked.

9.6 The Core Insight

In the process of trying to understand why different user types were behaving similar in this new created environment, the team realised that generating personas, user archetypes or any other type of segmentation of passengers wasn't useful to explain the observed behaviour. They landed in the core insight that is driving the way the

Fig. 9.8 Different user types classified in the familiar/unfamiliar trip spectrum

wayfinding and information system is designed: all trips, regardless of the user type, can be classified in familiar and unfamiliar trips.

However, the level of familiarity as shown on Fig. 9.8 is a spectrum for all passengers depending on their individual circumstances. However, if the system is designed for the extremes of that spectrum, the system will be usable and friendly for all. In the extreme of a familiar trip, real time—along with disruption—becomes the most relevant piece of information and it's required in a simple easy to read format and along the desired line of movement. On the other hand, in the extreme of an unfamiliar trip the passenger needs detailed information not only about the specific trip but the context of how the network operates.

With this new view of the world of the user of the public transport wayfinding system the user types could be seen and classified in a more accurate way that reflects the reality of what they live in the network. Figure 9.9 explains how different types of users are classified along the spectrum. Even though daily commuters will be familiar during their usual trip, they could fall in the unfamiliar side of the spectrum if their regular trip is disrupted or they are doing a different trip outside of the train line of their regular commute. Similarly, a tourist or visitor that commonly starts its interaction with the network in the extreme of the spectrum of the unfamiliar trip, after few interactions with the network they start to move closer to the familiar trip of the spectrum. In some cases, regular visitors could become a familiar user.

It is also relevant to highlight that regardless of the level of familiarity along the spectrum, all users will need to satisfy not only their functional but emotional needs as referred on Fig. 9.9. This means that all users will always require to feel certainty, trust and reassurance during their journey, therefore providing regular updates of real time information along the journey is important for familiar and unfamiliar trips. Similarly, unfamiliar trips require assurance with fixed signage along the journey,

Fig. 9.9 Emotional and functional need

especially when there are changes in the space they are navigating e.g. before and after the gate line—even if they are in the same location. During the detail design process, solutions should carefully consider the balance between reassurance and overcrowding of signs, through usability test.

9.7 The Core Insight in Action

As the expression goes, "the devil is in the detail". Even though the familiar/unfamiliar trips insight is logic and simple, the design team still developing and refining its application process to detail design of wayfinding solutions across the network. During three days of design iterations the high-level foundational principles for the development of a user-led wayfinding system was defined, but detailed design solutions are still an evolving piece of work that has required further testing and iteration.

As shown on Fig. 9.10, new wayfinding guidelines are responding to a purpose-led system instead of asset-led. Previous design guidelines were based on asset types without major consideration to the context of each specific location inhibiting innovation and flexibility in the introduction of new solutions. The development of new wayfinding guidelines based on purpose, delivering the right type of information at the right time in the journey before any asset is selected, will allow the introduction of improved and new solutions that will meet contemporary user expectations.

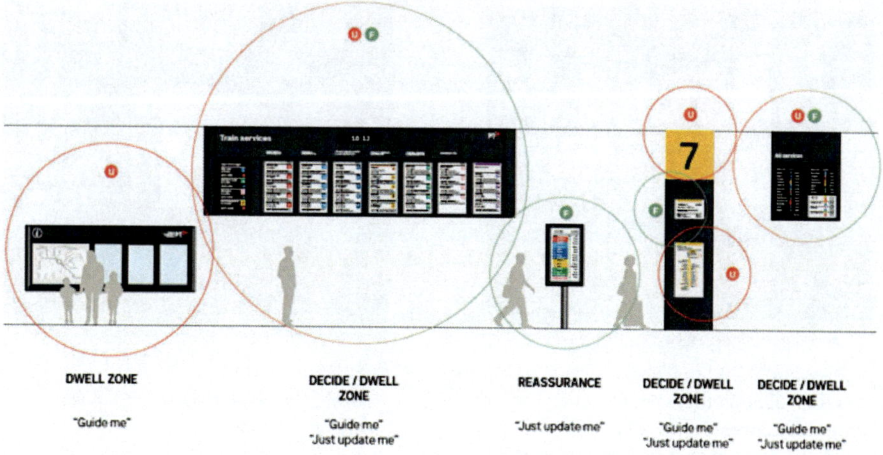

Fig. 9.10 Purpose-led versus asset-led solutions applying the familiar/unfamiliar trip insight

Under this concept the definition of which trip type will need to be served is essential before assets are defined. Even though the end-to-end solution will need to cater for both, some assets will need to be catered for one of the extremes of the spectrum when conflicting needs between familiar and unfamiliar trips arise—like location or level of detail required-, while others will need to serve both. For example, different combination and types of assets will be required for unfamiliar users that need dwelling spaces with a mix of fixed signage, network maps, ticketing and real time information to plan their journey in the same location instead of moving around the station concourse trying to put of pieces of the puzzle together.

Additionally, sometimes the same type of information will need to be delivered in separate pieces of assets with different content design. Figure 9.11 exemplifies how the same type of asset—digital displays—need to be used in different ways for familiar and unfamiliar trips. For familiar trips, the content design will need to provide a quick snapshot of service updates located in key points along the flow of passengers. On the other hand, unfamiliar trips will need different content design of the same real time information but with more detail, accompanied by network maps and other types of general network information in dwelling points to allow time for planning.

9.8 How Does It Look like in Practice Now at Flinders Street Station? A Before and After

As a result, the colour map was launched across the network in early 2017 and new colour-coded wayfinding screens were installed at Flinders Street in January 2019. The map was installed in all trains and key interchange stations across the network,

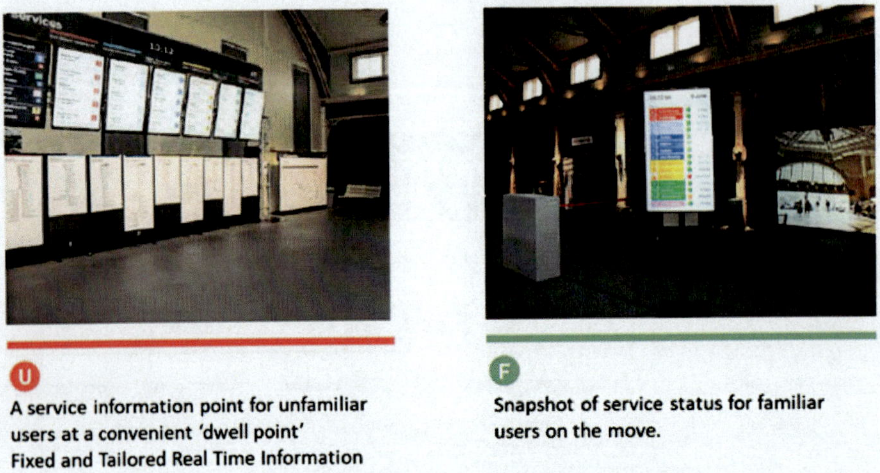

Fig. 9.11 Same asset with different purpose requires different content design

integrating new design features like the unification of metropolitan and regional train networks in one map and a grid map style system with station index ordered alphabetically to facilitate location of stations. Sometimes simple 'old ways' of doing things are the most cost-effective solution as proved during the testing sessions with unfamiliar users trying to find new stations across the network.

The extension of the colour system beyond the map across other wayfinding assets, specially the fixed ones, has been proved difficult to implement specially at interchange stations like Flinders Street. Even though most of the time the lines use the same platform, from time to time due to operational requirements, changes of platforms are required. This means that painting fixed assets (like platforms) of a specific colour will not always be accurate when services are 'transposed' to use a different platform from the regular one. Nonetheless, detailed design testing and iteration of potential digital solutions are being explored as well as operational changes, that could be applied to improve and facilitate not only station navigation but the entire journey as a whole with higher levels of consistency—proven also to be critical to reduce cognitive load.

However, its application on digital assets—known as Passenger Information Displays (PIDs)—was a viable option given its flexible nature to adapt to any operational changes. The final detail design we see today on the displays at Flinders Street Station was also tested with users. Even though, the principles of design follow a similar look and feel of the ones used at the mock-up station, the designs were refined through additional detailed testing to make sure the interface usability was clear and easy to use.

The displays not only follow the colour system of the map but applied the principles of the core insight to support familiar and unfamiliar journeys, including use

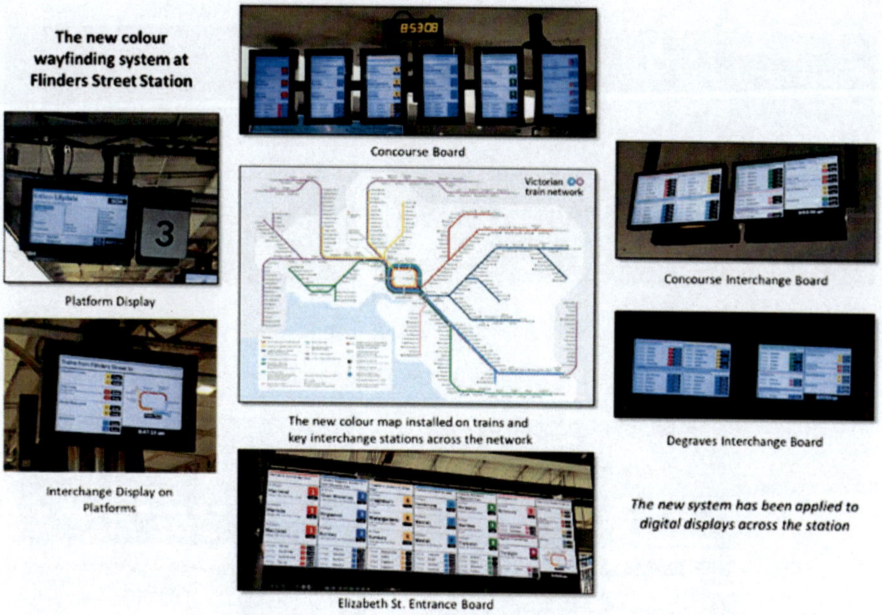

Fig. 9.12 The application of the new colour wayfinding system at flinders street station

cases like starting the journey at the station with the support of network boards at the concourse and main entrances, and interchange between platforms as shown on Fig. 9.12.

Most of the displays at the station haven't been replaced but its content redesigned and, in some cases, repositioned to serve better all types of users. For example, with the current configuration of Melbourne's train system that links all CBD stations through a 'loop', including a tunnel, and the direction on how trains get in and out the loop during the day changes, interchange information between them was found specifically hard to follow. With the new design that presents a visual representation of the loop in real time, it is much easier to read than the previous version that only presented stations in a list and at the bottom of the platform service information display (see Fig. 9.13).

A general comparison of the same assets before and after the colour system shows a clear difference as seen on Fig. 9.14. Not only the application of the new colours but detail design to enhance content readability was considered. Even though more work needs to be done around integration with fixed assets and information for unfamiliar trips, the system has already proven its effectiveness.

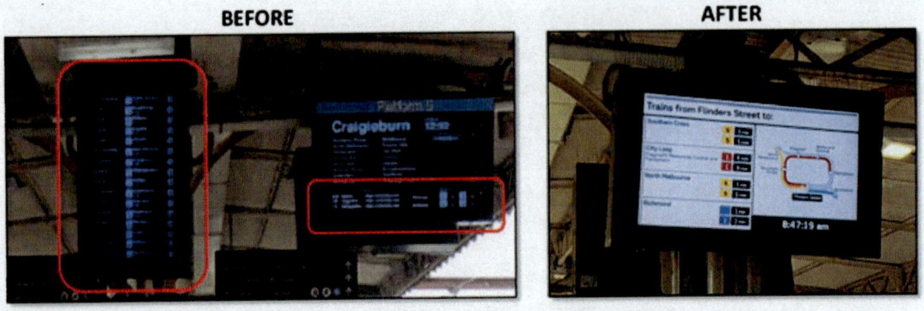

Fig. 9.13 Interchange displays at flinders street station—before and after

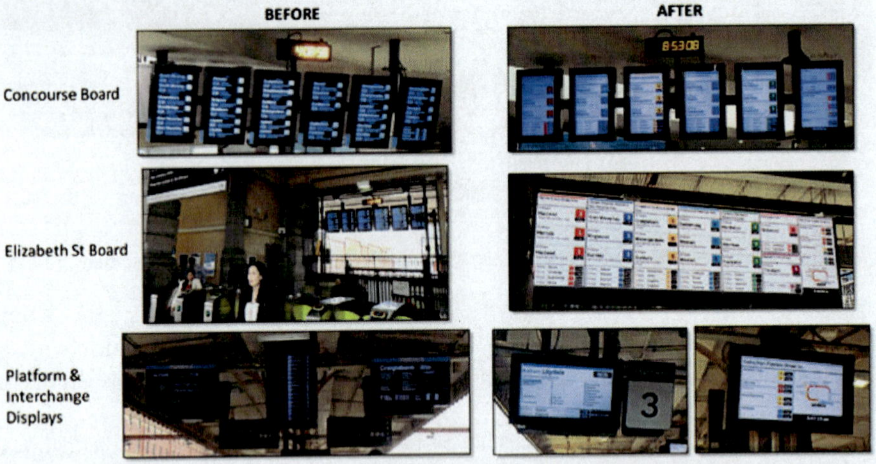

Fig. 9.14 Before and after the colour wayfinding system applied to displays at flinders street station

9.9 Conclusion—The Effects of the Project on the Internal Team and Other Pieces of Work

After the project finished the design team was interviewed to understand how the project affected the ways of working and the immediate outcome for the organisation has been the incorporation of usability test as a standard practice within the design team. Their reflections showed how getting involved directly with the user helped them to understand better the issues firsthand and debunk assumptions to quickly iterate their designs is fundamental to make their work better.

The extension of the familiar/unfamiliar trip principles to other aspects of the wayfinding system including detail design of visual elements, non-visual elements for vision impaired users, application to digital tools including journey planning and service design of customer facing channels is part of the fundamental pieces of work the teams will keep working on.

This piece of work has also allowed them to influence various elements of the end-to-end experience linked to wayfinding and information, beyond the look and feel. Development of frameworks and guidelines to influence environmental design to integrate wayfinding principles from early stages of station design, instead of coming at the end of the process, has been the fundamental shift towards a user-centred wayfinding system.

Acknowledgements Victoria Department of Transport.

Chapter 10
The Challenges of Designing the Rail Network Map/Diagram as the Critical, Missing, Transit Infrastructure for Mumbai

Mandar Rane, Jaikishan Patel, Kalagouda Patil, Aditi Kulkarni, Braj Bhushan, and Hriday Gami

Abstract It is intriguing how a population of 7 million passengers travelling daily [7] on the Mumbai Rail network manage to navigate in absence of a well-designed public information system. The authors inquire, why is it necessary to design a consolidated map/diagram for this high-density rail network (Suburban, Metro, Monorail) controlled by distinct public and private entities, running across multiple districts. Can this map be understood within a multicultural, multilingual context of Mumbai (India); with varying degrees of literacy and exposure among people? In response to these questions, authors designed a network map/diagram for Mumbai Rail network (Print), as an alternative to the current method of 'asking around' to access information. The designers of the map, have based their design on the DNA of the famous London Underground Diagram, LUD [4] drafted by Henry Beck [5] in 1933. Considering the context of Mumbai the designers have introduced new features in the map to enhance its utility. Intercept study (guerrilla usability testing with tasks, pre and post-task questions) method was employed with convenience sampling to test the designed map with 102 users in commute context. The insights, challenges, failures and design decisions about the use of station index and grid, advantages of pseudo-geographic representation, appropriate use of colours for the colour-blind, importance of closure and concerns of visual complexity with regards to information density, are captured in this chapter.

Keywords Transit maps · Mumbai rail map · London underground diagram · Information design · Communication design · Visual complexity

M. Rane (✉) · J. Patel · K. Patil · A. Kulkarni
IDC School of Design, Indian Institute of Technology Bombay, Bombay, India
e-mail: mrane@iitb.ac.in

B. Bhushan
Department of HSS, Indian Institute of Technology Kanpur, Kanpur, India

H. Gami
Department of Design, Indian Institute of Technology Guwahati, Guwahati, India

S. Coxon and R. Napper (eds.), *Advancing a Design Approach to Enriching Public Mobility*, Intelligent Systems Reference Library 198,
https://doi.org/10.1007/978-3-030-64722-3_10

10.1 Structure of the Chapter

First part of the chapter helps us understand the structure of the Mumbai Suburban Rail and the peculiarities of a high-density network. It identifies peculiar traits of commuters and passengers who travel on this network. Authors report their observation about why 'asking around' for information has become second nature, and eventually a habit of commuters and passengers. The second part of the paper focuses on the design of the Mumbai Rail Map (MRM). It discusses how the design graphically unifies the three types of train services (**Suburban, Metro** and **Monorail**), the design decisions implemented and the challenge to fit it in a limited space (as non-digital). After designing, an intercept study of the map in context (at the railway station) revealed findings around the discovery of introduced features and tasks performed.

10.2 Introduction: A Background of the Mumbai Rail Network

Indian railway is the legacy of the British Raj running its first passenger train between Bombay (Mumbai) and Thane on 16 April 1853 (Aklekar B. Rajendra) [1] The suburban rail network in Mumbai is spread over 465 km with 129 Stations running across multiple districts. It operates 2342 train services. It is divided into four rail lines namely, Central Railway (CR), Western Railway (WR), Harbour Railway (HR) and the Trans harbour (Fig. 10.1).

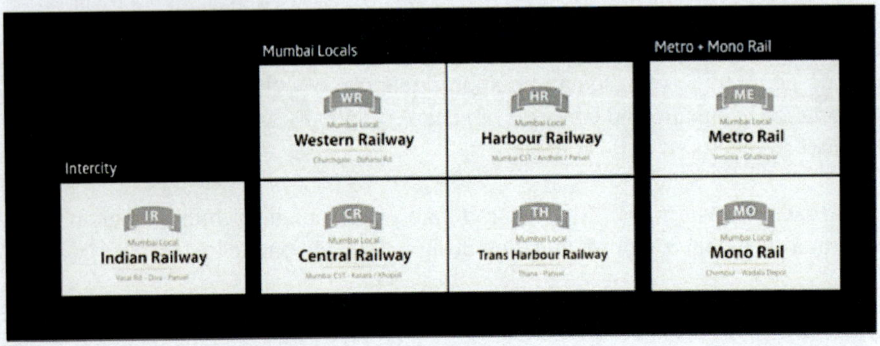

Fig. 10.1 Mumbai Rail network

10.3 High-Density Network and Its Concerns

Mumbai is an area of 634 square kilometres with a population of over 22 Million. The Mumbai Rail network has grown into a high-density transport network without a well-designed public information system. A large number of people commuting together in a disorganized manner on a network results in errors of navigation amongst novices who are new to the system.

Each subsequent train pours people onto the railway platform at an interval of three minutes. In the event of an emergency, the overcrowded platform can pose a safety risk. To get a glimpse of this high density the recently aired podcast, 'The science of the stampede' by Audiomatic, a conversation between Efa Hunt (UK) and Shibu Mani (Figs. 10.2 and 10.3) (India) describes the stipulated density per sq. m as follows:

- 2 people per sq. m—ideal (comfortable)
- 4 people per sq. m—slow-moving crowd
- 5 people per sq. m—Considered Maximum (High Density—should ideally be the limit)
- 12 to 14 people per sq. m in Mumbai Suburban trains.

When there are 7 people in a per sq. meter area, the crowd can begin to act like a fluid. (they have very less freedom of movement and can be pushed around in the direction of a single driving force). A density of 14 people per sq m, evades all the notions of personal space as seen in the Mumbai suburban train compartments (Fig. 10.3). Commuters unknowingly start formulating unsaid rules of navigation to exit and enter the train compartments. Therefore, novices who are unaware of these rules often have unpleasant experiences while travelling on this high-density network.

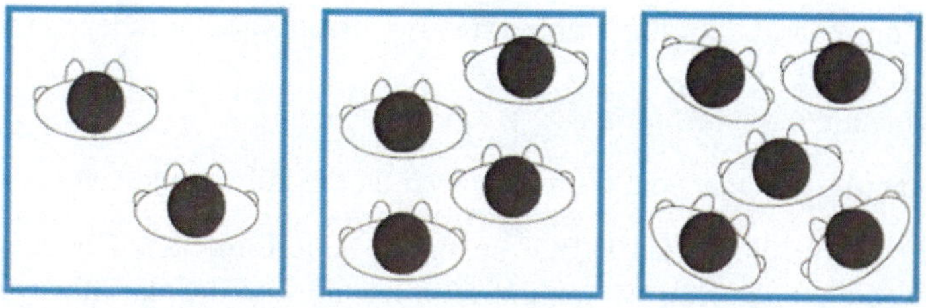

Fig. 10.2 Understanding definitions of the crowd—density per square meter

Fig. 10.3 Illustrating how 15 people are packed in a one square meter in the Mumbai train compartment towards explaining/experiencing high density for a passenger

10.4 Train Types, Destination Codes, Variety in Information Displays

As mentioned earlier (Fig. 10.1: about different train lines), the Suburban Mumbai rail has various train types as well. Train types include Slow, Semi-fast and Fast; which run on separate rail lines with train arrivals on same or independent railway platforms (Fig. 10.4).

Following are the type of trains:

1. **Slow train**—Alphabetical code displayed on the railway indicator: 'S'.

   A slow train is indicated by the uppercase letterform 'S'. This letter is also displayed on the railway station indicator installed on the platform as passenger information code and the on the façade of a train. Slow trains halt at all the stations towards their respective destinations. Confusion arises when the station code for terminal stations matches with the code for slow train. For example a train for a station named 'Karjat' also displays the letterform 'S'. When two 'S' are displayed on the station indicator it creates confusion amongst novice passengers who are not accustomed to the station codes. In some scenarios new displays are installed, they display the complete text, for example 'SLOW'.
2. **Semi-fast**—Alphabetical code displayed on the railway indicator: None (Fig. 10.5).

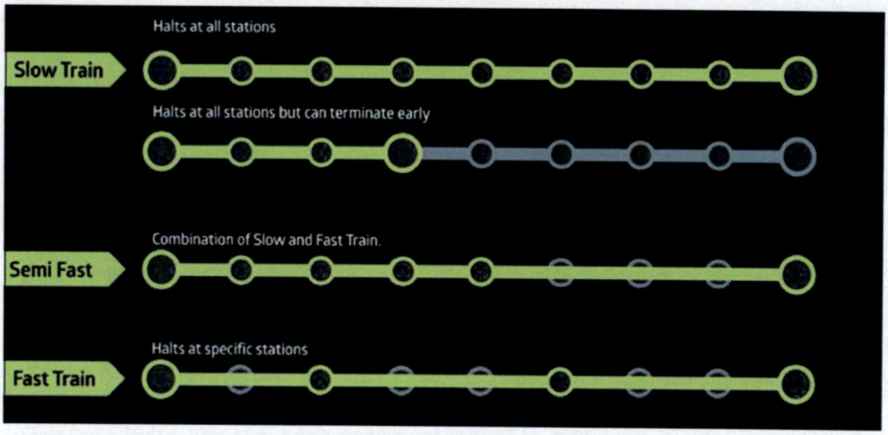

Fig. 10.4 Nuances of trains on the Mumbai Suburban Rail network

No code is defined for semi-fast trains. One has to pay attention to station announcements made verbally or ask fellow passengers for information. These types of trains initially/eventually halt at all stations, but skip a few stations in-between, halting at key junction stations.

3. **Fast trains**—Alphabetical code displayed on the railway indicator: F.

   These trains only halt at important junction stations and skip the in-between stations. If by mistake you board this train in crowd.

In a high density network, not being aware of this codification creates a lot of anxiety amongst passengers (Figs. 10.4 and 10.5).

10.5 Mumbai Suburban Trains Versus the Recently Launched Metro and Monorail

The functioning of the Metro is easier to understand for a novice because of its simplicity. Metro runs at a consistent time frequency between two trains towards fixed destination. In contrast, the Suburban Mumbai Rail has multiple train types running on a single network between several start and end destinations. Add to it, multiple routes, varied time intervals between trains, along with their respective alphabetical codes for terminal stations. This variety is difficult to grasp for a novice passenger, who needs guidance, as the trains are overcrowded and require quick decision-making and swift movements (considered skills of an expert commuter).

Fig. 10.5 Images of variety in station indicator displays at platforms with their station codes

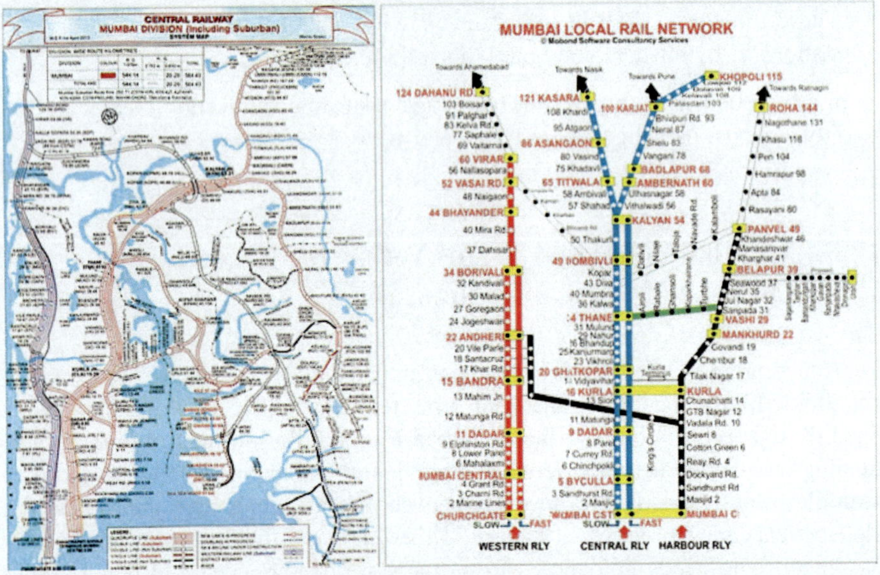

Fig. 10.6 (Left) A system network map of the Mumbai Suburban Rail network exists (official) and is on display at certain railway stations. (Right) A diagrammatic map of Mumbai Suburban Rail is available on the backside of official publication by the railway authorities

10.6 What Are the Current Mumbai Rail Network Maps Available to a Commuter for Navigation?

The three maps/diagrams (2 official and 1 unofficial) have the following issues (Figs. 10.6 and 10.7):

- None follow a grid and alphabetically ordered station index (conventional mechanisms used in maps to reduce search area and the scan time for the user when the network is complex).

Fig. 10.7 Arun Ganesh from the National Institute of Design (NID) designed the first schematic map for the Mumbai rail network (unofficial map) and is available online

- The map in (Fig. 10.6 - Left) is a system map ideally meant for train drivers, as it shows the exact representation of the railway tracks in kilometres and is therefore not useful to the commuter. It is displayed only at a few select stations.
- The official maps do not display station codes.
- Official maps do not display routes of other networks and points of transfers to other lines.
- Absence of a legend in official diagrammatic maps (Fig. 10.6 Right).
- The maps are not colour-blind friendly (use of the colour red and green for differentiating rail lines are to be perceived by the colour-blind as of the same tone (grey).
- There are 129 stations on the Suburban rail network, however currently only a few railway stations display the system maps at ticket counters (Fig. 10.6).

10.7 Information Seeking Habits of Passengers in the High-Density Mumbai Rail Network

In a crowded city like Mumbai, people take information availability for granted, often sourcing it easily from fellow commuters/passengers. Unfortunately, this has become their second-nature and now a habit. The lack of awareness about the need for a public information system has never been questioned and therefore its absence seems justified.

10.8 When People Can Ask Each Other and Find Their Way into the System, Why Do Authors Consider It as Unfortunate?

Authors have no reservations about people asking fellow passengers for travel information. But it has to be understood that the habit of 'asking around' is not out of choice. It has resulted from a lack of reliable sources of information in their environment. Over the period, this habit has manifested into a travel culture that forces passengers to coax fellow commuters to provide information about stations, platforms and train routes. Good passenger information is an essential ingredient of a successful public transport system: ill-informed travellers may not be able to identify services which best suit their needs, leading to poor perceptions and low use of public transport [2]. This can be considered true in the case of the Suburban Mumbai Rail network because the information system which can aid the user, exists but only partially.

10.9 Errors in Guidance on a High-Density Network Builds Peculiar Traits Amongst Passengers and Commuters

- Asking around becomes their second nature (habit formation).
- Reconfirmation is sought at every step, due to past experience of wrong guidance.
- Commuters and passengers form their personal mental model (Norman D. A) and eventually become experts.
- These so-called experts, guide novices, based on their established mental model.
- This interaction between novice and experts continues in a loop.
- Such interactions slow down the movement of the crowd which is a concern for high-density networks.
- In this context, the authors planned to provide an alternate and authentic source of information, enabling passengers, whether novice or expert, to make informed decisions about their journey. There was a need for a user-friendly map to provide a clear conceptual model of the complete rail network.

Fig. 10.8 Mumbai Rail Map Version 5 (English)—designed by communication design students of IDC School of Design, IIT Bombay, INDIA

10.10 About the Design of the Mumbai Rail Map (MRM)

The Mumbai Rail Map (MRM) was designed at the Industrial Design Centre (IDC) school of design, IIT Bombay, as part of the course work for students. Under the guidance of Prof. Mandar Rane two communication design students, Jaikishan Patel and Snehal Patil took up the challenge. The first version of the MRM was posted on Facebook to gauge public response. It was well received and people were quick to correct the minor factual errors in the map, especially regarding station names and their spellings, which were updated after verification with the railway authorities. After several iterations, we launched version 5 (Fig. 10.8) of the map in December of 2013.

10.11 Design Process

For a full case study of how we arrived at the current design refer to this article: https://mrane.com/portfolio/mumbai-railmap/ (Fig. 10.9).

10.12 Design of MRM Version 5 Had the Following Components: (Overview)

- **Instructions**: Graphic instructions on how to read the map.
- **Index**: Alphabetically ordered station index with adjacent colour and station codes.
- **Grid**: Division of the map into equal segments by a grid to reduce scan area and adding alphanumeric codes to the station index to ease the search.
- **Network and geography**—Graphical representation of the map: Simplified pseudo-geographic/diagrammatic graphical representation of the complete rail network (we adhered to the geography of Mumbai city)
  - Creating two separate adjacent line routes to represent the difference between fast and slow trains, a key feature of the Suburban rail system.
  - Depiction of travel time between stations by adding a number on the line. It helps to estimate travel time between two or multiple stations.
- At this point, we conducted an in-context intercept study for MRM version 5 (guerrilla usability testing with tasks, pre and post-task questions; explained in Sect. 5) and the following key design decisions were made to action findings from the testing study.
- **Border**—Insight of isolating the legend from the map: In version 5 of MRM, the legend was integrated with the map enclosed by a rectangular border. People were therefore not able to discover the legend without guidance. Revisions were made

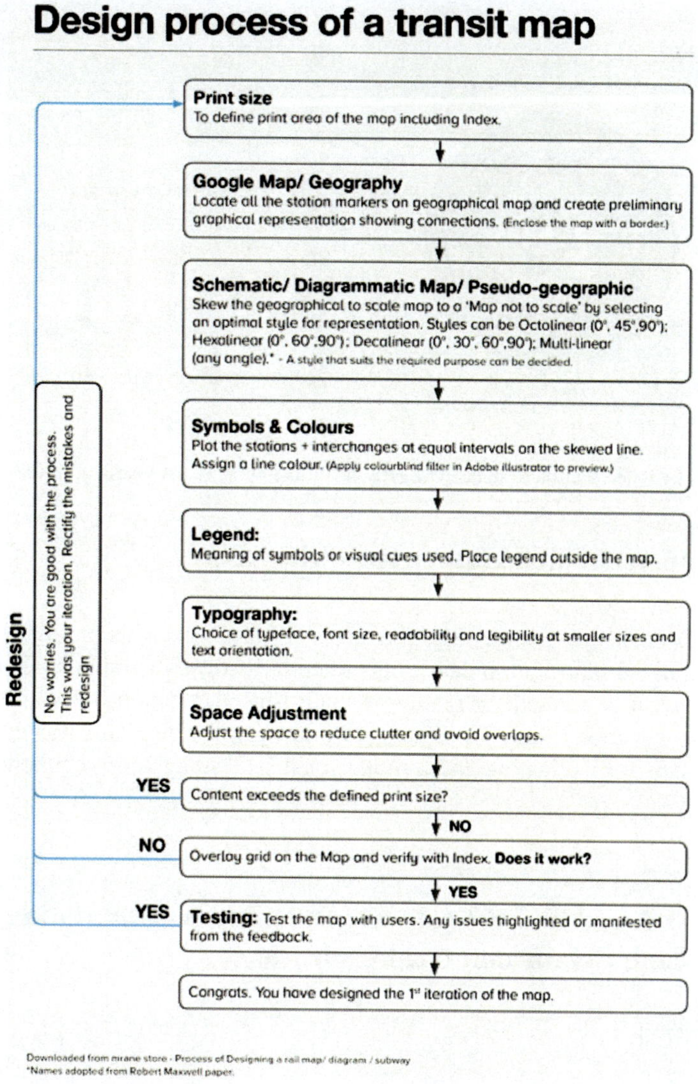

Fig. 10.9 Design process practised in the making of the MRM

to version 6 by isolating the legend by the gestalt principle of closure. **The area of the map should be enclosed with a distinct border announcing its limits.**

- **Colour**: Use of colour to differentiate the lines and the need for colour-blind friendly colours for lines.
- **Visual complexity**: Discussion on perceived visual complexity and its role in information processing.

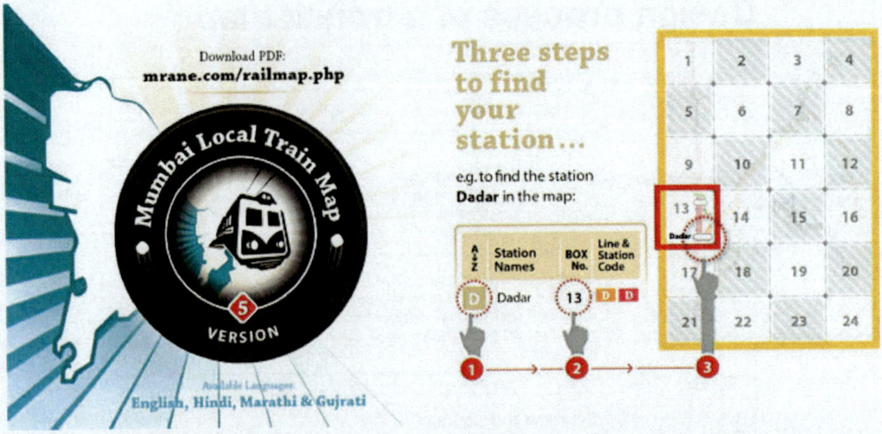

Fig. 10.10 Cropped section of the MRM version 5 illustrating visual instructions to find a station

10.13 Instructions: How to Read the Map

If a novice passenger plunges into the abyss of visual elements of a map to search for a station, the information density is bound to be overwhelming. To counter this people need to be trained for reading a map using the station index and the grid. MRM had a section titled 'Three steps to find your station'. This was done as use of print maps for navigation and training for map reading is not common in India (Fig. 10.10).

10.14 Index: Alphabetically Ordered Index with Station, Line Colour and Grid Codes

See Fig. 10.11.

10.15 Grid: Alphanumeric Grid Versus Number Inside the Box Grid

While testing the maps internally among design students, we received a suggestion to add a number on the map inside the box of the grid. Authors assumed, since map reading/training is not so common in India, placing a number in the box grid is better in comparison to an alphanumeric grid. Alphanumeric grid requires the user to visualize the imaginary lines of X, Y axis to plot the alphanumeric box on the

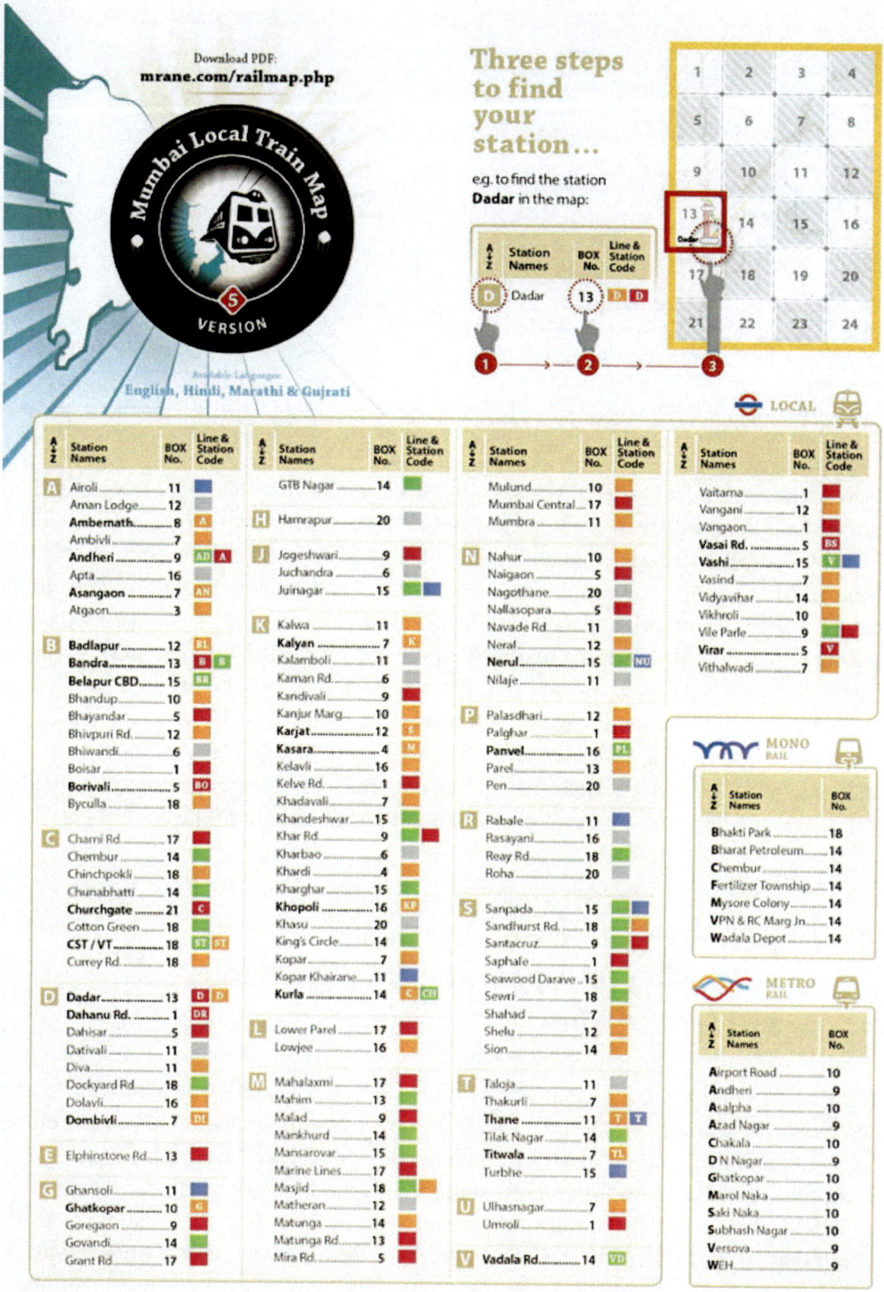

Fig. 10.11 (Displays the left half of the complete MRM). An alphabetically ordered search index of station names with line colour codes and station codes

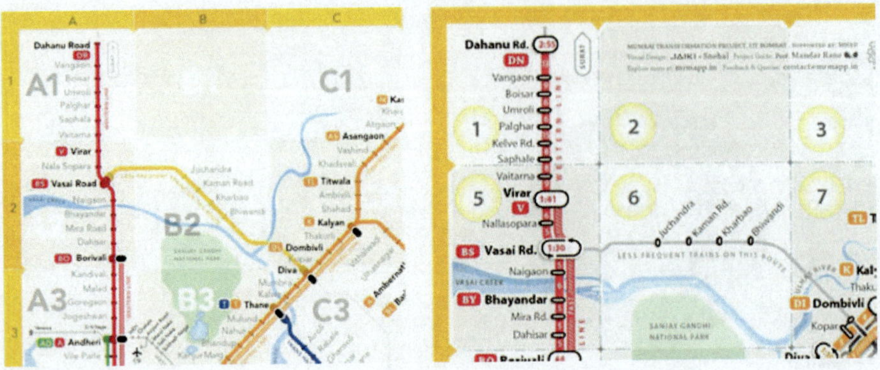

Fig. 10.12 The alphanumeric grid (Left), the number in the box grid (Right)

grid. However, during testing we found that only 55 out of 102 people were able to understand the number in the box grid without external help.

Authors later realized that their decision of introducing a number in the grid box for MRM against an alphanumeric grid, used worldwide, had a problem. In the future when the network expands and the number of lines increase, adjusting the grid number (see Fig. 10.12 right) inside the box becomes difficult. The number keeps on displacing itself inside the box into the available space as the information density of that particular box increases. When people are untouched by any conventions of the alphanumeric grid, is it advisable to introduce a new system of learning; is yet to be answered. However, adhering to conventions can reduce the cognitive load of the users, due its consistency worldwide and therefore its familiarity.

10.16 Network and Geography: Design of Pseudo-geographic Map, Station Markers and Lines Routes

Mr Beck's London underground diagram (LUD) is a visual system proven to ease navigation and journey planning [5]. It is diagrammatic and tries to reduce clutter by segregation of congested areas, by distorting geography. In reward it achieves a comprehensible and simplified representation of the network for the passenger.

MRM is also a diagrammatic representation of Mumbai which adheres to the geography of the location to its advantage. Since Mumbai city shares border with the Arabian sea, it gives a natural contour to the landmass. Designers used Mumbai's geography, by depiction of certain landmarks (bridges, architectural monuments) to their advantage as they felt it would be easier to build a mental model of the city for passengers, in comparison to their absence. More over the Mumbai Rail network is over-ground and not underground. Therefore, authors prefer to call it a pseudo-geographic diagrammatic representation.

Fig. 10.13 Station markers, line route thickness and travel time depiction by number

The station markers on the map are designed in the shape of a capsule, so they can extend to overlap on the fast and slow line routes.

The line routes were made thicker to accommodate the travel time (number) between two stations, one more unique feature of the MRM.

Creating two separate line routes for slow and fast trains to avoid confusion. This is a unique feature of the Mumbai Suburban Rail as fast and slow trains arrive at independent platforms. This must be communicated to the passenger visually (Figs. 10.13 and 10.14).

10.17 Border: Announcing a Visual Boundary for the Map

In MRM version 5, the legend appeared to be the part of the map geography, as it was included within the map. In several instances while testing, the users were not able to locate the legend when asked to explain the meaning of symbols in the map (Fig. 10.15).

In the revised (Version 6) of MRM the area of the map was enclosed by a thick boundary leading to a closure. This design decision gave the legend an independent identity and made its discovery easier.

Fig. 10.14 Design of the Mumbai Rail Map, Version 5

10.18 Colour: Design for Inclusion. Use of Colour and Its Functions

Understanding colour blindness—Colour-blindness is a visual impairment that people suffer from when the colour receptors known as cone cells, in their retina are missing or dysfunctional. There are several types of colour-blindness but the

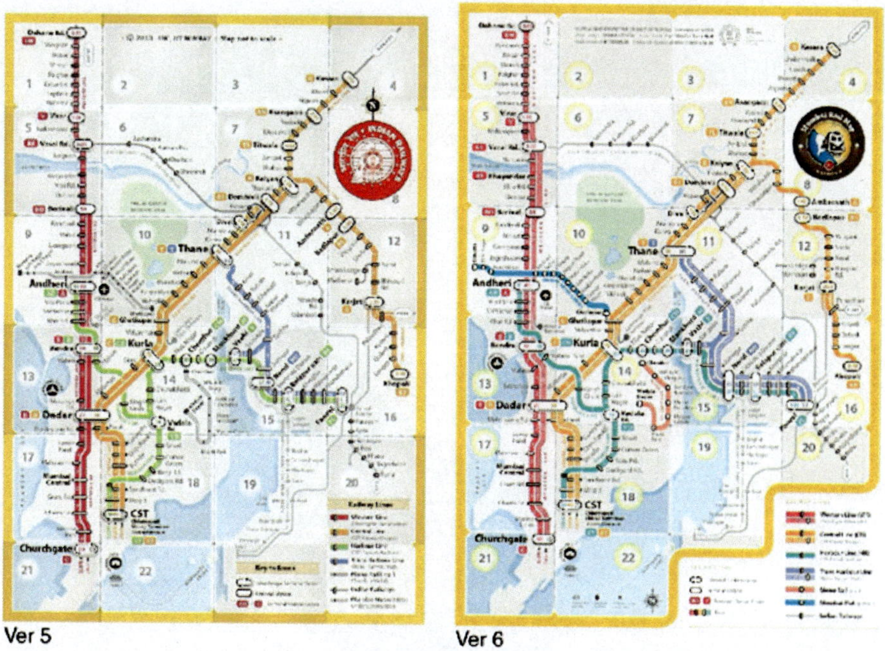

Fig. 10.15 The legend in version 6 of the MRM (right) is excluded from the map and is an independent entity. Notice how the legend in version 5 (left) appears as a part of the map geography making the discovery difficult for the user while searching for meaning of symbols

most commonly found is 'red-green colour-blindness'. Humans with normal colour vision, known as trichromats, process colour information through three types of cones responsible for absorbing long-wavelength (red channel), mid-wavelength (green channel) and short-wavelength (blue channel) (see Fig. 10.16).

Dichromats, on the other hand, see colours as a combination of only two of the three channels. In most cases, red-green colour-blindness is a congenital disorder, which means that it runs in the family (see Figs. 10.17 and 10.18).

10.18.1 There Are Three Types of Dichromats

1. **Protanopes**—Protanopes lack long-wavelength retinal cones, which translates into an inability to distinguish between the green-yellow-orange-red colours, also known as protanopia.
2. **Deuteranopes**—Deuteranopia is a lack of mid-wavelength retinal cones, which also affects their ability to differentiate between green-yellow-orange-red colours. Persons with deuteranopia experience the same distortion of colours as protanopes but without the dimming of colours that accompany the latter.

Fig. 10.16 Trichromats, persons with normal colour vision, possess the three channels red, green and blue (RGB) through which they process the colours of the visible spectrum

3. **Tritanopes**—Tritanopes suffers from a rare form of colour-blindness that results from insensitivity to blue light, causing confusion of greens and blues. The absence of short-wavelength retinal cones of persons with tritanopia imparts an inability to distinguish between blue-violet-indigo. They also experience a severe dimming of these frequencies along with an inability to distinguish between pink and yellow, and confusing purples with shades of reds.

Statistics suggest that 8% of men and 0.5% of women in the world are colour vision impaired [3]. Considering India's population, this is a significant number of people. As a society we have come to associate shades of red and green as opposite colours, signifying opposite meanings and even giving opposite instructions, as in the case of traffic signals.

Look, for instance, at websites for booking tickets online. The procedure is fairly simple until one comes to selecting the seats (see Fig. 10.19). The selected seats are shown in green and the unavailable ones in orange. This choice of colours makes the design inaccessible for individuals with colour-blindness.

Over repeated encounters and years of use, this has become a semantic convention. It appears to be the most intuitive choice to use green and red for indicating availability and non-availability, yes and no, correct and incorrect, go and stop, but we must begin to question the norm and be inclusive in our designs wherever possible. Most of the maps when designed use red and green to differentiate line routes.

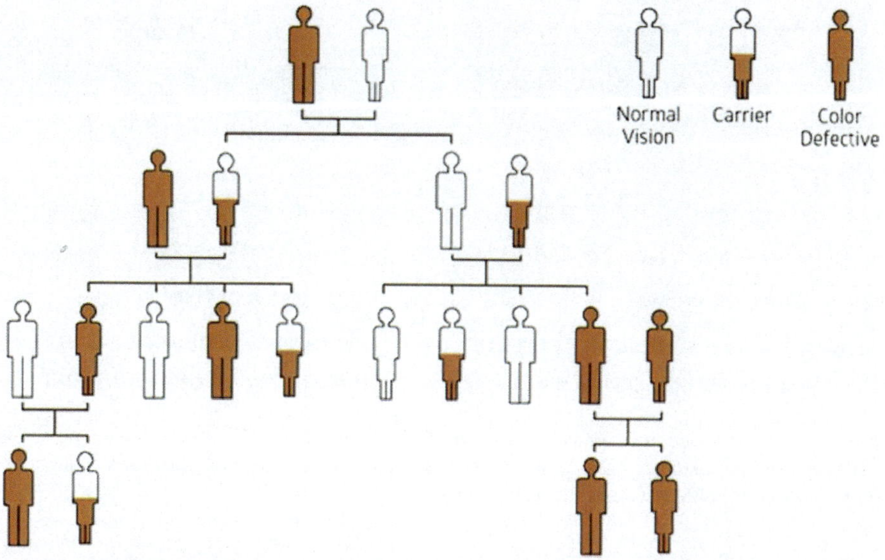

Fig. 10.17 Colour-blindness is, in most cases, a genetic disease inherited by the children from their parents. This means, if one or both of your parents is/are suffering from some type of colour vision deficiency, there is a certain chance that you or your children will have the same vision handicap The chance is strongly related to the type of colour-blindness

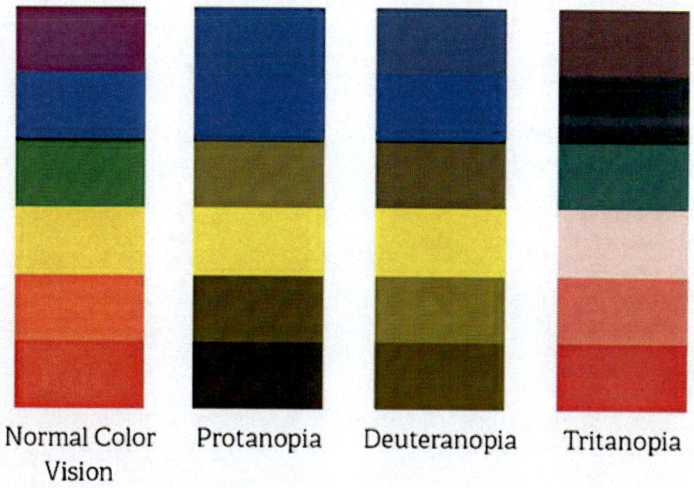

Fig. 10.18 Colour perceptions of dichromats as compared with trichromats. Notice how protanopes and deuteranopes see little difference between red, orange and green. Tritanopes, on the other hand, see yellow as pink and see little difference between violet and red

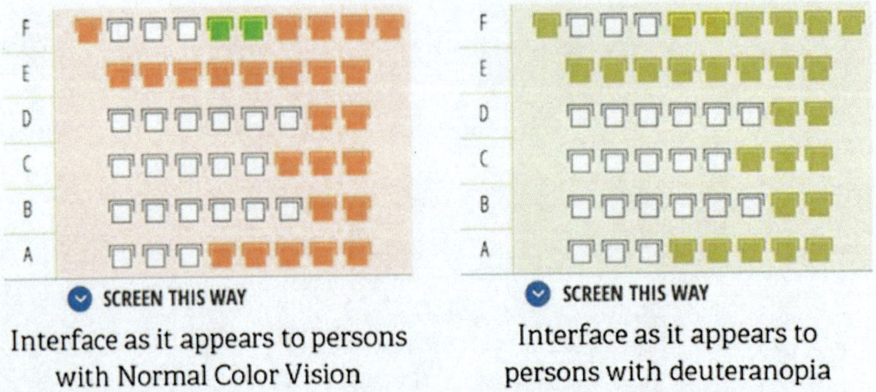

Fig. 10.19 Seat selection interfaces in a movie ticket booking website. Booked seats are shown in orange and the customer's selections in green. The interface can be confusing for a deuteranope who sees no difference between green and orange

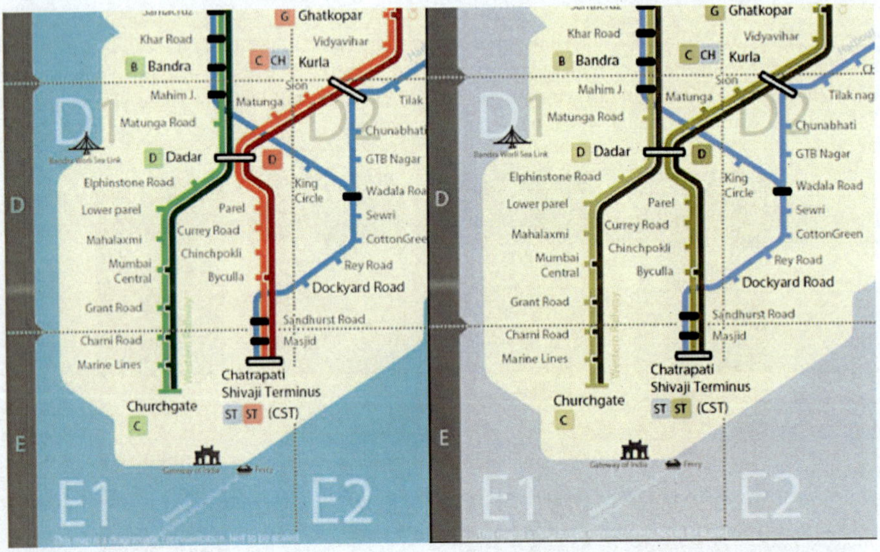

Fig. 10.20 (Left) An early iteration of the MRM that was not colour-blind friendly. (Right) Notice how the green and red lines appear to be the same colour when the map is simulated for deuteranopes

10.18.2 Designing for Individuals with Colour-Blindness (IWC)?

While designing the Mumbai Rail Map (MRM), designers were to assign colours for the line routes, a technique commonly used world over to ease map reading. The function of colour in a map is to create differentiation between line routes for quick identification. For the IWC, the choice of colours should appear different in tone even if they are not as distinguishable in hue.

If you see our initial iteration (see Fig. 10.20, shown), designers used red, green, blue and their shades for identifying the different line routes on the map. Alongside that image (right), you can see how the same map appears to deuteranopes. The colours appear dimmed with lack of distinguishable contrast. Now imagine correlating these lines with the legend and the index of the map. The hues of the red and green line routes appear same with only a slight difference in tone. This can be extremely problematic for persons with colour-blindness while planning their journey and can lead to serious errors.

In the following versions of the map, we changed the colour of the line routes to pink, orange, bluish-green, purple, turquoise and reddish-orange (see Fig. 10.21). We arrived at this colour palette by simulating the map using a plugin in the software Adobe Illustrator to see how it appears to protanopes and deuteranopes. This is typically managed by controlling the hue, saturation and tone of the colours used (Fig. 10.21).

The line routes were made thicker in the revised versions of the map such that colours can easily correlate while referring to the legend or the index. Maintaining a

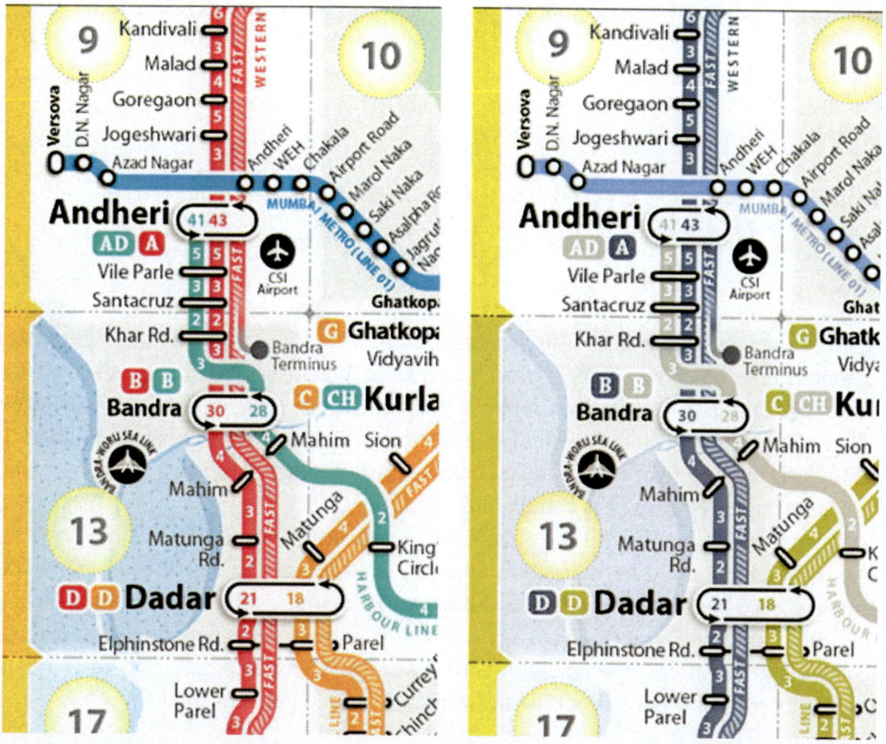

Fig. 10.21 The image of the map on the right shows how people with protanopia see the map. Had the lines been in pure hues, they would've appeared indistinguishable from each other

high contrast with the background is also important because colour vision deficiency can cause dimmed eyesight.

Even though the MRM is colour-blind friendly, it is not yet a complete solution. The next step would be to introduce a visual distinction between the line routes that is not based solely on colour but on line type (dotted or dashed) or overlapping visual pattern (implemented in MRM Version 6) or texture. This will make it clearly distinguishable for the individuals with colour-blindness. As we add redundancies to ensure usability for the IWC impairment would it add to the density of elements and therefore increase in its visual complexity.

10.19 Perceived Visual Complexity of the Map Versus Reluctance to Process Information

Often we are asked this question, "the Mumbai rail map appears too complex. The complexity here is referred to as its visual appearance. Visual complexity is conventionally defined as the level of detail or intricacy contained within an image [9]. It can be considered as an important trait of a map, as the novice usually gets overwhelmed by the variety and number of elements on the map. Such appearances lead to reluctance to process information. In other words, they become unapproachable and uninviting (Fig. 10.22).

10.19.1 The Perceived Visual Complexity and Underlying Order

The map/diagram (Fig. 10.22) appears like a vast ocean of visual and textual elements jostling for space. It creates a feeling of fear/anxiety for the novice traveller with a potential reluctance to process information as he has no idea of his location on the map. In his book 'Living with complexity', [8] mentions about the mess on Al Gore's office desk that is perfect for him but not for a stranger entering his cabin. There is an underlying order (of how things are arranged) to which Al Gore is familiar with, but the stranger entering his cabin is not.

If we draw an analogy to the mess on Al Gore's office desk towards the arrangement of elements in a map; then the map for a novice obviously appears complex. But as novice begins to travel on the rail network, he understands how the map is laid out (the underlying order). Over time, the visual complexity seems to disappear and the content becomes approachable. Isn't it strange? This is because, with frequent travel, his mental model about the network as a whole gets clarified.

Fig. 10.22 This is a part of a 22 × 32" print map of the New York Subway. Does it appear too complex? Can it be simplified?

10.19.2 But then a Question Arises, How to Make the Map Approachable for Novices Who Are Likely to Be Overwhelmed by Visual Complexity?

The visual complexity in a map is subject to how dense (number of line routes) is the network and what is the available space to fit it in. Designers have to operate within the constraints of available space. No doubt, that the visual complexity is overwhelming for novices as they are not familiar with the underlying order. But there is a way out.

Visual complexity can be tamed for novices with the tools of search which are built in within a map. These tools are standard conventions. Imagine searching the meaning of a word from a thick dictionary if you are not aware of the method of alphabetical search. Your prior knowledge of search by alphabetical order and the fact that it was taught to you makes it easier to search. Similarly, an alphabetically ordered index, alphanumeric grid and 'you are here' depiction are standard tools to read a map, subject to the fact that the audience is aware of these conventions or is trained to use them. Therefore the visual complexity for novices can be tackled, if your audience is trained to read a map. Training is an effective tool to reduce the reluctance to process information in context of novices.

10.20 MRM Study: Objectives

Following are the objectives of usability testing.

- Are participants able to comprehend the tools of search (index and grid) used in the map?
- Whether users are able to discover, understand and use features built into the map like visual language depicting 'fast' and 'slow' trains, travel time between stations.
- Whether participants are able to accomplish tasks/steps (example, A to B travel, with transfer at C) in trip planning?

10.21 Methodology

MRMs Version 5 in English, Marathi and Hindi languages (separate map/diagrams) were installed on the railway station foot-over-bridges (FOB) at Dadar (West) and CST stations in Mumbai for 6 days in May, 2014. Since we wanted to conduct the study in context, intercept study (guerrilla Usability testing with tasks, pre and post-task questions) method was employed. Convenience sampling was used to recruit participants while they were waiting on the FOB or passing by and willing to participate. Care was taken to recruit commuters/users across

- different socio-economic groups,
- genders,
- literacy levels,
- age groups,
- diverse mother tongues,
- engaged in various professions,
- knowledge of the rail network (novice, a regular commuter, superuser).

12 volunteers (6 at each station), helped capture over 102 study response.

Fig. 10.23 The MRM being tested by students of Industrial Design Centre (IDC) at the foot over bridge at the CST (Central railway) Station in May 2014

10.22 Usability Testing Protocol

- MRM was installed on the FOB (with the permission from railway authorities).
- Student volunteers gave a brief introduction of themselves and the MRM usability test to commuters who agreed to be a part of the testing.
- Demographic data was captured, commuters' responses to tasks from the test were captured, commuters are thanked for their time and given a copy of the portable MRM (Fig. 10.23).

10.23 Usability Testing Tasks

Task 1: Identify your destination (desired railway station) on the MRM.
Task 2: How long (time) would it take you to get there, from here (Dadar/CST)?
Task 3: Which sort of train would you have to catch to go to a particular station?
Task 4: Plan your trip from A to B
(involved an understanding of slow and fast train routes and knowledge of how to change-over from one line to another between western, central and harbour lines).
Task 5: What is the fastest route to go from A to B
(required knowledge of using other modes such as Metro and Monorail along with the Suburban train network.

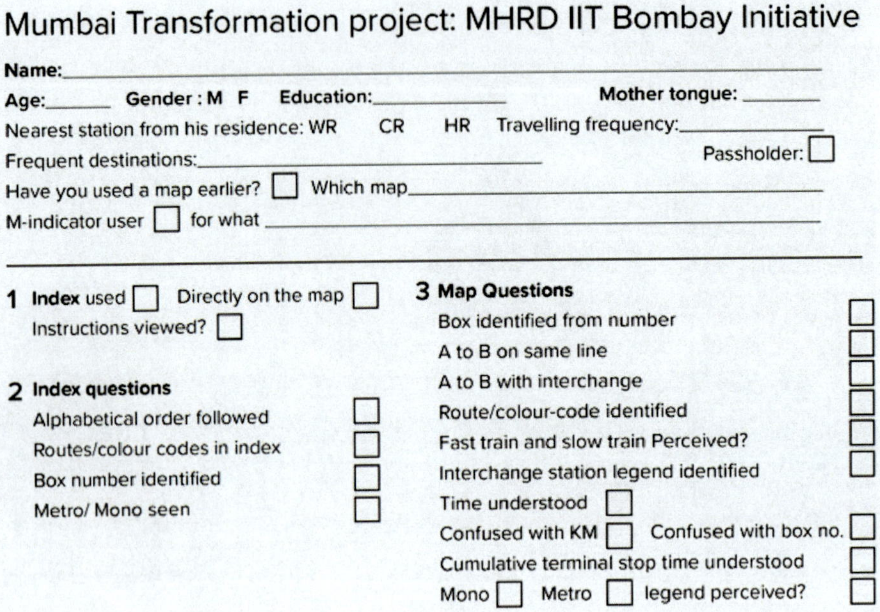

Mumbai Transformation project: MHRD IIT Bombay Initiative

Name:

Age: **Gender : M F** **Education:** **Mother tongue:**

Nearest station from his residence: WR CR HR Travelling frequency:

Frequent destinations: Passholder: ☐

Have you used a map earlier? ☐ Which map

M-indicator user ☐ for what

---

1 Index used ☐ Directly on the map ☐

Instructions viewed? ☐

2 Index questions

Alphabetical order followed ☐

Routes/colour codes in index ☐

Box number identified ☐

Metro/ Mono seen ☐

3 Map Questions

Box identified from number ☐

A to B on same line ☐

A to B with interchange ☐

Route/colour-code identified ☐

Fast train and slow train Perceived? ☐

Interchange station legend identified ☐

Time understood ☐

Confused with KM ☐ Confused with box no. ☐

Cumulative terminal stop time understood ☐

Mono ☐ Metro ☐ legend perceived? ☐

Fig. 10.24 This form helps capture demographic information as well as responses to tasks

10.24 Format for Capturing Data

See Fig. 10.24. The picture in (Fig. 10.24) illustrated the format used for capturing data. The area above the horizontal line captures the demographic data, whereas the bottom part documents the task success.

10.25 Findings/Feedback Received from the Audience

Mumbai Rail Map is designed as a 'you are here' map as well as a portable print map. Here are a few of the findings (Figs. 10.25 and 10.26).

Awareness of map reading as a skill was overall low

- 63% of commuters interviewed, had never used a map before.
  Implication:
- Map reading training is needed at the elementary school level with reinforcements through repeated encounters with maps in the everyday environment.
- Though infrequently used, maps may be needed in stressful or anxiety-causing situations such as navigating during time bound travel or emergency evacuation.

Discovery of search through index and grid approach was poor

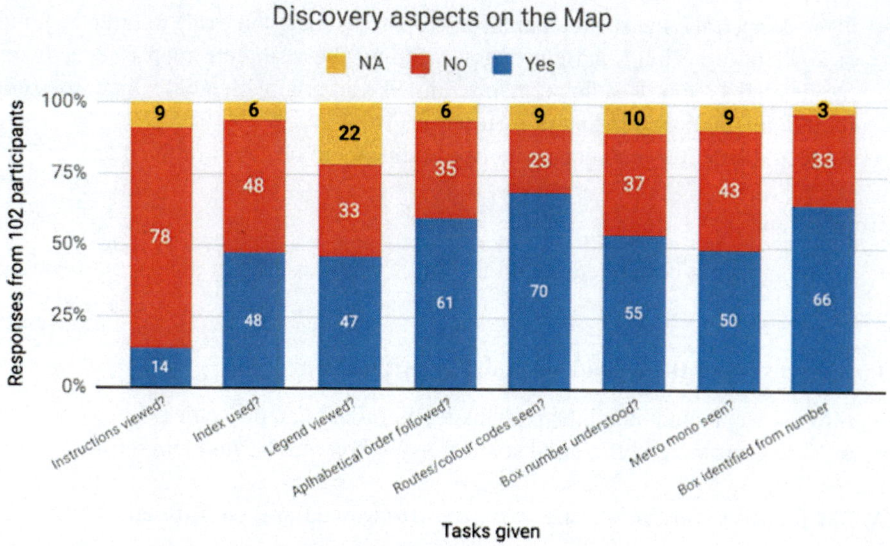

Fig. 10.25 Graph displaying the count (number of people) able to discover information on the map

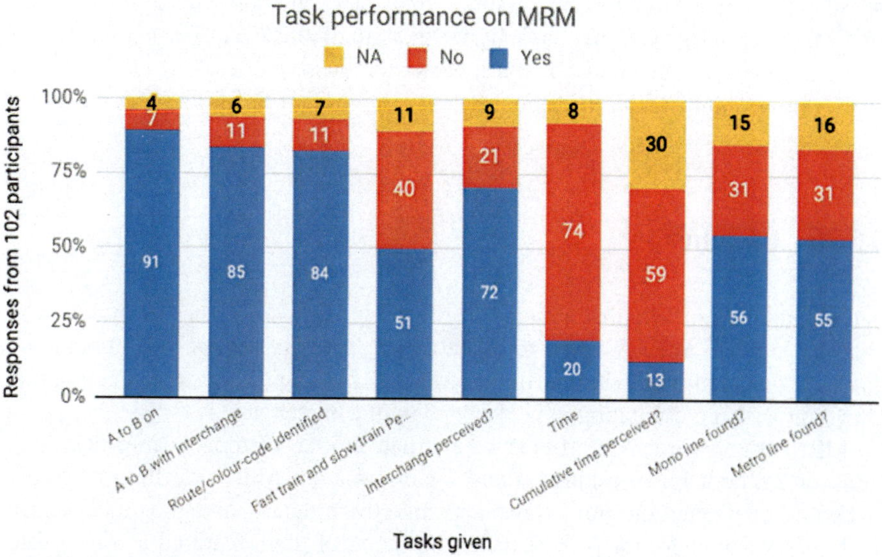

Fig. 10.26 Task performance on MRM. Displays the count (the number of people) able/not able to perform the assigned tasks

- 73% skipped the instructions and search tools (index and grid) to directly jump straight into the map and ended up scanning the complete map to search for the desired station. Reading a map seems like an attainable goal which does not require instructions/reading of instructions [6]
- At first, the map legend goes unnoticed (appears to be a part of the map).

Implication:

- Searching for a desired station in the entire map resulted in anxiety and wasted time.

Post discovery of the legend, organically or due to probing

- 60% of users understood the alphabetically ordered station index.
- 66% of users were able to find stations using box number/numeric grid.

MRM features were moderate to poorly discovered and comprehended

- 51% of commuters perceived the fast and slow lines.
- 70% of users could find station code and line route colour.
- 80% Commuters mistook 'travel time' for distance in kilometres or station codes. In the map, it is explained visually using a small clock icon located at stations, such as Churchgate and CST. But it went unnoticed.
- Few users confused Metro-Mono routes with the Diva-Vasai shuttle route as both were grey in colour.

10.26 Conclusion

Suburban trains in Mumbai along with Metro + Monorail form a complex, high-density network. We attempted to make this complexity accessible to commuters through a printed map within a limited space. This map is a pseudo-geographical diagrammatic representation to be decoded by the passengers.

MRM was designed to provide an alternate source of reliable information (in addition to the habit of asking around for information while travelling on trains). When we conducted the study travel apps like (M-Indicator and google) were used by expert commuters for A to B travel, but none of them featured a consolidated network map. MRM aim was to provide a standard conceptual diagram of the all rail routes on the suburban network, similar to LUD. Authors suggest that if maps like MRM are made omnipresent then commuters can form error-free mental models.

After testing; backlit (you are here) MRMs of 5 by 4 feet in three languages (English, Hindi and Marathi) have been installed at two Suburban railway stations in Mumbai. The authorities warned us that the maps might be vandalised (if placed at an easy-to-read height). However, MRMs have been in use for 5 years now with minor pen scribbles at the bottom. During our visit for installation of MRM people were quite intrigued by 'these huge artworks' and were keen on locating themselves on map. MRM's true impact can be investigated only when we are able to implement it at scale in conjunction with local authorities across all 129 train stations. MRM could be realised as a missing part of a well-designed information system that Mumbai needs.

References

1. Rajendra, A.B.: Halt Station India. Rupa Publication India, New Delhi (2014)
2. Balcombe, R.: The Demand for Public Transport: A Practical Guide. TRL Report TRL 593, p. 91 (2004)
3. Colour-blind Population. Colblindor. 28 April 2006. Website https://www.color-blindness.com. Accessed on 23 June 2015
4. Walker, J.A.: Article: the London underground diagram. In: Triggs, T. (ed.) Communicating Design: Essays in Visual Communication, B.T. Batsford (1995)
5. Garland, K.: Mr. Beck's Underground Map: A History Capital Transport Pub. Reprint ed. (November 30, 1994)
6. Keller-Cohen, D., Meader, B.I., Mann, D.W.: Redesigning a telephone bill. Inform. Des. J. **6**, 45–66 (1990)
7. Mumbai DNA (Newspaper Article): When can we travel like humans? (2013) Viewed 20 November 2013. http://www.dnaindia.com/mumbai/report-when-can-we-travel-like-humans-1891154#
8. Norman, D.A.: Living with Complexity, p. 5. MIT Press, Massachusetts (2011)
9. Snodgrass, J.G., Vanderwart, M.: A standardized set of 260 pictures. Norms for name agreement, image agreement, familiarity and visual complexity. J. Exp. Psychol. Human Learn. Mem. **6**, 174–215 (1980)

Chapter 11
Developing a Design Research Approach to Inter-city Rail Carriage Interiors

Sylvain Jolivet and Selby Coxon

Abstract Historically, rail had been the first great arbiter of travel over significant distances that could be experienced by the common man. A nineteenth century technology that overcame distances in journey times unobtainable by the contemporary means of that century. Since then the travelling public has fallen out of love with the mode. Contrast the romanticism of travelling by luxury sleeping coach to the crowded standing room only of a commuter service. The late twentieth century has been dominated by increased accessibility to air travel and the associated culture that manifests itself in airport termini, aircraft interiors and associated accessories. However, High Speed Rail (HSR) is enjoying a resurgent interest in global national transport planning agendas. Economic and sustainability metrics indicate benefits over alternative transportation methods and over comparable geographical distances (1). Recent developments in rail carriage interior design both leverage the 'flavour' of the airline aesthetic with a new confident form language that celebrates the space to move about, amenity, comfort and psychological wellbeing that places the mode at a competitive advantage. This chapter examines the critical issues and contemporary approach as exemplified by the French National Railway (SNCF) to create a human centred resurgently confident approach to the future of rail.

Keywords User experience · High speed rail · Comfort · Sensory experience · Emotional wellbeing

11.1 Introduction

Rail passenger travel has been with us since the Stockton to Darlington railway opened in 1829. It covered 25 miles of track at, what was for the time, the dizzying

---

S. Jolivet (✉)
SNCF, La Defense, Paris, France
e-mail: Sylvain.jolivet@sncf.fr

S. Coxon
Faculty of Art Design and Architecture, Monash University, Melbourne, VIC, Australia
e-mail: selby.coxon@monash.edu.au

S. Coxon and R. Napper (eds.), *Advancing a Design Approach to Enriching Public Mobility*, Intelligent Systems Reference Library 198,
https://doi.org/10.1007/978-3-030-64722-3_11

speed of 50 kilometres per hour. Despite the cycles of boom and bust over the intervening two centuries, rail has been an enduring mode of transport especially in the sweet spot between too great a distance to be comfortable by car and too short a distance to be efficient to fly. High speed rail (HSR) is a cost effective and environmentally friendly alternative to roads and airports [1]. HSR can be brought into city centres and so can engage easily with a variety of other modes such as metro trains, buses, taxis bicycles and pedestrianisation.

The attractiveness of any mode of transportation is enhanced by how quickly it can transport passengers from point A to point B. Since this cannot be done in an instant then passengers will inevitably 'experience' the moments of their journey with a mixture of either joy and enthusiasm, or misery and indignation. Travelling any significant distance invites the overcoming of inconveniences such as sustenance and biological needs. The quest for comfort at some level is something deeply human.

Until the advent of the motor age, the railways had no real competitor for the movement of people across country. Early train interiors were little more than conveyances to accommodate the maximum density of passengers. This perspective changed when European royalty embraced rail as the convenient means to traverse their nations while enjoying palatial interiors that reflected their status. Notably the grand interiors of the imperial train carriage built for Napoleon III by the Compagnie de Chemin de Fer de L'Est. The widely published images of these royal carriages may well have influenced contemporary George Mortimer Pullman in his pioneering work in making transcontinental travel in the United States more comfortable with the creation of sleeper cars. The Pullman name became synonymous with innovative interiors that recognised 'place making' and in particular a superior level of comfort while travelling by rail [2]. These carriage designs delivered such innovations as the experience of relaxing in a reclining chair, while enjoying a rapidly passing landscape and consuming the best food and drink available.

Pullman had come from a carpentry background and the interior of his 'Pioneer' train carriages reflected the prevailing architectural taste of the time, somewhere between palatial and bordello. Rather than something to be endured, Pullman's taste conveyed the potential pleasure of travelling comfortably. His eclecticism and sumptuousness remained influential throughout the second half of the nineteenth century. Growing health concerns and a better understanding of disease transmission, made more obvious the potential harbours of dirt, dust and bacteria in such elaborate interior designs and thus ushered in a more rational approach to interiors. Forward thinking design philosophies of the Deutscher Werkbund and later the Bauhaus advocated a pared back aesthetic that emphasised elegant geometry with restrained but highly crafted details.

The culture of long-distance rail travel peaked in the decades that straddle the 1930's to 1950's, before both air travel and the motor-car really took hold of the popular imagination. Rail maintained its allure through the significant contributions of Industrial Designers such as Walter Dorwin Teague, Norman Bel Geddes, Henry Dreyfuss, Paul Cret and Raymond Loewy. Locomotive exterior designs exemplified modernity and speed through styling affections such as streamlining. Carriage interiors also orchestrated the same generous swept rounded ends, tapered cross-sections, domed roof linings and tear drop forms. Materials included tubular steel, large mirrored surfaces to amplify the perception of width, wood veneers and simple geometric textile designs. This was a machine aesthetic that transcended the affections of class and responded to Raymond Loewy's ethos of MAYA; most advanced and yet acceptable. The approach pushed the boundaries of an 'acceptable' aesthetic in the commercial world.

By the 1960's it was becoming clear that competition from airlines would require the rail industry to re-think their services and designs. Services became faster as demonstrated by the Japanese Shinkansen 'bullet train' and carriage interiors mirrored the sleek interior pastel colour palates as experienced within an aircraft. No longer aping architectural affections but competing with airlines for a chic identity and a compelling mix of technology, brand and design.

11.2 SNCF

The Société nationale des chemins de fer français (SNCF), is France's national rail company, formed in 1938 from an amalgamation of six regional railway companies. It is a highly successful organisation (as measured by the fortune 500) with global business interests in transportation through subsidiary operators. The company has a highly embedded design culture which has engaged with notable French designers; in particular Roger Tallon, who was the leading creative force behind Corail cars (COmfort by RAIL), the TGV Atlantique, the TGV Duplex, projects such as the Meteor (line 14 of the Paris metro today) and the Montmartre funicular for the RATP, the VAL (automatic light vehicle) designed by Matra (OrlyVal and CDG Val) and the Mexico City metro.

" Industrial design is neither an art nor a mode of expression, but a systematic approach able to be applied to any design problem.".

Roger Tallon.

HSR operates in an increasingly competitive operational environment, not least since the European continent opened up to enable different national operators to run services across borders creating direct competition on routes between capital cities. Service quality is a significant contributor to mode attractiveness. Predicting that customer dissatisfaction would limit business growth, SNCF created a strategy that placed the concept of 'user experience' at the heart of their research and design program. Drawing particular inspiration from Olivier Le Goff's thesis, "The invention of comfort: a new social form", which identifies comfort as a key component in the appreciation of different modes of transport.

11.3 The Concept of Comfort

The presence of other people, as experienced while travelling by rail as well as being beholden to a fixed timetable, has long made the personal car a more attractive alternative for travel. Long distance journeys mitigated some of these perceptions since travelling by rail avoided fatigue as well as giving the passenger a generally faster journey time while having the opportunity to eat and sleep without stopping. The key element that these amenities brought together was the idea that they could be experienced 'comfortably'. Human factors literature often speaks of comfort as a relative and changing quality that can only be defined as the absence of the antonym; discomfort [3]. Embodied in avoiding discomfort are the qualities of; temperature, lighting, air quality, sounds, vibrations, olfactory, sense of security and tactile support amongst other nuances. The extent to which these are important to a passenger may change throughout the course of the journey.

There is a growing body of research that addresses the effects of travel on emotional wellbeing [4]. Passengers evaluate their journey experiences to include not only utility (cost, time, punctuality, frequency, safety etc), but their moment by moment moods as discrete events punctuate their journey; such as noise, crowding, the view from the window, positive engagement with fellow passengers etc [5]. To persuade customers that train travel is a better alternative to the greater polluting and energy-consuming personal modes of transport, the features and sites of public transport must appeal to the customer and respond according to sensory as well as functional expectations. Indeed, transforming travel into an improved sensory experience is the main focus of transport design. These qualities are conceptualised best in a customer satisfaction pyramid as seen in Fig. 11.2. At the base of the pyramid are the service expectations, such as cleanliness, punctuality, safety etc. These things are satisfactory. Above the line are 'delighters', experiences of the journey that make it pleasurable, engaging and memorable for all the right reasons. One might characterise the role of design to elevate the passenger experience from the zone of expectations into the zone of delight.

11.4 A Passenger Focused Approach to Comfortable Carriage Design

Traditional Industrial Design approaches to manufactured goods, it could be argued, have clumsily married the available manufacturing techniques with the functional intent of the object. Human factors research and increasing sophistication in manufacture have improved the connection between making and using. However, nuances of human perception and a strong business imperative to deliver an exemplary experience demand more sophisticated methods.

SNCF adopt a mixed methods approach to collating, analysing and interpreting data from their customers. These include questionnaires, open ended interviews,

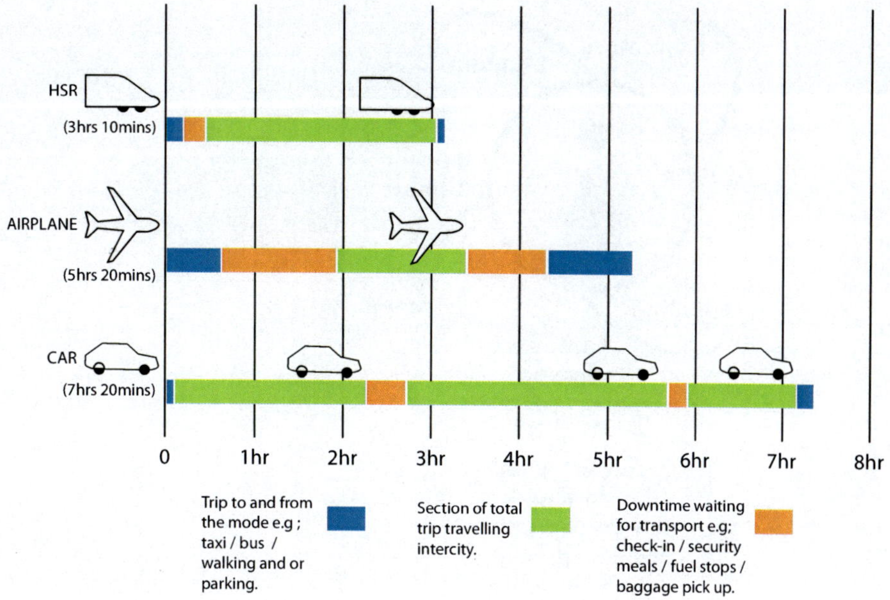

Fig. 11.1 Speculated comparison of intercity modes over the same distance. In this example, from central San Francisco to Los Angeles. Diagram adapted from [1]

Fig. 11.2 Customer satisfaction pyramid. Authors image

surveys and observations. Psycholinguistic ways of interpreting data are frequently used. This is a technique that unpacks the link between the concepts of language and words and how they are interpreted differently by different people. By understanding what passengers say about their trips the key comfort and sensory elements of their experience can be identified.

On board comfort is subjective with complex interactions between different factors as Fig. 11.3 shows the elements of perceived comfort are entangled, personal and can change over time.

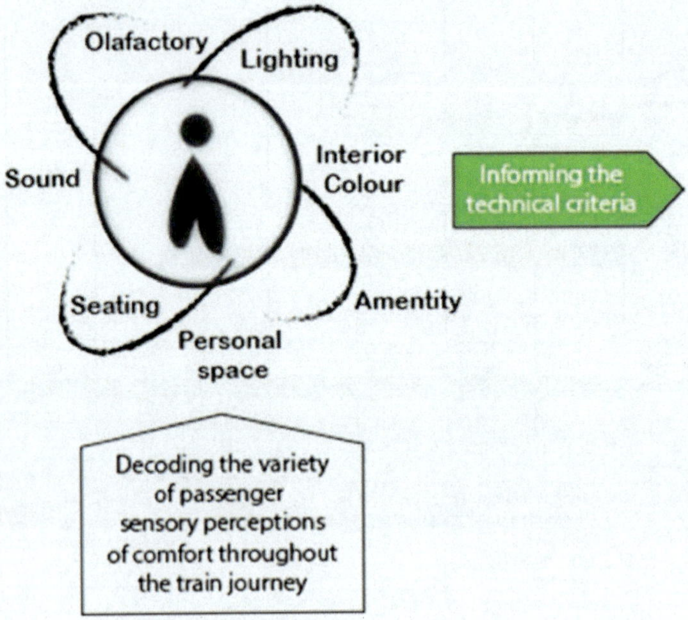

Fig. 11.3 Entangled qualities of passenger comfort. Authors image

Fig. 11.4 Building technical specifications

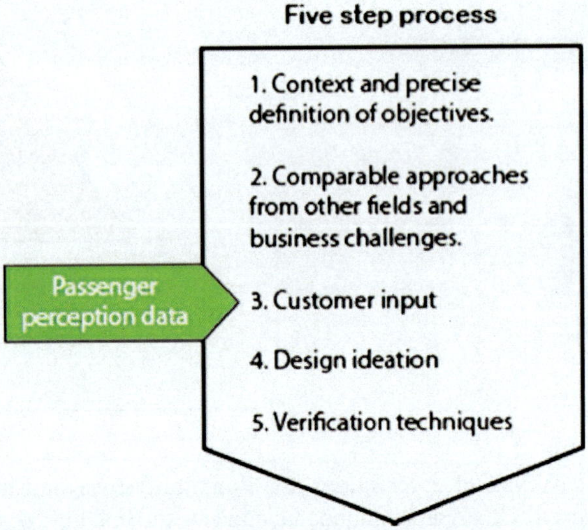

Passenger feedback supports the creation of draft requirements for designing and developing the interior elements; seating, décor, sound etc. To pull together conflicting passenger perceptions the 'design' of the carriage needs to capture a central idea; a clear concept to be responded to. For HSR the idea projection is that unlike airline passengers, customers of rail are not strapped into their seats. They

Fig. 11.5 Digital simulation of the Concept Train at Gare St Lazare (Paris)- Sylvain JOLIVET for AREP

Fig. 11.6 Concept Chair models. © RCP Design Global

are free to move about and expect to enjoy a variety of activities unavailable or restricted in the environs of an aircraft or a car. Emerging from this core idea, "free to move about" is the design specifications for all the elements within the carriage. This includes developing techniques to maximise space, colour schemes, lighting, seat dimensions and accessibility to gangways, and amenities such as toilets, luggage and refreshment carriages. The ability to carry on business activities through the use of mobile phones and portable computer devices has been one of the more remarkable changes in passenger behavioural expectations in recent decades.

The engineering of the constituent elements of the carriage interior are developed in response to the passenger data. So, seating solutions that cater for the wide variety of body shapes and the minimising of vibrations and unpleasant noise are evolved through this part of the development process. Virtual reality testing of developing concepts is playing a greater role in design decision making. Unlike full size mock-ups, the creation of virtual environments which can be experienced by a wide range of stakeholders is much quicker and cheaper to produce. Experiences of spaces and

Fig. 11.7 Dyapos postural comfort test rig. © Cédric GALLAIS

features can be examined up to a point and then the building of physical prototypes at pre-production point remain necessary, but the overall costs of development are diminished.

This conceptual exploratory framework has given rise to approaches combining sociology, anthropology, engineering and art. At the heart of the process, the designer listens, translates, integrates, and, via 2D and 3D projections, models and prototypes, then exhibits the future of products and potential creative directions to test on sample groups. Promising results mean that commercial operations have come to consider this form of prospective design as indispensable to the creation of transport solutions. Many transport solutions have emerged from this method: from Concept Train Ile de France 2002, for example, came the NAT (Nouvelle Automotrice Transilien), introduced into commercial service in 2013; from Concept Chair TGV 2010 came the first- and second-class chairs of the TGV L'Océane, introduced into commercial

Fig. 11.8 3D model of the interior design of the concept NGTE © Sylvain JOLIVET

service at the end of 2016; from Concept of Toilets 2009 came the main colour scheme, ambient lighting and the design of future TGV toilets including those used in TGV L'Océane and in the fleet TGV inOui.

The following images exemplify carriage interior concepts developed through the deep inspection of human experience on passenger rail and then re-engineered to meet those expectations.

11.5 TGV L'Océane

The Train à Grande Vitesse (TGV) is the flagship operation of SNCF in France. The concept was conceived during a period of the mid-1960's that was particularly ambitious in technological development and included the Ariane rocket and Concorde supersonic airliner. Unlike those two, the TGV remains in place and has gone from strength to strength breaking the speed record for HSR in 2007.

From 2012 a central project in SNCF's research has been in the re-vitalisation of the TGV brand and in particular the L'Océane, flagship service. This line runs between Paris and Bordeaux. Central to the strategy of creating 'place' during a train journey is the careful blending of utility with comfort and an aesthetic that reinforces brand value. This is especially important for the business passenger whose sense of place while travelling is predicated upon a satisfactory place to work rather than just to relax and watch the world go by.

The designers at the heart of the process began to develop a strong travel concept formed of products and spaces consistent with these objectives, that is, a vision of

Fig. 11.9 View of the model sink.© reserved rights

high-speed design which was democratic, accessible and egalitarian, always comfortable and with on-board digital connectivity. Drawing from the private sector's standards of comfort and privacy, the idea was to create not just a means of movement but a space and time between departure and destination that prioritised well-being, quiet, and reassurance. Keeping in mind that the duration of trips undertaken on this train would be between two and five hours.

The first-class seat included a fold down desk entirely equipped for business customers (including a mini additional tray table, USB ports, 220 V outlets, an individual book light, and desk lighting with manually adjustable dimmer), as well as a turning base that allows nine out of ten users to face the direction of travel. A vanity mirror is also embedded in the back of the seat, strengthening the relationship between user and object. Both first and second-class spaces have restyled LED-lighting that is as natural and relaxing as possible, while also being less energy-consuming. The aesthetic of the light bands obscures the source itself to offer ambient lighting. A low light on each pier (the structural element between each window) punctuates the space, enhancing the domestic and warm ambiance of the setting. Finally, the lighting

Fig. 11.10 View of the TGV L'Océane 1st-class carriage. © Pierre Antoine PLUQUET

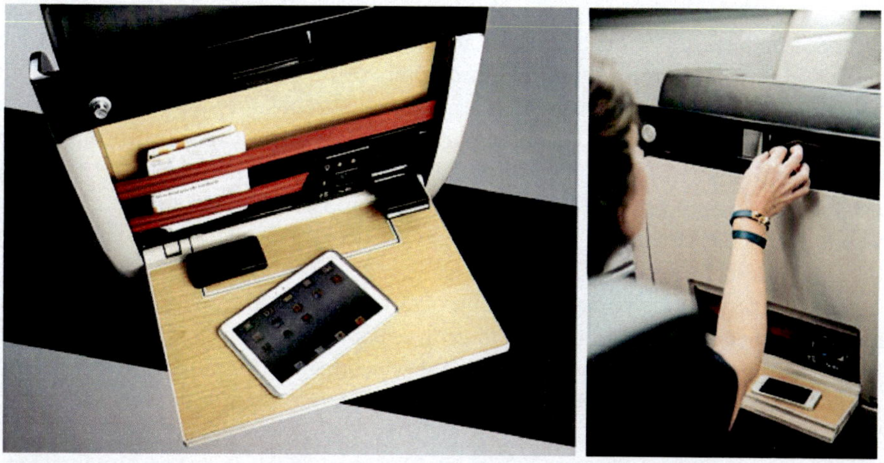

Fig. 11.11 Left; 1st-class seat nomad desk. © Jean-François STILLING. Right; mini additional tray table and vanity mirror © Florent DARTHOUT

has three different modes to adapt to different times of day; normal during the day, reduced in the morning and evening, and a night setting where only the seat markings and individual sources remain on.

The second objective was to markedly improve travel comfort for families and leisure customers who use the TGV less often. Here, the key focus was fluidity

Fig. 11.12 TGV L'Océane—view of 2nd class carriage. © Pierre Antoine PLUQUET and 2nd class seat mini desk. © Florent DARTHOUT

of space and optimisation of seat room. Intensive efforts were made to improve second-class seat design, slimming the structure of the seat to create more leg room and reducing hardware to ensure that seat mass met regulatory requirements and to increase seated capacity. Each seat is equipped with an innovative synchro-tilt system (patented by SNCF) for base and back, a USB port per person, one 220 V outlet between two, a mini desk, a cupholder and a compartment for overhead luggage. Finally, at the back of certain second-class lower-deck carriages, a generous 5-person bench seat transforms the space into a true living area for group travel.

During the journey, passengers can move from one end of the train to the other and make their way to the bar in the centre. For that reason, the bar remains the keystone of the journey. As well as a food service, the bar symbolises the possibility of travelling differently, chatting with a group of friends or family, or even listening to musicians play during the concerts organised by the events team of SNCF-Voyages. The interior architecture, with a different table arrangement and perch seats towards the centre of the carriage, makes the space convivial; the bar is restyled as a warm and contemporary living area. Features such as the table lamps and piers designed by Ionna Vautrin and materials like light oak, white Corian and leather brighten the space and encourage passengers to make it their own. The counter, with a real-wood top, a pared-back design, and a bright-red overhanging canopy, is completely open to stage the barista and invite a friendly rapport between customer and server. To the right of the counter is a refrigerated display case for fresh, ready-to-serve products. The dining car as a whole is divided into two sections: a standing-room side for on-the-go dining with a long, high table complete with two divides to group customers,

Fig. 11.13 View of the TGV L'Océane bar. © Florent DARTHOUT

and a seated side made up of three four-person booths, elevated to offer a view of the passing landscape out the window. Finally, each booth includes a Corian cantilever table and an iconic lamp also designed by Ionna Vautrin.

11.6 Conclusion

Long distance rail is seeing a resurgence in popularity as an alternative to flying. Intercity services especially in the European context address door to door time constraints as well as sustainability concerns that airlines struggle to match. Where there are perceived disadvantages to travelling by rail, an earnest attention to carriage interior design has been addressed to create the trappings of a comfortable enjoyable experience. Unlike other competing transport modes, HSR enables passengers to enjoy the freedom to move about the train and visit dining cars, bars and cafés during their journey.

Fig. 11.14 The iconic TGV L'Océane lamp appearing to be made from a single lustrous surface engender the sleek dynamic associated with high speed—design Ionna Vautrin. © Florent DARTHOUT

The increasingly sophisticated array of amenities introduced into successive iterations of HSR cater overwhelmingly for the business market where the ability to work and hold discussions with colleagues is a vital differentiator to the airline experience. There are a wide variety of qualities that form the opinion of a rail passenger's journey. Techniques such as organising these qualities in a customer satisfaction pyramid demonstrate that passengers have basic expectations of a service as provided by good engineering and that what delights and excites a traveller beyond these satisfiers can be demonstrated by good design.

To create this experience, significant rail operators such as SNCF develop in house design expertise that transcended a singular Industrial Design manufacturing approach and embraced a variety of human centred research skills; psychology, anthropology, sociology and psycholinguistics to design interiors that engendered a sense of place and positive well-being. It is anticipated that the business case for HSR will only become stronger, with both ecological and experiential benefits of travelling directly between commercial centres. It is likely, as shown in China in recent years, that speeds for HSR will become even faster. Design will be under increased pressure to be the significant differentiator in an increasingly competitive global market.

Acknowledgements The authors wish to thank SNCF.

References

1. Terplan, E. (2011) Beyond the Tracks: The Potential of High Speed Rail to Re-shape California's Growth San Francisco's Planning and Urban Research Association Policy Paper Accessed 20th April 2020
2. Votolato, G.: Transport Design: A Travel History. Reaktion Books, London (2007)
3. Vink, P.: Comfort and Design Principles and Good Practice. CRC Press, New York (2005)
4. Redman, L., Friman, M., Garling, T., Hartig, T.: Quality attributes of public transport that attracts car users. Transp. Policy **25**, 119–127 (2013)
5. Garling, T., Ettema, D., Connolly, F., Friman, M., Olsson, L. (2020) Review and assessment of self-reports of travel-related emotional wellbeing. Journal of Transport and Health, 17, Elsevier
6. Eds Hayashi, Y. Ram, KE. Bhanle, S. (2019) Handbook on High-Speed Rail and Quality of Life. Asian Development Bank

Chapter 12
Data Visualisation Prototypes for Problem Definition in the Design of Accessible Transport Solutions

Vincent Moug and Indae Hwang

Data visualisation prototype viewer: http://gapgeometrydatavis prototype.surge.sh/

Abstract Public transport is a challenging environment for both passenger and operator. The operation of large networks carries expectation of timely delivery of passenger-oriented services. With numerous stakeholders such as government, private industry, and the general public the expectation around the primary concept of mobility within transport networks often differs. This chapter examines how data visualisation has been engaged to better understand an enduring mobility challenge of Australian metropolitan rail networks, the universally accepted accessibility issue widely known as 'the gap'. With increasing global requirements for disability compliance and anti-discrimination within public transport, data visualisation is applied within a broader design project. Engaged at the problem-definition stage this data-driven approach provides investigators greater insight and visual transparency towards understanding the impact this issue has on mobility for all users of public transport.

Keywords Accessibility · Prototype · Design process · Data visualisation · DSAPT

V. Moug (✉)
Queensland College of Art (QCA) Griffith University, 226 Grey Street, South Brisbane 4101, QLD, Australia
e-mail: v.moug@griffith.edu.au

I. Hwang
Department of Design, Monash University, 900 Dandenong Road, Caulfield 3162, VIC, Australia
e-mail: indae.hwang@monash.edu

S. Coxon and R. Napper (eds.), *Advancing a Design Approach to Enriching Public Mobility*, Intelligent Systems Reference Library 198,
https://doi.org/10.1007/978-3-030-64722-3_12

12.1 Introduction

12.1.1 The Problem of Accessibility in Public Transport

Public transport (PT) presents a complex system of interconnected products, services, and user-experiences. Operating at a macro scale throughout cities, regions, and countries, and then at a micro-scale enabling individuals with everyday commuter activities. PT provides alternative modes of transport where passengers can forgo responsibilities tied to navigating traffic and other road users by engaging with a more detached mode of travelling. The PT system is made up of conveyances; i.e. buses, trains, and trams, and adjoining infrastructure; i.e. roads, stations and bus stops, which aim to provide amenity to the broadest spectrum of the population. As contemporary societal values shift towards a more empathetic and inclusive norm, the historical legacy and resilience of these complex systems are now being challenged [1]. The social model of disability, how the built environment and system prevent inclusion of certain demographics in the population, specifically those passengers with disabilities, is now under intense review [2].

The Disability Standards for Accessible Public Transport (DSAPT) legislation in Australia provides a contemporary framework for evaluating, and challenging, the existing PT environment [3]. Aimed at the provision of inclusive transport systems across Australia and aligned with inclusive aspirations of the Federal Disability Discrimination Act (DDA) [4]. The DSAPT documents wide-ranging instances where PT systems are to adhere to specific spatial dimensions, product configurations, and or user-centric operational requirements. For example; a wheelchair-accessible toilet must be without airlock, provide minimal manoeuvrable floor space, and throughout the design of amenities provide amenable interfaces for all users.

Given the scale, interconnectedness, and inherited physical conditions of PT systems, the researchers engaged in design activity to examine a specific DSAPT requirement. One applied to the rail environment. The DSAPT clause of specific interest is referenced below. An early graphic visualisation is also presented to provide a relatable scale to the measurements, as defined within the clause (Fig. 12.1).

8.2 When boarding devices must be provided

(1) A manual or power assisted boarding device must be available at any accessible entrance to a conveyance that has:

(a) a vertical rise or gap exceeding 12 mm (**AS/NZS3856.1 (1998) Clause 2.1.7 (f)**); or

(b) a horizontal gap exceeding 40 mm (**AS/NZS3856.1 (1998) Clause 2.1.8 (g)**).

Source: Disability Standards for Accessible Public Transport, 2002, p.23

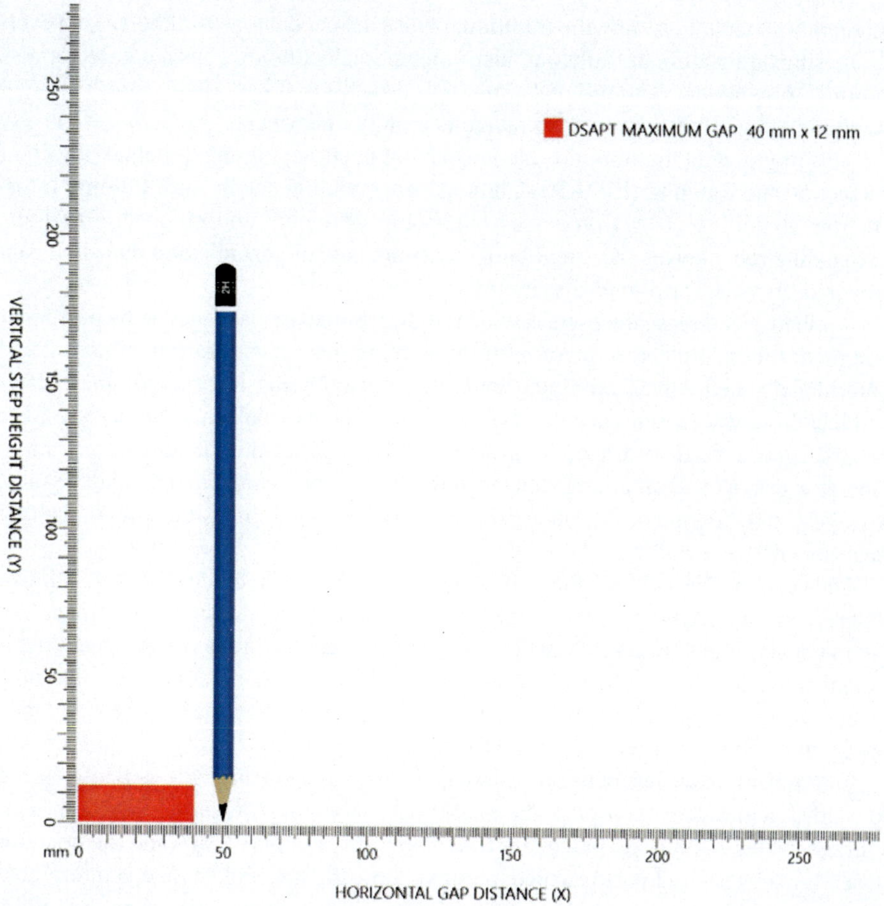

Fig. 12.1 DSAPT minimum compliance of 40 mm × 12 mm for boarding/alighting gaps on public transport *Source* Moug 2017 p. 49. *Source* Disability standards for Accessible Public Transport, 2002, p. 23

12.2 The Link Between Data and Mobility

12.2.1 Data in the Public Transport Environment

PT networks combine physical assets of varying operational lifespan. A large portion of a network's assets is understandably static and low-tech. These being the station platform and rail track. Taking a build and observe approach, most assets are observed intermittently and require manual onsite evaluations, in order to determine levels of degradation and compliance with standards. PT conveyances on the other hand are not static or low-tech. Forgoing the traditional guise as mechanical workhorses, an ever-increasing digitisation of trains and instrumentation technology has increased opportunities to generate *dynamic* real-time digital datasets. From diagnostic to spatial mapping a digital information exchange is rapidly underway.

This digital data disruption is also occurring in surveying equipment, with Light Detection and Ranging (LIDAR) scanning, drone mounted tech, and Building Information Modelling (BIM) processes enhancing designers and engineer's toolkits. Expanding repositories of virtual built environments, of periodic and real-time data now readily exist for digital distribution.

Applied use of these data sets is wide ranging. Some are packaged to be passenger oriented, for example coordination of timetabling, passenger load distribution, and intermodal scheduling. Combining with digital wayfinding information these data-led insights are accessed via enabled devices such as personal smartphones via apps, or passenger information display systems (PIDS). Other digital datasets are more operator oriented. Containing data valuable for asset management, maintenance, lifecycle analysis, and background scheduling information. Important for controlling real time network traffic.

Data capture has become the norm, infiltrating the public context. Administrative interest in understanding service performance and patronage of PT has meant an increasingly large amount of data can and is collected. Data that may reveal operational and behavioural attributes of both assets and passengers alike. Data that may provide useful insights into understanding and improving the PT experience and may assist in resolving or reviewing DSAPT compliance issues.

Interlinked relations between passenger, environment and PT experience are provided within the datasets. The real-time scheduling of services becomes of intrinsic value to both the passenger (how long am I going to wait) and the operator (meeting the required performance metrics). How the interlinked data is interpreted and presented to each group requires careful consideration, ensuring the value of the information to each group is prioritised.

From the perspective of advancing design research around mobility, datasets can be leveraged to help bridge junctures between *quantitative* knowledge gathering exercises, and a *qualitative* reflective exercise. It is within the design process that engaging data visualisation presents an exciting opportunity; for novel insights or patterns to emerge and to inform problem-solving or problem-defining design activities. In this context, it contributes to an understanding of improving accessibility at the platform-to-train interface, and therefore an individual sense of mobility.

12.2.2 Mobility in the Public Transport Environment

Before we shift the focus to data driven design activity it is important to familiarize with the primary concept of mobility. Mobility is the capacity to move freely. For passengers and operator's within PT this is as a primary objective. Moving freely throughout the network system irrespective of whether it is the passenger or conveyance. Optimistically, a positive PT experience occurs where passengers move throughout the network, self-directed and independently. Where passengers encounter timely arrival and departure, independent boarding and alighting of the train, and at convenient locations. All in a comfortable and safe environment. Mobility as a concept underpins the design activity focused on accessibility, with accessibility a key benchmark of the DSAPT. So how does one determine accessibility?

In leveraging digital data, the designer's cognition of the problem is empowered throughout the early stages of design process, that of discovery and problem definition. Converging the digital knowledge component with the physical component, knowns and previously unknowns, the data can influence decision making. Ultimately aiming to define, improve and determine accessibility, and to inform policy targeting the PT environment. The DSAPT recognises that PT in Australia is not one-hundred percent accessible. Significant progress has been made in both awareness of the issue and remediation of some assets; independent access is not available to all passengers. Especially those with disabilities who are otherwise readily independent in other aspects of everyday life.

As we turn our attention back to the platform-to-train interface the DSAPT policy has indicated acceptable physical parameters for accessibility. The *'gap'* is to be no more than 40 mm in the horizontal distance and 12 mm in vertical step height as shown earlier (Fig. 12.1). This aims to accommodate wheeled mobility devices moving between the platform edge, known as the coping, and the tread step of the train carriage. The origins of these exact measurements are layered but in summary are derived from multiple, interlinked Australian standards with reference to Australian building codes, vehicular designs, and mobility aids such as tethered hoists [5]. The base question remains, how does the DSAPT recommendation relate to the real-world conditions? This is where employing data visualisation as a method for design discovery and problem definition is engaged.

12.3 Data Visualisation

12.3.1 Definitions and Taxonomies

Data visualisation is commonly associated with digital tools. Computer-supported, primarily visual, and engaging interactive methods of discovery, explanation, and decision making [6]. Analog arrangements of visual information remain a useful tool, often in earlier conceptual drafting activities, however, the application of digital tools allows an effective and efficient way to interpret and understand substantially larger amounts of data. Often in more simplified compositions. Engaging digital tools and techniques enable translating and interpreting large datasets, revealing attributes such as patterns, anomalies, and peripheral insights—taking advantage of vision as the primary human sense. Composing the visualisation, to order the data, is an important first step in this process.

A taxonomy of data visualisation 'types' exist. These 'types' are described as 1, 2, 3-dimensional data, temporal, multi-dimensional, tree, and network data. Each 'type' prompts different approaches to composition, end-user interaction, and visual communication outcomes [7].

1d, 2d, 3d datasets denote shifts between linear (often alpha-numeric attributes are grouped i.e. in tables) to planar or map-based data (2d), and then to volume-based representations of real-world objects (3d). *Temporal* describes data over a period. *Multi-dimensional* configures relational and statistical datasets where value attributes are translated to dimensional representations. *Tree* hierarchies describes a parent-child relationship, analogous to a genealogy tree. *Network* captures a complex relationship between data elements. The relationship is displayed as "node" or "lines" between elements.

These 'types' of data influence the structure, interactivity, and visual elements of the visualisation prototypes. From an end-user perspective the integration of affordances, as described by Shneiderman, can be implemented to support both navigation and understanding of the data by the user [7]. User tasks and improved understanding can be achieved through the following strategies;

- Overview: Gain an overview of the entire collection.
- *Zoom:* Zoom in on items of interest.
- *Filter:* Filter out uninteresting items.
- *Details-on-demand:* Select an item or group and get the details when needed.
- *Relate:* View relationships among items.
- *History:* Keep a history of actions to support undo, replay, and progressive refinement.
- *Extract:* Allow extraction of sub-collections and query parameters.

These are relatable and useful prompts and tasks that underpin many data visualisation tools.

12.3.2 Designing Data Visualisation

At its core data visualisation is to provide accurate representation alongside useful information. As designers we construct pathways to meaning through the organisation of data-driven visual elements. Often engaging a hierarchal process or step-by-step approach to organising data. This at times includes strategies of mapping, selection, presentation, interactivity, human factors (usability and accessibility), and evaluation [8]. Aiming to ensure data visualisations maintain purpose. Enabling viewers to analyse and understand data better and encourage the exploration of complex data in engaging and compelling ways. An emphasis on conciseness and effectiveness in visualising data is distilling visual elements to easily comprehend the information. To limit the amount of 'Chart Junk' such as pattern overuse, heavy linework/grids, and colour palettes and vectors for artistic appeal [9]. Tufte suggests when communicating complex datasets that visualisations be; substance based, not distort data, concise, coherent, several levels of detail, clear purpose, and closely integrated with a statistical and verbal description of the dataset [9].

12.4 A Strategy Towards Prototype Development

With the broader project investigating platform gap mitigation strategies and data 'types' better understood, design activity began. A method of *data visualisation* is engaged to identify patterns or insights around 'the gap' geometry and problem distribution. Leveraging existing industry data provided a defined *quantitative* geometry range and datum that underpinned preliminary design activity. The data represents a metropolitan rail network. It is acknowledged that some subsets of the data used are now outdated due to progressive development and re-development of some stations within the broader network. The dataset remains valid, in that it constitutes a 'point in time' that provided the designers with a locus from which to build the prototypes. It is anticipated that future datasets can be substituted into the data visualisation to provide contemporary and temporal data visualisation outcomes.

The next steps identify how the designers define and visualise relational and statistical information. Determining user interactions and attributes for selection to manipulate and display the *multi-dimensional data* [7].

12.4.1 Designing Through the Data

As designers engage in problem-solving activities various methods of '*sense-making*' are applied to better understand and group information; a process of problem definition. Often a broad holistic design approach is used that combines both blue-sky hypotheses alongside empirical evidence, in this study the data or 'evidence' is being prioritised. A design rationale of how to interpret and synthesise this evidence is then applied from the designer's knowledge base. It is here qualitative and quantitative attributes that emerge are studied through a design research process [10]. Aiming to increase the designer's understanding and frame the problem scenario, towards defining it.

An accepted investigative approach to problem definition, and problem-solving activity, is the method of Who, What, Where, When, Why and How. Commonly known as the Five W's and How (5W1H), an approach used by many disciplines, both within and outside of design-led practice but where investigative methods are required. With the originating data source quantitative in nature, the data sets are presented in tabulated, numerical datasets. Logging specific values under specific columns. It became apparent the importance of data visualisation was communication to a broader audience and revealing attributes obscured by the original data format.

12.4.2 Data Communication Strategies

As proposed, the use of data visualisation is to amplify cognition. To decipher data, configure it for interpretation, to enable insight, all of which can either be exclusive to the data or applied to further design activity [6, 8]. Data visualisation is rarely an unguided activity. Within this body of work a preliminary exercise first considered the potential of a data driven approach to informing design outcomes. It was further supported by a series of visual data arrangements. The collaborating designer's intent towards prototyping these arrangements were threefold; (i) make the data **accessible** (ii) make the data **relatable** (iii) **learnings** from the data would aid further design development and allow reflection on any novel insights.

12.4.3 Preliminary Exercise—Decoding Through Data

At the initial phase of the decision-making process, the design task began with a simple reflection exercise. Designers actively engage in reflective practice and in this example it was reflecting on the 'existing' outcomes of previous data visualisations attempts, not to be confused with methods of *reflective-in-action* where reflection happens within a specific design activity [11]. Here the task of framing a new data visualisation approach began with a reflection on the earlier engagement with the data. The prior data investigations and outcomes simply showed distribution of (X) width and (Y) height values of the gap geometry. Providing a direct correlation in gap geometry size relative to the DSAPT requirement. This can be viewed within a composite of static overlays which involved three different stations within the same network, with data captured at 0.5–0.7 m intervals along the three platforms (Fig. 12.2).

As the designers reflected on the early visualisations and a clear operational discord was observed, the broader narrative of the impact of this operational discord was missing. The challenge became how to provide holistic and discrete data insights through data visualisation. To enable greater empathy and understanding of the impact on the affected demographic.

Shifting from earlier attention on representation and reflection, a more in-depth data enquiry using the 5W1H approach was initiated. Building on the earlier organisational processes centred on improving cognition through data structures [8]. It was quickly acknowledged by the designers that although each descriptor lends itself to multiple interpretations the 5W1H attributes are applied and defined as.

- *Who—is the data for?* The data is to inform the designer's own process. Forming part of an internal design exercise within the Mobility Design Lab at Monash University and within the lead author's Ph.D. research work. It was recognised early as a valuable tool for communicating the problem to a broader group; including peers, non-technical audience, and industry/government bodies.

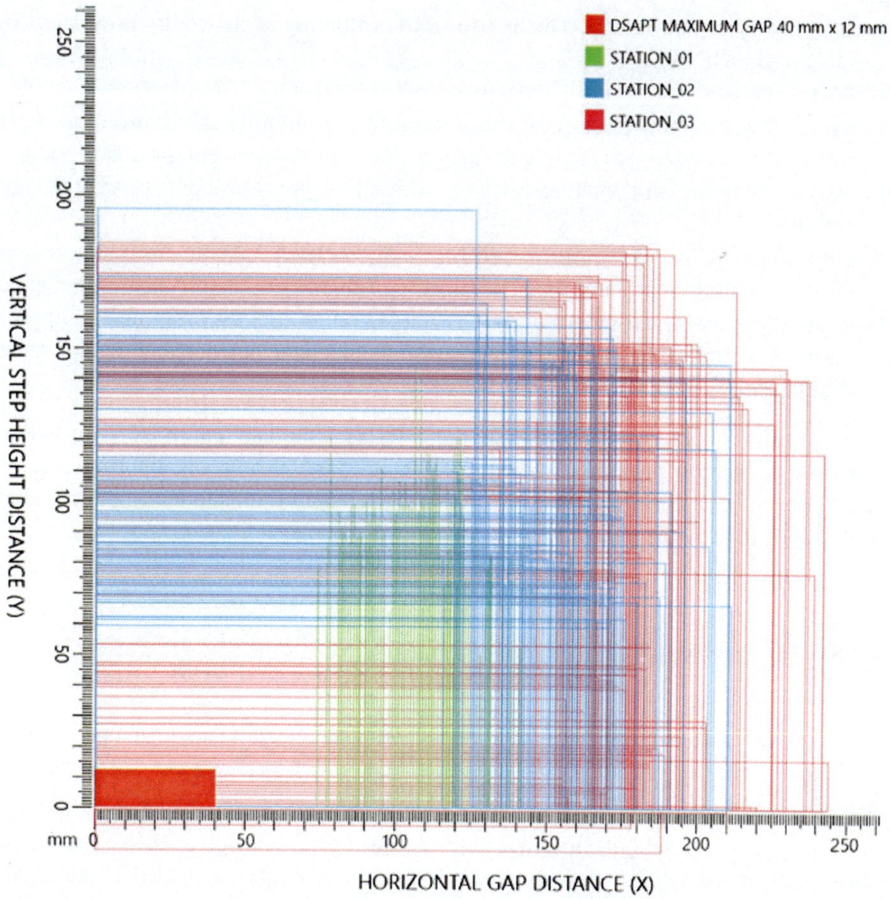

Fig. 12.2 Comparison of THREE different platform stepping data and DSAPT minimum compliance of 40 mm × 12 mm. Image reconstructed from Moug 2017 pp. 67–68

- *What—information do we have and what does it represent?* Datasets contain quantitative data of gap geometry i.e. measurement and location data of a metropolitan rail network. Providing initial parameters to build the data visualisation prototypes.
- *Where—is this data going to be used and/or displayed?* Visualisations are to be used as an investigation and communication tool within a Ph.D. course of study. As digital visualisations they are to be displayed as an interactive user interface (UI) across digital desktop devices, and potentially handheld.
- *When—is this data collected and when might it be useful?* The data is collected through periodic network surveys.[1] It is a useful data asset from which to test and validate visualisation prototypes to reveal underlying networks conditions,

[1]Primary survey data is from 2007 and it is acknowledged that subsets of this data has been superseded through iterative upgrades within the respective network.

specifically gap geometry. This is useful in evaluating accessibility provisions in view of the DSAPT.

- *Why—is this data important and why represent it visually?* Data contains quantitative evidence of network conditions. Highlighting mobility issues around access for disabled passengers. It is anticipated that visual representation can reach a broader audience and could directly or indirectly support problem-solving design activities.
- *How—can we use this data to inform design activity?* Aside from supporting broader communication of the accessibility problem and the physical impedance to mobility; the development of data visualisation prototypes presented an opportunity to reframe the information to support physical prototyping and product design specifications.

The initial reflections and engaging the 5W1H method set parameters for which a data visualisation prototype would seek to address. Prompting internal discussions between collaborating designers, from both 3D product and interaction design professions in providing the above 'briefing' statements.

12.5 Data Visualisation Prototypes

12.5.1 Part 1—Making Data Accessible

The term 'accessible' has a dual meaning in relation to the design problem. It remains in the physical sense about enriching mobility through the design of accessible transport solutions. Maintaining the primary aspiration of the broader project and challenge—designing improved cross-gap access. Accessible now extends to explore the accessibility of information contained within digital datasets, and how this can be reconfigured to engage a broader audience and to support future applied design activity.

The dataset in tabulated format (Fig. 12.3) captures specific information within the rail network environment. Containing in excess of thirty thousand rows of information extracted from more than one hundred stations, it is an exhaustive list.

ID	Easting	Northing	StationName	PlatformNumber	ProfileLocation	X_Measurement	Y_Measurement
159	331102.75	5807149.2	001	1	0.61	1608	1062
158	331102.62	5807148.25	001	1	1.57	1638	1078
157	331102.49	5807147.29	001	1	2.54	1603	1070
156	331102.35	5807146.21	001	1	3.62	1600	1065
155	331102.22	5807145.22	001	1	4.61	1635	1081
154	331102.09	5807144.24	001	1	5.6	1626	1086
153	331101.96	5807143.25	001	1	6.6	1596	1069
152	331101.83	5807142.24	001	1	7.62	1626	1084
151	331101.7	5807141.27	001	1	8.6	1591	1075
150	331101.57	5807140.28	001	1	9.6	1626	1086

Fig. 12.3 Tabular array of underlying dataset (partial selection). Data source provided Moug 2017

Recognising also it is difficult to navigate and draw meaning outside the immediate sightline.

Valuable information is contained within the dataset. This includes TWO key columns of ***X_Measurement*** and **Y_Measurement**. These record the distance from the track centreline (X) and the height above the top of rail (Y) (Fig. 12.4).

The tabulated data is not illegible by any means; however, the format has been prescribed for the intent of data acquisition rather than the communication of data. Originally captured as asset identification and management data, it is directed towards problem identification with the intended audience predominantly engineers. This is a key distinction, where the purpose is to now compose the *same* data into a format that is driven by design discovery and of problem solving. In doing so reinterpreting how numerical data can be perceived and making it more widely accessible.

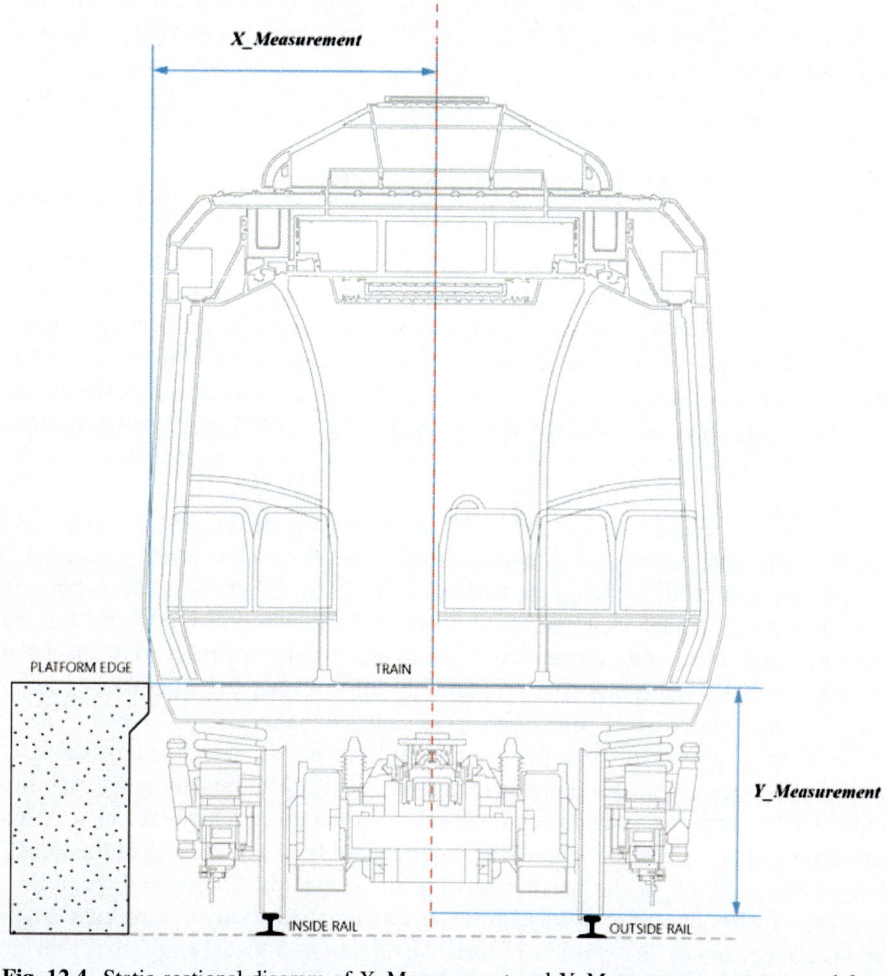

Fig. 12.4 Static sectional diagram of X_Measurement and Y_Measurement reconstructed from Moug 2017 p. 122

Considering the *Who* and *What* statements in relation to the tabulated dataset, the development of the first prototype interface is described. A data organisational workflow was first applied. Dynamic list elements enabled users to navigate and select specific *stations* and *platforms* for review. These were considered the top-level criteria and presented a simple and well understood UI element. Initiated by the top-level selection criteria from the underlying tabular data *three* interlinked visualisations exhibit a dynamic response to the selection criteria (Fig. 12.5).

In earlier prototype iterations once the top-level criteria were selected a slider bar element enabled users to dynamically change the locale position along the station. A range typically between 0 and 160 m (Fig. 12.6). In doing so a combination of dynamic graphics and text-based information was instantaneously updated throughout the interface. Presented in a single screen layout the three interrelated visualisations sought to accurately convey the dataset. Each focusing on a specific element of the gap geometry data. Allowing the viewer to access geometry information about the station and train, the gap, and the remedial requirements to comply with the DSAPT.

12.5.2 *Part 2—Making the Data* Relatable

In making the data more accessible, the next task was considering how to effectively translate the data for improved audience cognition. Compositing relevant graphic imagery and crafting the prototypes to enable new insights or patterns to emerge. How could the numerical data be made useful to the viewer and to the project. This was about making the underlying data *relatable*. As the earlier composite of the pencil (Fig. 12.1) is used to give an identifiable scale to 'the gap' geometry, then giving real-world identity to the dataset provides context and greater clarity for the audience.

In earlier prototypes some of the visual elements were used as internal indicators for the immediate design team, i.e. the compass was inserted to reveal the station's geographical orientation within the network (Fig. 12.6). Other elements such as the train in top orthogonal view indicated both the track side position of the derived platform edge line, while reinforcing the curvature along some stations. Decision making around the original dataset, hierarchy of information, what user-tasks to integrate, and avoidance of 'junk' visual elements continued.

To complement the top orthogonal platform visualisation, and aided by the slider element, the second visualisation provided a graphical indicator on the gap variability along the platform. This was presented in an earlier iterative prototype using a closed circular composition (Fig. 12.7). Using a perfect circle as an indicator of the 'ideal' height and width set by the specific network rail standards. The second radial line, identified by its non-uniform shape provides a visual indication on respective *height* and *width* deviations, in a positive or negative direction. A visual gauge to reveal the issue of compliance with DSAPT clause 8.2.

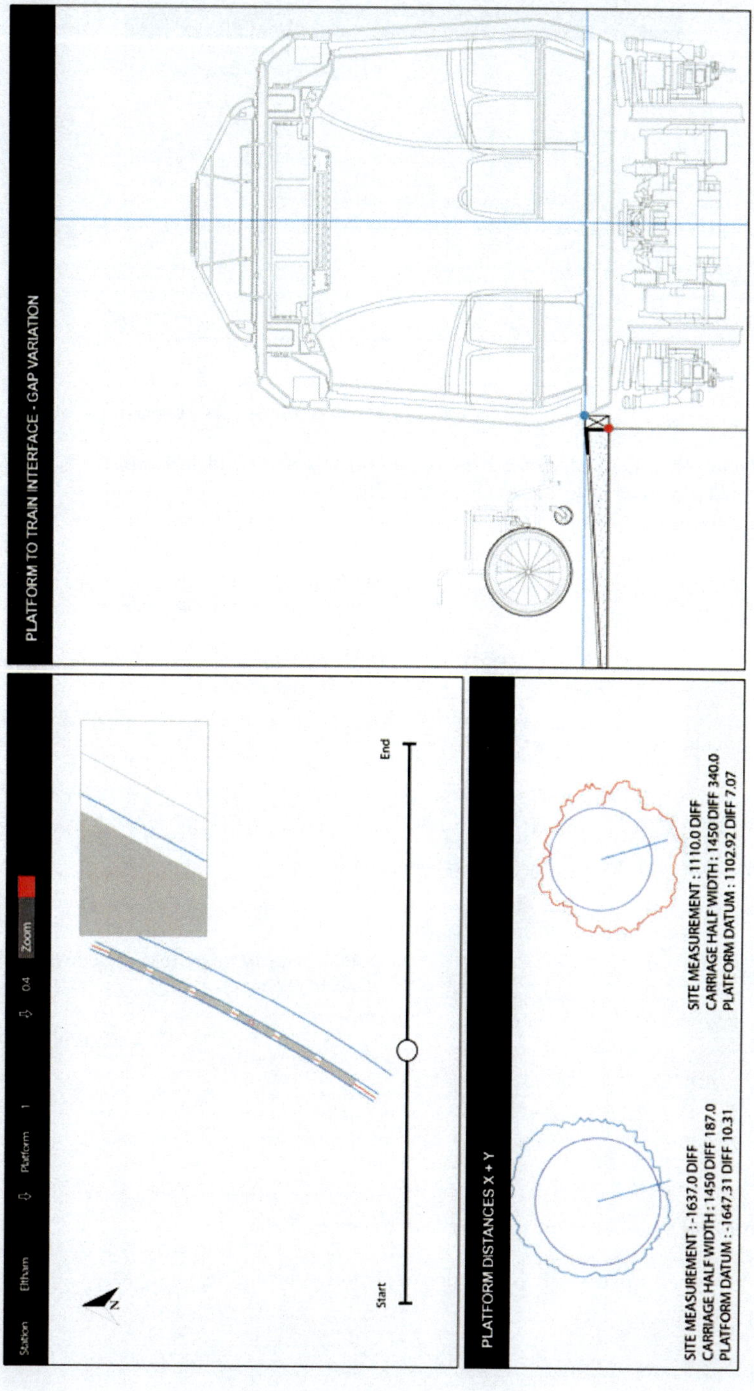

Fig. 12.5 Prototype 1 screen grab of early interlinked data visualisation

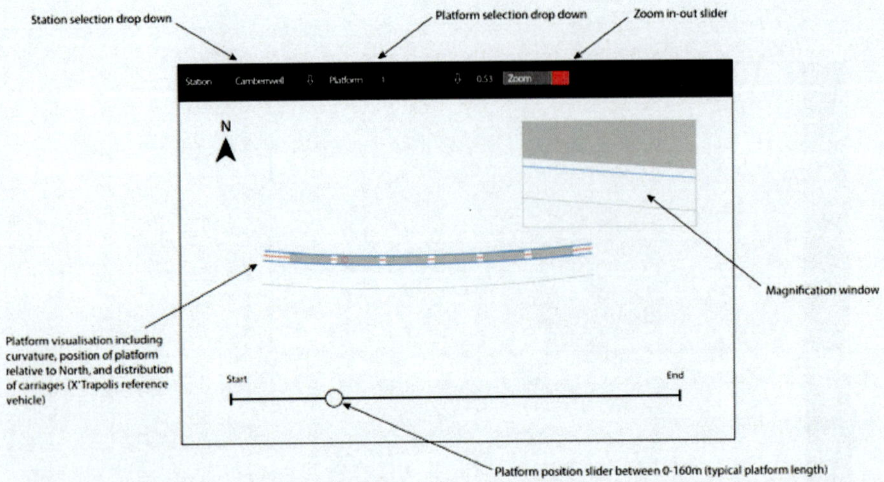

Fig. 12.6 Prototype 1 Early dynamic data visualisation prototype of orthogonal layout. With selection list and slider element from Moug 2017 p. 167

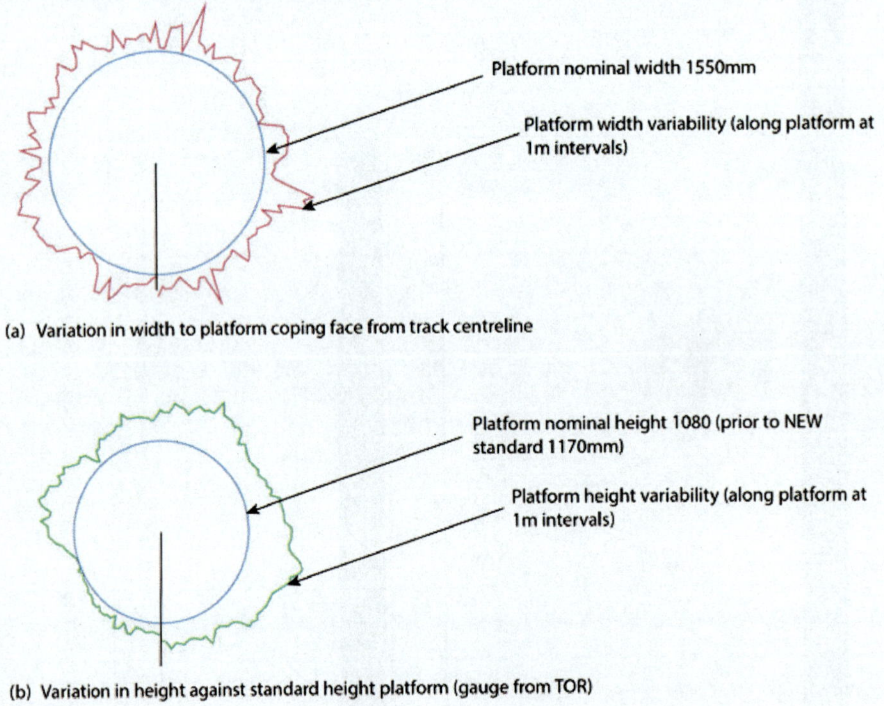

Fig. 12.7 Prototype 1 Conceptual visualisation of platform stepping data from Moug 2017 p. 164

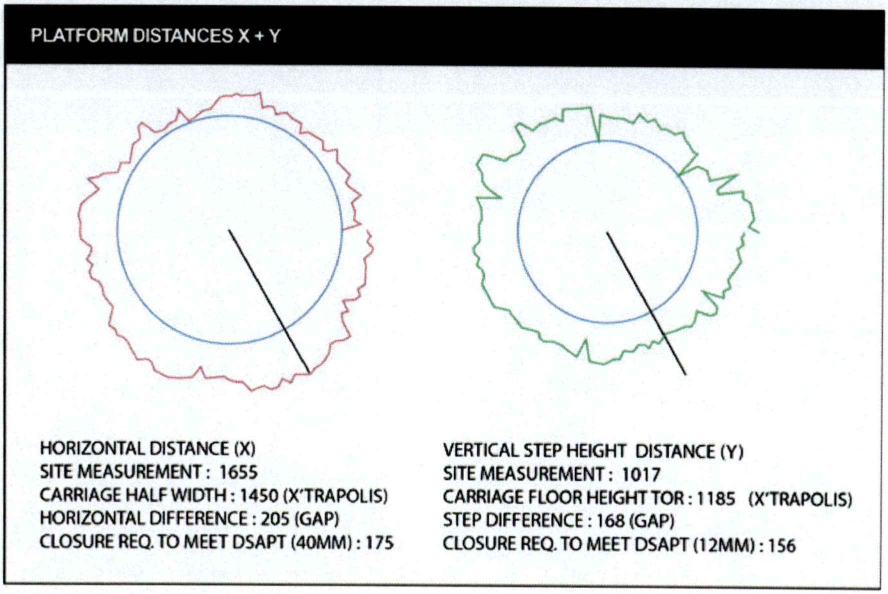

Fig. 12.8 Prototype 1 Visualisation of platform stepping data to validate proposed design envelope from Moug 2017 p. 170

Additional dynamic text-based information was included in the data visualisation platform, further indicated by the visual position of the 'needle'. The presented text notation describes not only the deviations in the distance but also crucial information such as (a) the distance between the platform and a specified carriage design; and (b) the actual gap size in accounting for the maximum allowance of the DSAPT (Fig. 12.8). With the scale of the overall problem communicated visually on a micro level the inclusion of numerical information, the actual dimensions, was a valuable addition.

Lastly, in order to convey the overarching project intent, the inclusion of graphical elements in the wheelchair, cross section of the train, and the platform provides a relatable image for viewer comprehension (Fig. 12.9). Drawing the focus away from a numerical *X* and *Y* alignment problem, the addition of a manual wheelchair sought to contextualise the impact of misalignment. The composition of train and wheelchair further assists to create empathy with the user group and the problem, and like earlier visualisations with the pencil (Fig. 12.1) it provides a recognisable reference from which to gauge the impact of the problem.

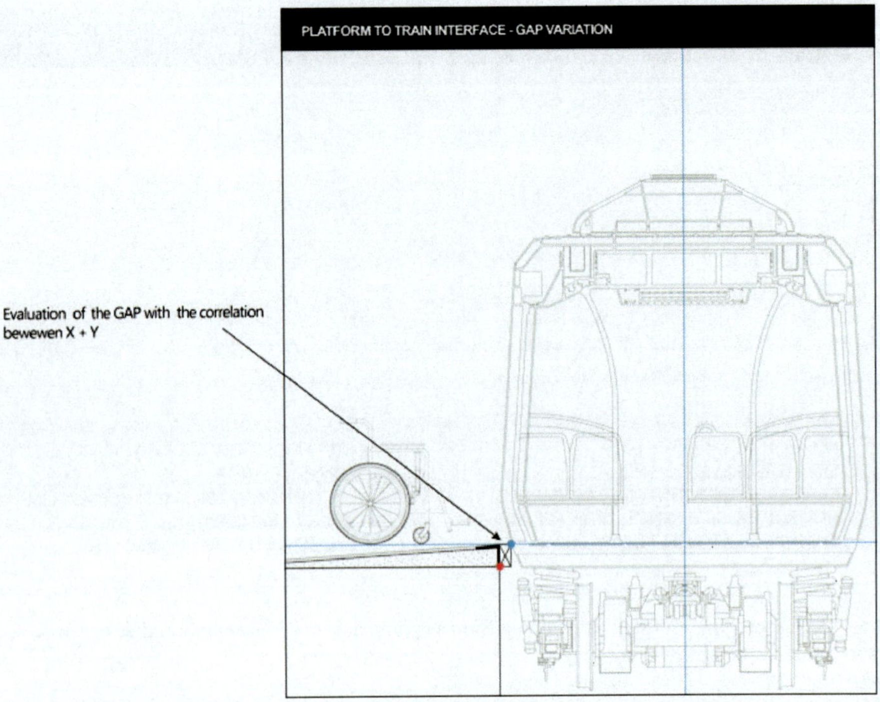

Fig. 12.9 Prototype 1 Visualisation of platform train interface to validate proposed design envelope from Moug 2017 p. 171

*12.5.3 Part 3—***Learnings** *from the Data Visualisation Process*

The primary objective of the data visualisation was about re-presenting quantitative attributes within the data in a visually compelling and informative way. This was achieved by making the data *accessible* and *relatable.* A secondary objective was the *learnings* revealed through development and resolution of the data visualisation process. How can these prototypes inform future design activity in demonstrating problematic conditions and uncovering new insights or opportunities? Then applying this to the broader development project around accessibility and mobility in PT.

12.5.4 Iterative Development

The data visualisation prototypes shifted from a desktop application-based (Fig. 12.5) to a browser-based configuration (Fig. 12.10) several subtle changes were made in line with accepted organisational methods [7]. A shift from slider to scroll wheel navigation (B) (Fig. 12.11) in the top view provided the user with the ability to zoom

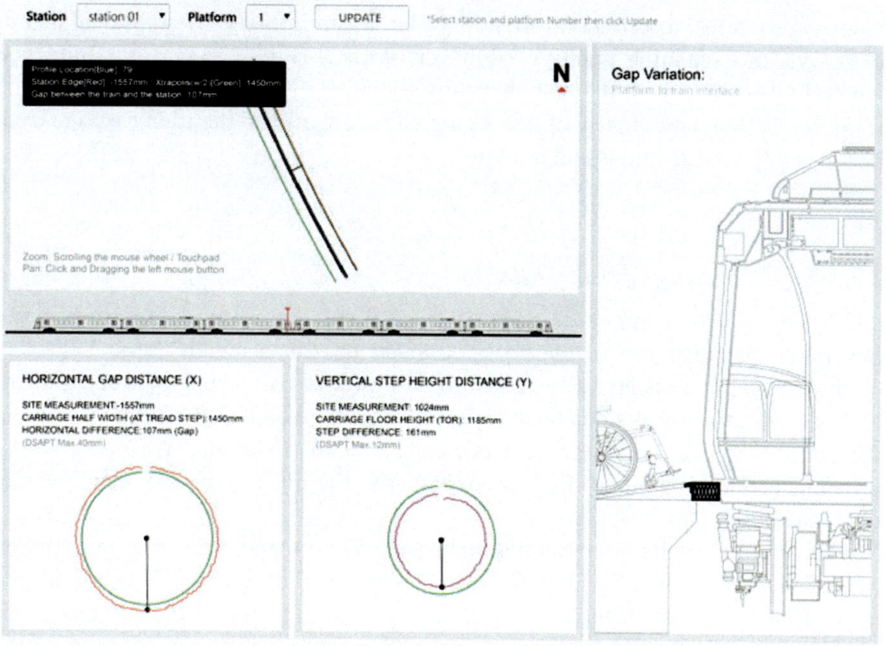

Fig. 12.10 Prototype 2 A browser-based data visualisation prototype tool for problem discovery

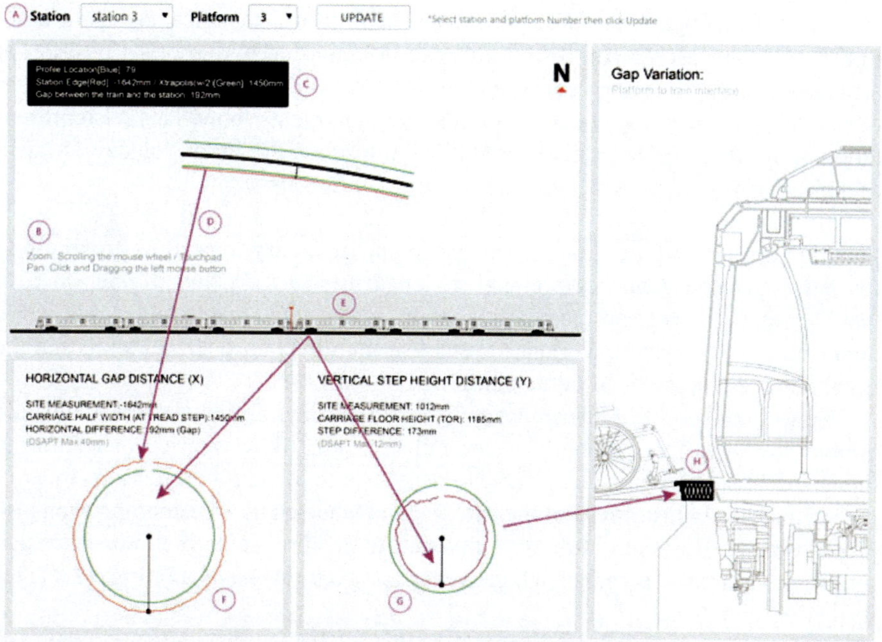

Fig. 12.11 Prototype 2 iterative interface and information organisation

to areas of interest to extract information. This was reinforced by additional details of the various scanning points via a hover box (C) (Fig. 12.11). Continuity was applied throughout with the colour coding provided clearer relational information (D) (Fig. 12.11); particularly of the coping edge (green) and the train carriage (red) between gap visualisations and top view.

12.5.5 Learnings from Data

Several key insights or *'learnings'* were drawn from this design process.

In developing a visual analysis tool, it empowered non-technical users with the ability to access information about a specific boarding point or platform, throughout the entire network. This became a powerful tool for individual reasoning of the problem and how it may affect their personal mobility or enrich their general understanding of the problem.

Given the data collection was undertaken at 0.5–0.7 m intervals an approximation of the longitudinal profile of the platform coping could be derived. This is achieved through interpolation of the individual survey points. It provided a representative profile of the coping edge to then compare the relative position of the edge of the conveyance to the platform. The juxtaposition between straight and undulating lines also highlighted possible inaccuracies or outliers, within the data collection method.

The addition of geolocation data into the visualisation also indicated that the baseline X and Y measurement of 'the gap' can be impacted by the offset position of articulated carriages against a curved platform. Consequently, the gap can be greater or smaller than the perpendicular centreline measurements show. This misrepresentation of information is a critical one and will impact on further design development of a physical prototype which needs to take this into consideration.

The addition of a side orthogonal view of a six-car train set (E) (Fig. 12.11) provided detailed information regarding the gap geometry located in proximity to the passenger entry/exit doors along the length of the train. Identifying where it would potentially be easiest to board the train. It was noted that to reflect real-world application it would be important to establish the direction of travel and align the train to the appropriate end of the platform.

The sectional profile with composited graphic overlays provided strong indication around cross-gap inaccessibility of the rail network. As recognised in relation to the DSAPT requirements. It was able to demonstrate this through reinforcement of the vector-based illustrations of the carriage and wheelchair. The configuration also provides an early visualisation tool in which to position some preliminary design interventions as shown with an indicative *bridging arrangement* (H) (Fig. 12.11) for further development.

As the user re-selected data inputs (A) (Fig. 12.11) i.e. the station and the platform, a repository of data sets was able to be exported in Scalable Vector Graphics (SVG) allowing for further comparative analysis. From here station curvatures were able to be observed, gap geometry at theoretical door positions interrogated, and patterns emerged of the overall network conditions and accessibility issues compared to DSAPT parameters (Fig. 12.12a–d).

12.6 Discussion of Next Steps

As the project develops within the iterative design process of prototyping, the current browser-based configuration provides a 'first' look at data driven problem definition. Moving from static spreadsheet to a dynamic analysis tool, it now provides clarity to the problem and new opportunity for reflection, all within a user-friendly visualisation tool. It also identifies and confirms the need for further accessibility-oriented design research. As the original dataset has shifted in the digital sense, in doing so it has provided an improved cognitive shift in a visual sense, provide overall clarity and comprehension of the data.

Some preliminary discussions on configurations of the interface have instigated small changes and sought to provide further clarity to the problem and how it is represented. A proposed re-configuration of circular to linear arrangements (Fig. 12.11 (F) + (G)) for the X *and* Y 'dials' is one obvious change moving forward. Feedback from early user engagement indicating the circular arrangement, in reflection, could be best suited as a possible navigational interface for smaller screens i.e. smartphones. These small shifts in design cognition complement and inform the development of the data visualisation as a multi-dimensional communication tool.

Further analysis of user engagement is intended, seeking to observe and integrate features identified earlier, such as historical changes, and the extraction of site-specific information to provide a usable repository for information on projects outside the immediate visualisation. As the project further shifts from a problem *discovery and define* stage towards a physical 3D development stage the supporting visualisation, and the underlying dataset may too shift from dynamic 2D representation to underpin more engaging 3D data visualisations. The translation of data from a numerical dataset, to a graphical interface, to a future volumetric model will be leveraged for future product design engineering applications.

12.7 Reflections on Prototypes

This design activity was less about technical skills development or innovative interface design. It sought to consolidate knowledge of a difficult situational problem. By applying an investigative framework in the 5W1H approach and then the organisation of data into relatable data structures consistent with well-understood practice. The outcome communicated to a broader audience an engaging experience and clearer

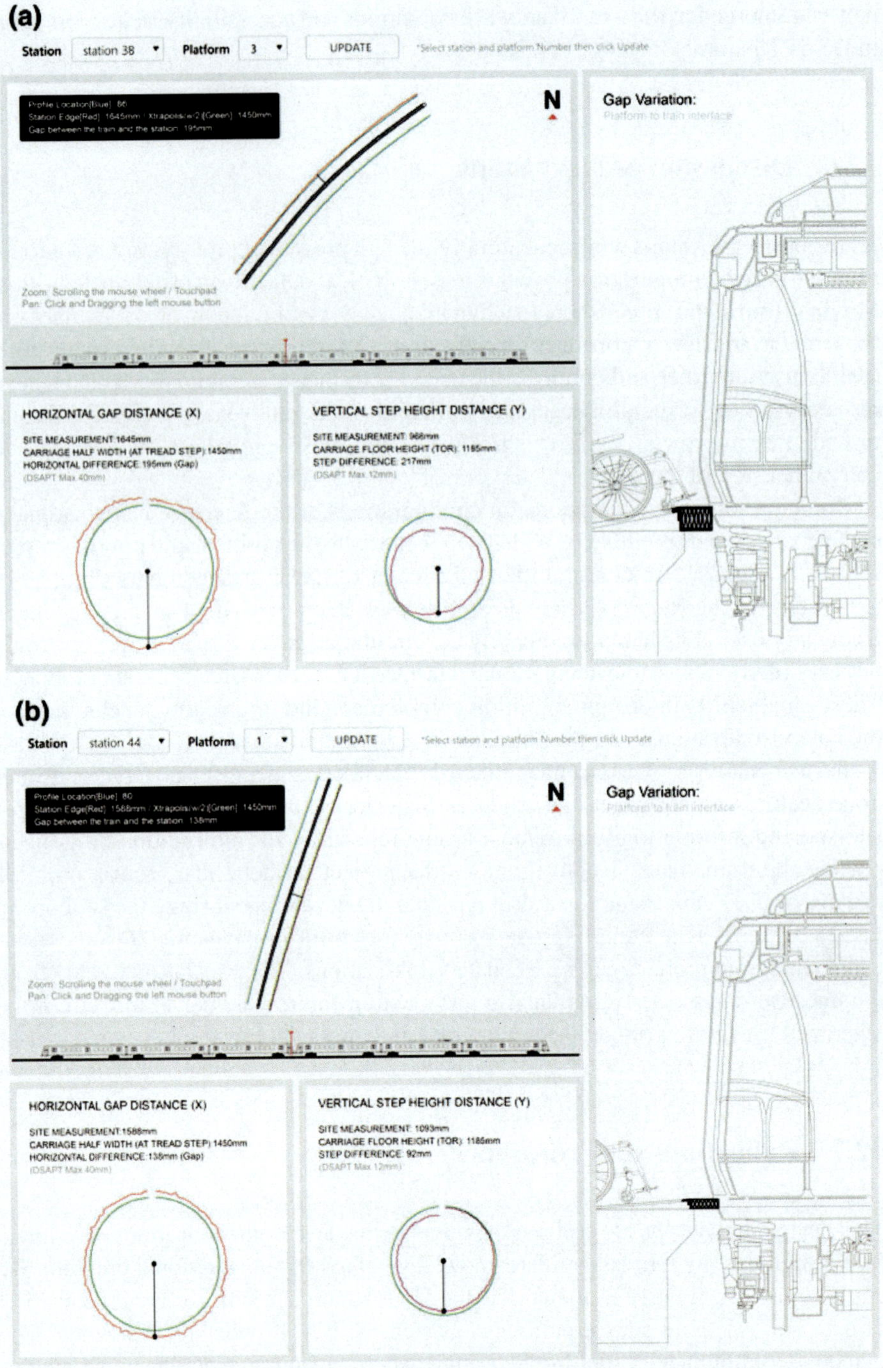

Fig. 12.12 **a** and **b** Station 38: Platform 1 (top) and Station 44: Platform 1 (bottom). **c** and **d** Station 53: Platform 1 (top) and Station 76: Platform 1 (bottom)

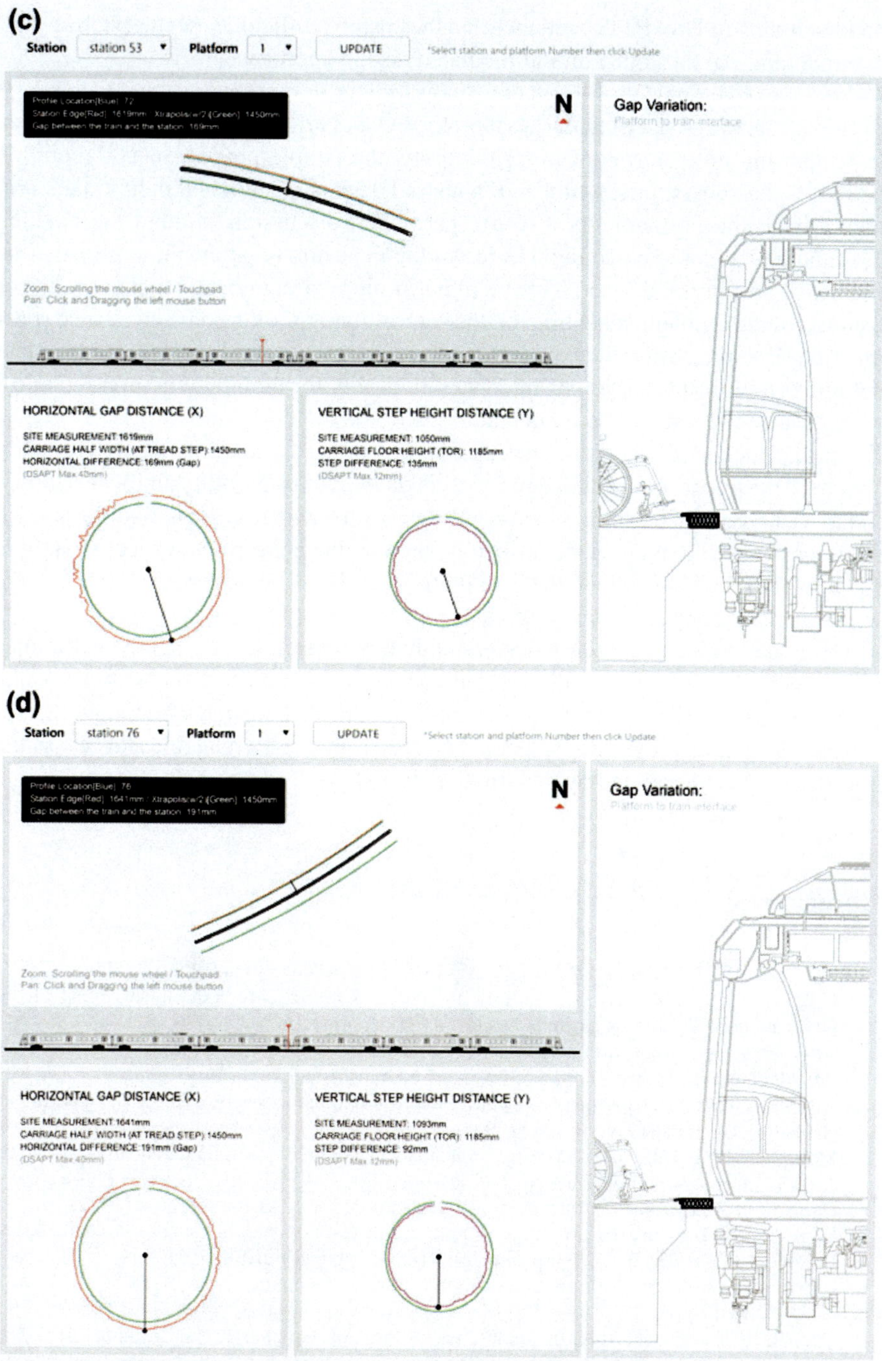

Fig. 12.12 (continued)

understanding of DSAPT compliance and the ensuing difficulties of the existing built environment. All through a digital mediated design approach.

Iterative development of the data visualisation prototypes was shown to successfully package large datasets into exploratory visual arrangements to aid cognition. In supporting other forms of design activity, the simple process of collaboration between researchers; interaction and industrial designers; enriched individual learnings around the problem. As a result, peripheral discussions around the existing data and how it could be configured led to the inclusion of geolocation information contained within the dataset. This resulted in more direct observations around the location of the problem depending on the station configuration and conveyance specification. Demonstrating the value of data visualisation as a discovery and problem definition tool for professionals interested in design for mobility and non-technical users, whose interest is more personal.

Further inclusion of graphic representations of key system components readily conveyed the impact of the built environment on individual mobility and accessibility. A data visualisation interface gave users the ability to navigate, assess, and understand the issues at both a micro scale, a specific area on the platform at a specific station, and at macro scale—i.e. how do the conditions at the users' routine station compare with other stations across the network?

Using data visualisation as a design method to unpack and enrich the communication of digital datasets has allowed a broader understanding of accessibility in the context of the design research. In this application it demonstrates how digital tools can enrich audience cognition of technical datasets through a relatable and accessible interface, one underpinned by quantitative data.

References

1. Meadows, D. H.: Thinking in systems: A primer. Chelsea Green Publishing (2008)
2. Grönvik, L.: Definitions of Disability in Social Sciences: Methodological Perspectives. Department of Sociology, Uppsala Doctor of Philosophy (2007)
3. Attorney-General's Department: Disability Standards for Accessible Public Transport 2002 (DSAPT), p. 80. Canberra Office of Legislative Drafting and Publishing (2002)
4. Attorney-General's Department: Disability Discrimination Act (DDA). A. -G. S. Department. Canberra, Australian Government (1992)
5. Moug, V.: MIND THE GAP—a design strategy for accessibility compliance and user autonomy at the platform-to-train interface, p. 248. Design, Monash University. Doctor of Philosophy
6. Card, S., et al.: Readings in Information Visualization: Using Vision to Think (1999)
7. Shneiderman, B.: The eyes have it: a task by data type taxonomy for information visualizations. In: Proceedings 1996 IEEE Symposium on Visual Languages. IEEE (1996)
8. Chittaro, L.: Visualizing information on mobile devices. Computer (Long Beach, Calif.) **39**(3), 40–45 (2006)
9. Tufte, E.R.: The visual display of quantitative information. Graphics Press, Cheshire, CT (2001)
10. Creswell, J.W.: Research design: qualitative and quantitative approaches. Sage Publications, Thousand Oaks, California
11. Dorst, K., Dijkhuis, J.: Comparing paradigms for describing design activity. In: Cross, N., Christiaans, H., Dorst, K. (eds.) Analyzing Design Activity, pp. 253–270. Wiley, Chichester, New York (1996)

Chapter 13
Mitigating Rail Vandalism Through Design Intervention

Amy Killen

Abstract The chapter summarises a research project undertaken to explore how the design of vulnerable rail assets might be better adapted to reduce incidents of vandalism within the context of Melbourne's rail network. The design approach utilised the novel application of linking rail assets, users and operators through both physical and system-based interventions. Ultimately, the project's outcomes demonstrate how the proactive redesign of vandalism-affected assets can assist operators in developing more targeted, proactive and effective solutions.

Keywords Vandalism · Graffiti · Public transport · Rail environment · Design intervention

13.1 Introduction

The chapter positions itself within Melbourne's public transport system, which includes an integrated network of trams, trains and buses. The rail network consists of 15 lines and 222 stations, with services transporting approximately 450,000 customers each day [1]. With regards to governance, Metro Trains Melbourne (MTM) is responsible for the delivery of public transport services and the management of rail assets, including land, railway tracks, infrastructure and rolling stock [2]. Operationally, they work under an agreement with the Victorian Government in order to meet specified performance targets with regards to reliable and efficient service delivery. As stated within MTM's Customer Compensation Code, "while we always aim to meet and exceed our performance targets, on occasion, certain circumstances may mean we fall short" [3]. One such circumstance includes that of vandalism within the rail network and is the subject of the design research project discussed within this chapter.

A. Killen (✉)
Faculty of Art Design and Architecture, Monash University, Melbourne, VIC, Australia
e-mail: amy.killen@monash.edu.au

S. Coxon and R. Napper (eds.), *Advancing a Design Approach to Enriching Public Mobility*, Intelligent Systems Reference Library 198,
https://doi.org/10.1007/978-3-030-64722-3_13

Rail vandalism can be observed as causing a variety of social, time and monetary impacts on rail operations. In 2019, MTM estimated that $10 million was spent annually to tackle vandalism across the rail network [4]. In the context of Melbourne's rail network, it is ultimately the public that financially sustains the transport system via taxes and fares and, thus, bears the monetary burden associated with such behaviour. When observing vandalism-related expenditure within a wider operational context, expenses are not simply limited to the cleaning, repair or replacement of rolling stock and infrastructure. Costs can also be attributed to the additional security measures required, such as the implementation of CCTV, the installation of fences and the employment of security staff. Administering fines and arresting perpetrators again increase the resources required, both human and capital.

From a revenue perspective, significant losses can result from vandalism-related disruptions. In order to receive incentive payments from the Department of Transport (DoT), MTM must meet performance targets as noted. Specifically, these targets relate to service delivery and punctuality. Service delivery is measured as a percentage of timetabled trains delivered, while punctuality refers to the percentage of services on time. "On time" is defined as those arriving at their destination no later than four minutes and 59 s after the scheduled arrival. As explained by PTV, "incentive payments [and penalties] are based on a detailed measure of the passenger impact of service disruption types. These include delays, cancellations, short services and a number of other disruptions" [5]. Should such delays and/or cancellations result in customers changing their mode of transport, operators may experience lost revenue in addition to any DoT-imposed penalties. The nature of this franchise agreement emphasises the operational and financial impacts vandalism can have on operators, especially those forms which impede MTM from achieving its performance targets.

Socially, such incidents can be observed as contributing to the perception that there is no ownership or control of the network, while also contributing to an overall negative journey experience among commuters [6, 7]. This, in turn, may discourage future utilisation of the network while fostering a sense of distrust towards rail operators. Nathan Glazer, a prominent sociologist, summarises this sentiment, stating that "the proliferation of graffiti, even when not obscene, confronts the subway rider with the inescapable knowledge that the environment he must endure for an hour or more a day is uncontrolled and uncontrollable" [8]. A study conducted by Mueller, Thompson and Hirsch [9] supports this view, finding that of 1800 surveyed train users, five out of nine free-text responses in relation to travel-related issues were connected to incidents of vandalism. Interventions to reduce such behaviour within rail environments, therefore, have far-reaching operational, financial and social benefits affecting both operators and the public.

13.2 Offenders

When discussing vandalism behaviour and culture, questions surround the characteristics of perpetrators involved. In a study by Taylor et al. [10], the Western Australian

Police Information Management System database was used to extract the three-year "offending histories" of 798 graffitists. From the recorded data, 65.5% were male and 11.3% were female (for 23.2%, the gender was not recorded). The mean age was 15.86 years, with the largest group (64.4%) being adolescents at the time of their first reported incident of graffiti. The Crime Prevention Division of the New South Wales Attorney-General's Department's (CPD) 2009 study [11] found similar data when interviewing a number of individuals involved in graffiti-based offences. Of the 47 participants who recorded their age, 16.6 years was the average, with the majority (37) also male. Furthermore, it has been estimated by the British Transport Police that the peak time for this activity is between 4 and 7 pm, which directly correlates with the school timetables of young vandals [7]. Research conducted in 2002 by Halsey and Young [12] among South Australian graffitists found that the majority became involved in graffiti through friends or acquaintances. Younger vandals were seen to engage at the highest levels in "tagging", while older vandals engaged most in creating paintings. These paintings are known by vandals as "pieces", with the term short for "masterpiece" (ibid.). CPD (2009) notes that those engaging in tagging often wish to become accepted as pieces—those who create pieces—by their peers. As many vandals do not have the required skills to progress to creating pieces, they will either continue tagging or stop creating graffiti altogether. The most reported vandalism tools in order of prevalence included spray paint, marker pens and etching tools, followed by a number of lesser-used implements. Half of the participants interviewed had been engaging in graffiti for three or more years, which aligns with Halsey and Young's South Australian research reporting the average length of participation in graffiti culture to be 3.92 years.

13.3 Terminology

Prior to exploring vandalism's effects within the rail environment, it is beneficial to understand the differences between the terminology "vandalism" and "graffiti". A widely referenced definition of vandalism is given by van Vliet [13] as the "wilful damage to or destruction of property owned by others". Graffiti, however, involves the unauthorised defacement of public or private property [14–17]. In the public consciousness, negative opinion is often found in relation to undecipherable tags, which many consider a social crime or an act of youthful vandalism [12, 16]. Tags can be explained as a "stylised personal reference" [18] such as a word, symbol or picture. In defence of graffiti culture, the researchers Garner et al. [19] maintain that tagging should not be viewed simply as an act of vandalism but, rather, as a reflection of the society which has created them. Halsey and Young [16] also argue that graffitists view their work as actions that transform otherwise uninteresting public areas and not as acts of common vandalism. As discussed by Allen [20], "there is some debate as to whether or not graffiti ought to be considered more broadly such that, for example, art exhibits of legal spray paint art would also be considered graffiti". Sanctioned graffiti, which is often well received by the public, further confuses the relationship

Fig. 13.1 Graffiti as vandalism versus graffiti as art (author's own images)

between graffiti and vandalism while also creating a conceptual distinction between "graffiti as crime" and "graffiti as cultural expression" [21, 22].

Figure 13.1 illustrates graffiti as vandalism compared with graffiti as art.

Although graffiti on rail-owned property can be considered a certain type of vandalism, there has been limited research studying whether those committing acts of graffiti also commit more destructive acts such as seat slashing or window etching [23]. What differentiates a graffitist from a vandal is not well documented and this can often come down to the opinion of the individual viewing the effects and under what conditions. Where vandalism may destroy or render property worthless, graffiti often does not degrade the overall function of property. For example, graffiti on a wall does not impede its overall function and it is, therefore, often the consequences of such behaviour that determine whether acts of graffiti should be considered vandalism. Graffiti is, therefore, best understood as a complex interaction between society's infrastructures and human behaviour while being considered within its environmental context. Although graffiti can be labelled as a form of vandalism, the connection is problematic, with further research required in this field of study. However, for the purposes of this chapter, graffiti will be considered under the umbrella of vandalism when discussing deterrent and removal methods in the context of the rail environment. The "rail environment" will be defined in general terms as any combination of the following; "the area located in, below and above a railway corridor, and the area located on, below and above the 25-m-wide strip of land running along each side of a railway corridor" [24].

13.4 Deterrence Methods

Before determining where design could contribute to easing the burden vandalism places upon rail operators, it was important to analyse current mitigation techniques. Upon review of prominent international and local cases in relation to reducing vandalism, with a focus on rail environments, deterrence measures largely fall within six key areas: maintenance, materials and design, technology, authority and law, community involvement, and holistic approaches. A selection of the most common methods in each of these identified categories have been detailed in Fig. 13.2. Approaches embedded in a holistic framework, partnered with a core understanding of underlying motivations and culture, were seen to yield the most effective results in reducing vandalism. Such approaches also need to be tailored towards long-term results rather than quick fixes in order to be financially viable solutions. Furthermore, successfully reducing vandalism can be seen as involving preventive, immediate and proactive responses. Socially, enlisting wider community support helps to address the issue from multiple angles, while the ability to engage both first-time and repeat offenders helps break the cycle of behaviour. The introduction of innovative materiality and design choices, coupled with smart technology systems, can also greatly assist in developing targeted responses.

Beyond necessary understandings in relation to the culture and motivations of vandalism, ten prominent strategic lessons emerged from reviewing the available literature. Such lessons should, where possible, be incorporated into future abatement programs. These lessons are:

1. Tailor investments towards long-term solutions rather than quick fixes.
2. Embed rapid-removal programs into a holistic framework for more effective results.
3. Invest in innovative vandal-proof materials and designs.
4. Implement CPTED principles where possible when designing rail environments.

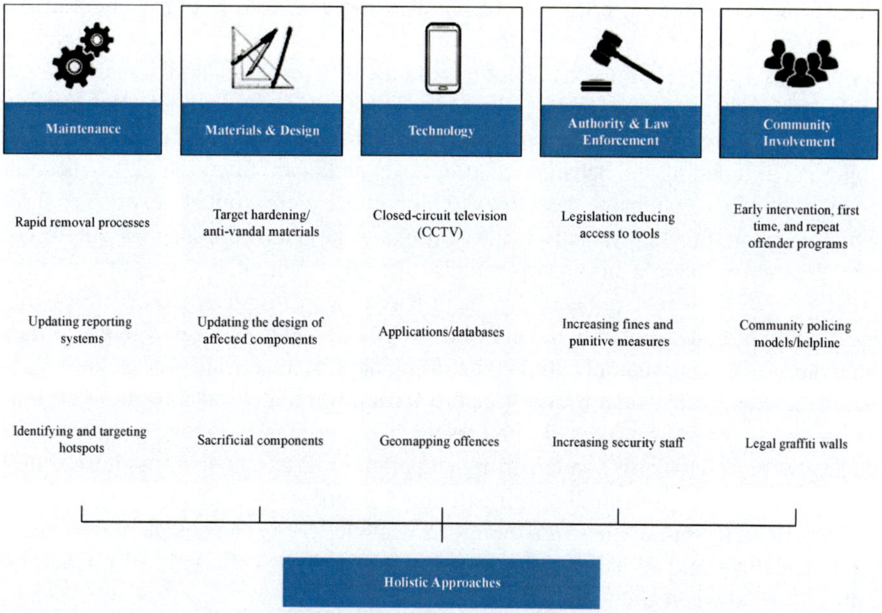

Fig. 13.2 Prominent vandalism deterrent approaches

5. Implement intelligent technology systems to assist with collection, analysis and dissemination of information, with such systems also able to support reporting and maintenance processes.
6. Include partnership approaches between transport authorities and outside organisations such as schools and councils to reduce problem-displacement issues.
7. Isolate breakdowns between rail-operated security and law enforcement groups.
8. Incorporate creativity when engaging with young people, as law enforcement and situational crime- prevention measures are not sufficient.
9. Enlist community support to assist in establishing ownership of the rail environment.
10. Enforce effective and consistent measures with regards to the criminal justice system.

When considering the above lessons, questions arose regarding what additional research was required to better inform future prevention, intervention and reactive strategies in the context of Melbourne's rail environment. This included identifying gaps as well as potential opportunities surrounding current approaches as explored further below.

13.5 The Gaps in Knowledge

The review of common mitigation techniques and available MTM data confirmed two prominent gaps in knowledge surrounding existing vandalism-deterrence approaches. Firstly, mitigation techniques often aim to protect a large scope of targets in the affected area, rather than a narrow focus such as a single asset. Protecting an at-risk asset or set of assets may provide potential for a more targeted and sustained approach in reducing the detrimental effects of vandalism. Such at-risk assets could be identified by considering their links to high-consequence outcomes such as those which result in train delays and/or cancellations. A targeted approach was, therefore, hypothesised to have a high cost–benefit ratio, with potential assets examined in greater depth in Sect. 1.6. Secondly, there was a lack of understanding of how this issue could be addressed with the use of a design intervention or set of interventions. Innovations in this area appeared to be dominated by materiality or technological advancements, with research largely centred around architectural changes to the built environment. There was, thus, limited research exploring how physical interventions could assist in deterring vandalism to vulnerable rail assets such as those within the rolling stock or surrounding infrastructure. In order to better guide the project's development, the broad question, therefore, became, "How can design practice mitigate vandalism on rail assets most at risk with respect to the level of operational impact caused when such damage occurs?".

While discussing the research findings within this chapter, the term "greatest operational impact" (GOI) will be used. Clarity is, therefore, required with regards to this term. As noted in the Introduction, the effective operation of services is measured using performance targets set by the Victorian Government with regards to service delivery and punctuality [25]. These targets ultimately measure delays and/or cancellations to rail services. From a financial perspective, disruptions can lead to penalties imposed by the DoT, with these penalties dependent on factors such as number of passengers affected, line, direction, time of day and length of disruption. Beyond monetary and time impacts, delays and cancellations also negatively affect commuters by contributing to a negative journey experience as discussed [6, 7]. To summarise, forms of vandalism resulting in delays and/or cancellations can be seen as those causing the greatest operational impact on the effective operation of services. Acts of vandalism are relative to particular assets such as those within the rolling stock, stations and/or the surrounding rail environment, as will be outlined below.

13.6 Identifying Melbourne's "At-Risk" Rail Asset

Identifying the rail asset at risk of causing the greatest operational impact when vandalised was explored through the data collection and analysis stage of the project. This stage included both qualitative and quantitative research methods as shown in Fig. 13.3.

Through the semi-structured group interview, the term "critical fault" was first introduced—these are faults that result in the removal of a service. A prominent critical fault, as highlighted by MTM, includes types of vandalism that affect the windscreen. Particular reference was made to forms that impinge on the driver's line of sight or eliminates their ability to operate the train safely. Therefore, even small amounts of graffiti or other forms of vandalism to the windscreen can result in significant disruptions to service. When referring to the incident response data, such types were also highly reported—with graffiti represented most heavily. Upon identifying the windscreen as the at-risk asset, it was important to determine what makes this component susceptible to vandalism. Although the potential to disrupt services

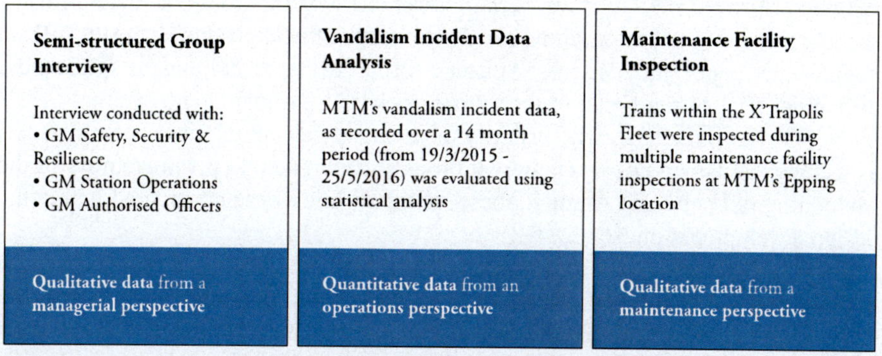

Fig. 13.3 Research methods used to identify Melbourne's "at-risk" rail asset

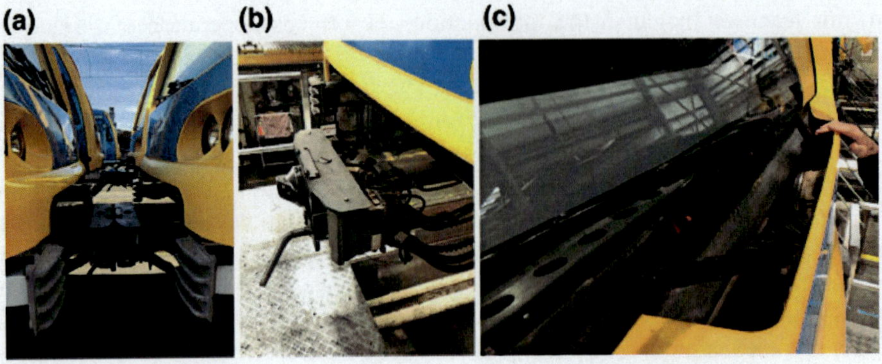

Fig. 13.4 **a–c** Rail coupler and windscreen wiper cavity (author's own images)

when damaged makes this asset a key target of vandals, this question focused on the attributes of or relating to the windscreen that could be influenced or redesigned. Due to the nature of this question, susceptibility was assessed using qualitative methods. Through the semi-structured group interview and maintenance facility inspection, vulnerabilities in this area were largely observed as being linked to accessibility, materiality and surveillance. With regards to accessibility, the windscreen is largely an external target, with access possible through various points within the rail environment. For example, rolling stock situated within rail depots is particularly vulnerable overnight, as identified during the interview. Accessibility also relates to the ability to make contact with the windscreen itself, with various components within the driver's cabin affording areas to stand/crouch and grip. Of particular note, the coupler provides a flat, accessible surface to stand or crouch on, while the open cavity surrounding the windscreen wipers affords a secure level of grip as shown in Fig. 13.4a–c.

Furthermore, materiality proved an important constraint, with the windscreen made using glass and covered with a protective and removable plastic film [26]. Although such films provide a high level of protection, the process is still reactive and provides little deterrence to actual vandalism behaviour. Lastly, security within rail depots proved ineffective in certain areas due to difficulties encountered when seeking to monitor all areas of the site. For example, the High Capacity Metro Train maintenance depot spreads across a distance of 118 hectares, helping to explain these difficulties [27]. Surveillance of the rolling stock is also limited, with external CCTV not yet in widespread use. The inability to effectively detect the presence of vandals as well as the damage of assets in real time proved to be a key vulnerability of the Melbourne rail network. As such, these susceptibilities were considered within the design development phase.

In order to move from data collection and analysis into the design phase, the main findings that emerged needed to be translated into possible design directions. This task was guided through the formulation of a research hypothesis: "Vandalism affecting the windscreen causes the greatest operational impact and the introduction of suitable design interventions will assist in reducing this impact". Such interventions needed to consider the context in which the windscreen sits, that is, the driver's cabin, and how components within the cabin might assist in restricting access to the windscreen while also protecting the driver's ability to operate the train.

13.7 Design Intervention

To better direct the design phase, it was important to consider the insights gained when exploring best practices in deterrence approaches. As noted, it was concluded that holistic approaches are most effective when developing mitigation techniques for rail vandalism. These approaches incorporate deterrence measures in the areas of maintenance, materials and design, technology, authority and law, and community involvement. The design work has, therefore, sought to adopt a multifaceted approach in relation to protecting the windscreen by incorporating these five deterrence areas where possible. When referencing key lessons learned from the literature in relation to these areas, the design interventions aimed to: streamline the maintenance processes of operators, incorporate innovative materiality and design decisions, integrate intelligent technology, improve security measures and encourage community involvement. In the interest of clarity, this approach was distilled into three primary design objectives:

1. Protect the identified assets through physical design interventions.
2. Streamline the maintenance and security processes of operators.
3. Involve the users through intelligent technology systems.

In an attempt to meet these objectives, concept ideation was divided into two streams and included both physical and system-based interventions. Due to the nature of the project, the physical interventions took prominence and will be discussed in greater depth.

During the initial concept generation phase, sketches were produced to explore a range of possible solutions. When first referencing development of the physical interventions, such concepts included both those with an immediate focus, also known as the retrofit solutions, as well as those with a future focus, or the redesign solutions. These concepts referenced both the windscreen and the components within the driver's cabin as they relate to protecting the windscreen (example concept ideation sketches are shown in Fig. 13.5 opposite). In developing these concepts, the designs sought to alter the reactive approach commonly taken by rail operators to more effective, proactive responses. Consequently, such strategies have attempted to address the vulnerabilities identified within the data collection and analysis stage regarding accessibility and detection. Accessibility has been addressed by limiting proximity to the windscreen, including reducing physical affordances with regards to stance and grip. As explained by Hartson [28], "a physical affordance is a design feature that helps, aids, supports, facilitates, or enables physically doing something". Should vandals be able to access the windscreen, the next initiative sought to reduce the

Fig. 13.5 Example sketches demonstrating ideation of physical interventions

length of time they could remain unnoticed by security. This included detecting the vandals' presence while also alerting both vandals and operators to this detection. Although reducing the severity of the damage was considered, should vandals be able to access the windscreen and remain undisturbed by the detection/alert systems, measures often involve the implementation of sacrificial components such as nano-coating and protective window films. However, the sourcing and application of these measures are heavily researched and addressed by MTM. The focus, thus, remained on reducing vandals' ability to access the windscreen as well as reducing the time they remain undetected.

When developing concepts, the security afforded can be viewed in a number of different ways. Although the level of security is easy to assess from a technical standpoint, evaluating it from an aesthetic viewpoint proves more difficult and requires further explanation. While measures implemented through target-hardening approaches are often intended to transmit feelings of safety, the perception of such approaches may induce feelings of anxiety and fear for the public by drawing attention to the idea that their security is at risk [29]. Furthermore, overt deterrence measures may be viewed by potential offenders as a direct challenge while also identifying certain assets as targets by highlighting their presence [ibid.].

Questions, therefore, surround the level of visibility deterrence measures should deploy, including the message they project to potential offenders. On the other end of the spectrum, the application of invisible security measures can be seen as an alternative to highly visible target-hardening approaches. When considering concerns regarding public acceptability, Coaffee, O'Hare and Hawkesworth [ibid.] contend that there is increasing pressure to adopt design features that are less obtrusive without compromising aesthetics, going on to emphasise that invisible security interventions are "potentially more aesthetically pleasing, and offering a redress to the many drawbacks of more obvious and obtrusive security interventions in the built environment". However, it should be noted that the same researchers conclude that such measures should be treated with a "healthy degree of scepticism". For example, they may potentially present the area as insecure, inviting criticism from the public. In the context of counter-terrorism measures, questions have also been raised as to whether such approaches will lead to the implementation of more invasive technologies [30]. Although an extreme example, such concerns are still worth considering when developing vandalism mitigation techniques. Upon reflection, there is a delicate balance to strike when evaluating the aesthetics of the proposed design interventions. However, the answer may lie in the application of approaches known as "stealthy security measures" [29]. These measures are visible to the public and potential offenders, but are not identified as being primarily for security. Such approaches were, therefore, implemented during concept development. Figure 13.6 describes the aesthetics of security.

Before receiving stakeholder evaluation, the inclusion of retrofitted and/or redesign-based directions required reflection with regards to MTM operational considerations. When first reviewing the immediate-focused, retrofitted responses, directions were largely centred around protective cowls with detection/alert sensors where suitable. In reviewing the sacrificial or removable cowls proposed in reference to the windscreen, the driver's window, and lights, their ongoing removal and installation would require greater levels of maintenance when compared to permanent cowls

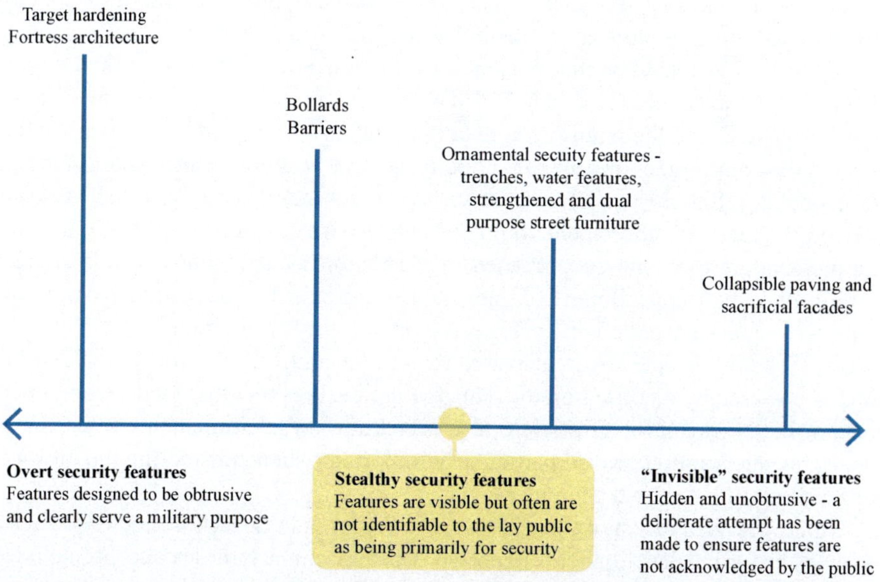

Fig. 13.6 The aesthetics of security (adapted from Coaffee et al. [29])

and, therefore, did not adequately meet primary design objective two: "Streamline the maintenance and security processes of operators". In reference to applying protective cowls to vulnerable door areas, the process of replacing problematic manual locks with more automated processes proved a more appropriate direction. Similarly, a protective cowl with regards to mirrors (an area held for balance by vandals when accessing the windscreen) seemed an outdated response when considering the potential benefits of implementing CCTV. The remaining concepts, however, appeared viable solutions worthy of stakeholder evaluation:

1a. Protective coupler cowl with detection/alert sensors
2a. Protective windscreen wiper cowl that reduces the ability to hold onto rolling stock.

The next stage within the refinement process included presenting the chosen concepts to stakeholders, who assessed their suitability and recommended modifications based on evaluation criteria in relation to MTM asset management and maintenance frameworks. In synthesising MTM's feedback, design revisions were made to the proposed interventions. When comparing the design outcomes to the design objectives, the intended concepts began to demonstrate a positive answer to the research hypothesis. To assist in further validating these directions, prototyping techniques were implemented and included the creation of a proof-of-concept prototype. By utilising a 6-axis CNC robot arm, the design interventions and exterior shell (the yellow fiberglass component attached to the front of the driver's cabin) were fabricated from expanded polystyrene (EPS), as shown in Fig. 13.7a–f. By creating a 1:1 scale prototype, the design interventions could be more fully realised, assisting with both evaluation of their efficacy as well as the communication of important details and features to MTM. This scale also assisted in providing proof-of-concept by demonstrating the concepts' feasibility and practical potential.

Upon conclusion of prototyping and evaluation processes, two refined retrofit interventions were presented to MTM. The first intervention, the protective coupler cowl, saw access to the windscreen greatly reduced. The angled design was trialled through user testing, confirming its effectiveness in limiting balance when standing or crouching. Should vandals access the cowl, sensors were introduced to detect weight and alert both the vandal, through a visual alarm, as well as operators, through an incident alert system. As proposed to MTM, this system could include either a mobile or desktop app tailored to organisation requirements. The second intervention, a protective, automated cowl, encloses the open cavity surrounding the windscreen wipers. The retractable cover, activated by remote control, slides on a moulded rail and is powered by a housed motor. This feature ensures that the windscreen wipers cannot be leveraged for grip while also protecting these components when not in use. This intervention proved particularly important when considering the physical constraints afforded by the coupler cowl.

Future-focused redesign concepts were also presented to MTM should they wish to invest in further development. To provide a brief summary, these concepts included redesigning the shell component to include additional security features (such as live CCTV) as well as introducing automated cabin door locks.

Fig. 13.7 **a–f** 1:1 scale prototype fabricated using EPS and CNC machining (author's own images)

A selection of final detailed renders, as presented to MTM, can also be seen in Figs. 13.8a–i. These renders are in relation to the immediate-focussed retrofit interventions only.

Lastly, in order to provide a holistic solution, a system-based intervention was also developed, as guided by design objective three: "Involve the users through smart technology systems". This objective lent itself to the creation of a community reporting app, as also encouraged during the MTM semi-structured group interview. At present, the connection between users, mobile apps and assets appears limited within the context of public transport operations in Melbourne. Upon review, many

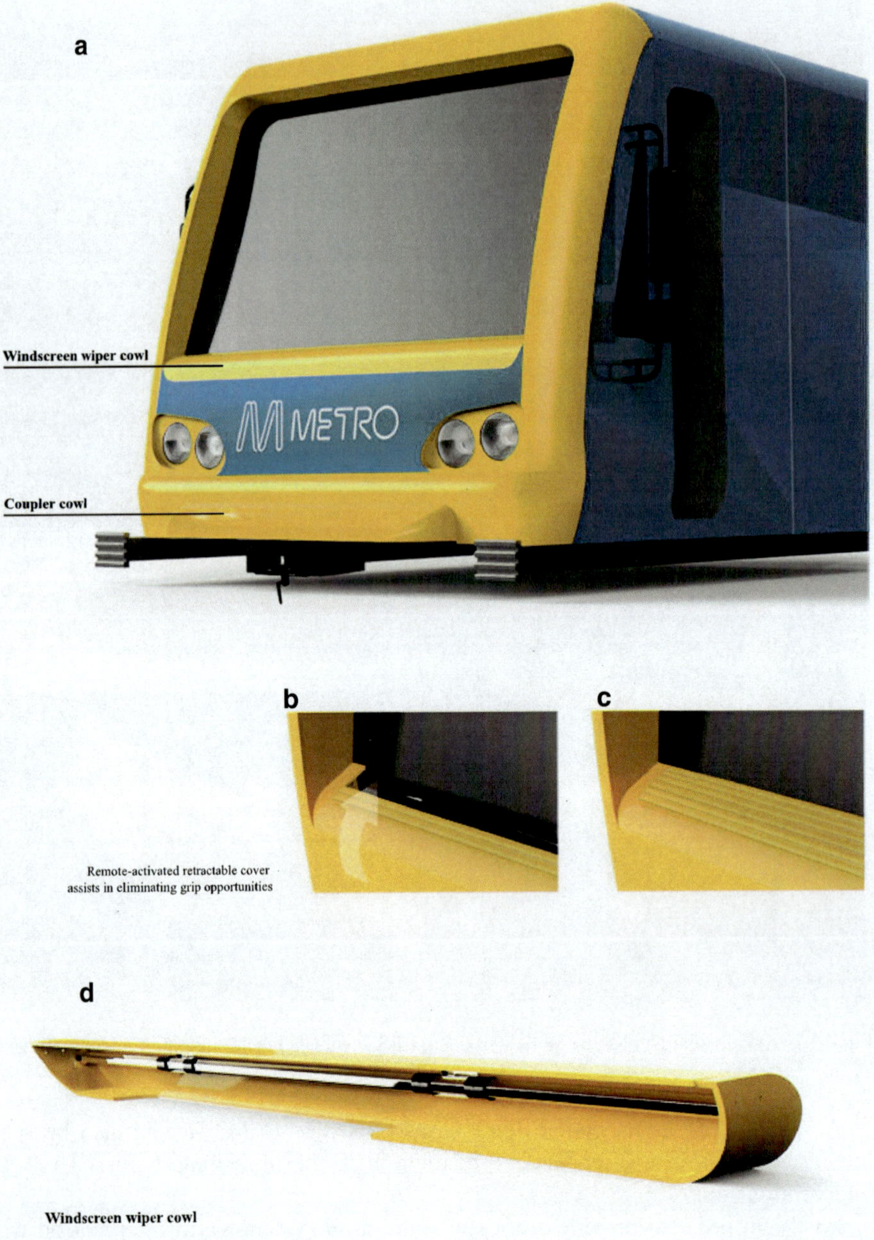

Fig. 13.8 a–i Physical retrofit interventions

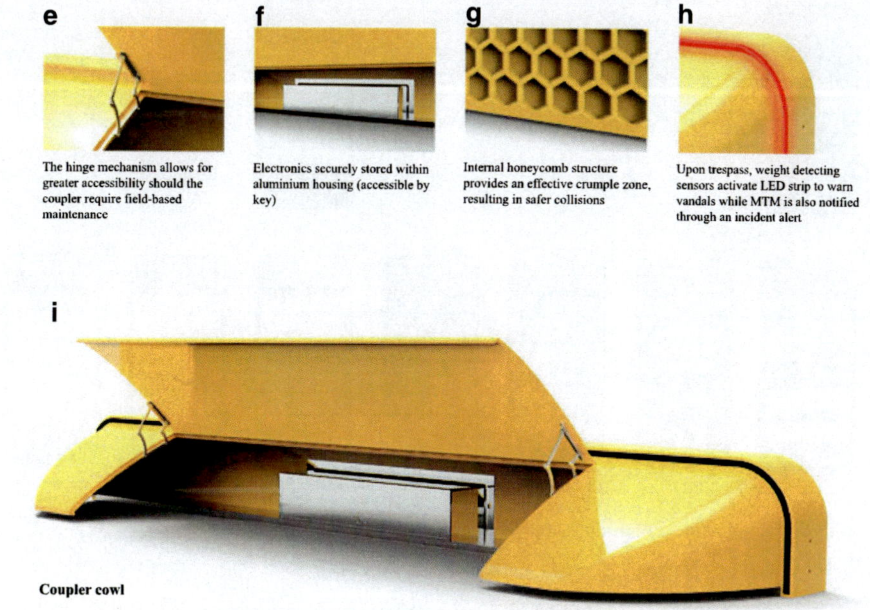

Fig. 13.8 (continued)

apps within this space are centred around providing users with information to navigate the network and its services, rather than allowing active participation from an operations perspective.

Upon review of common user scenarios, functional requirements were formulated:

1. The ability to lodge an incident of vandalism
2. The ability to view progress updates
3. The ability to contact an MTM representative if required.

The first of these functionality requirements can be broken down further into process steps, including the ability to:

1. Report the time of the incident
2. Report the line affected
3. Report the type of vandalism observed
4. Report the assets affected
5. Attach a photo of the incident if required
6. Enter additional free text information if required.

An example of the app's interface is shown in Fig. 13.9, with complete workflows presented to MTM. The next phase would see software engineers coding the application, followed by usability testing to ensure acceptable functionality.

Guided by the design process, the interventions have sought to integrate the assets, users and operators within a holistic mitigation framework. By introducing

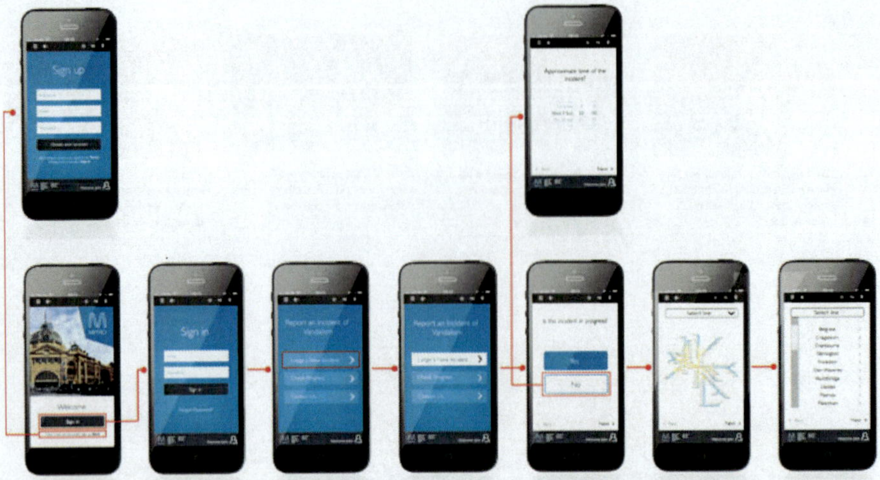

Fig. 13.9 "Lodge a new incident" application workflow

the primary interventions—the physical retrofits—the windscreen's vulnerabilities were addressed by limiting accessibility and reducing the time an offender goes undetected. When considering the costs associated with fabricating and integrating new components, it is worth revisiting the level of impact caused through types of vandalism targeting the windscreen. By totalling the 164 incidents recorded in relation to this at-risk asset, service was delayed for 2181 total minutes and 1,677,157 total passenger weighted minutes (PWM) from 19/3/2015 to 25/5/2016. In analysing the 124 incidents caused by graffiti alone, as largely targeted through the retrofitted interventions, the service was delayed for 1700 total minutes and 1,407,457 total PWM. As explained, disruptions can also result in penalties issued by the DoT dependent on factors such as the route, the direction of the service and the time of day. These disruptions were previously multiplied by the PWM when determining the operational impact and resulting penalties. At the time of writing, bonuses and penalties were calculated under the Enhanced Operational Performance Regime. This regime still "weights" disruptions based on the number of passengers using a train and compares this information to contractual service targets when establishing incentive payments or penalties [31]. Introducing the proposed interventions may, therefore, assist in reducing the delays experienced, thus also reducing the subsequent DoT-imposed penalties. Should MTM wish to invest in redesigning the shell as outlined, the ability to vandalise the windscreen will be further reduced. Lastly, the system-based intervention—the community reporting app—contributes to the multifaceted framework recommended by the literature and also encouraged by MTM. As explained, information entered via mobile apps has the potential to link to operator-run database systems, which can then analyse and disseminate important information regarding incidents. By empowering the community through such technology, a sense of ownership over the rail network is fostered while also adding an extra layer of surveillance that vandals must consider before committing an offence. By implementing better detection and reporting systems through smart technology, maintenance and security processes can also be streamlined.

13.8 Conclusions

Through the creation of the design outcomes, the overall question investigated through the design research project—"How can design practice mitigate vandalism on rail assets most at risk with respect to the level of operational impact caused when such damage occurs?"—was answered. As demonstrated, design practice can be used to mitigate vandalism affecting the windscreen—the most at-risk asset—by introducing both physical and system-based interventions, supported by lessons from the literature. As emphasised in the relevant theory, holistic approaches are highlighted as increasing the likelihood of success [32]. Their effectiveness can be further increased when incorporating both preventive and immediate, proactive responses. By taking a coordinated approach when tackling rail vandalism, the project acknowledged that this behaviour is a multifaceted and complex issue. As suggested by Bandaranaike [18] when discussing Ferri's [33] research, "while the physical environment may determine the place of crime, the social environment explains it". It, thus, proved essential not just to look at redesigns in the context of the rolling stock, but also to consider the wider context of rail vandalism. Interventions, therefore, sought to integrate the assets, the users and the operators within a proactive and holistic mitigation framework.

By reducing vandalism on at-risk assets, the reliability and efficiency of services are also increased, which, in turn, assists MTM in reducing maintenance efforts while also meeting the Victorian Government performance targets. Additionally, redesigning vulnerable rail components would largely benefit the public, among both rail users and non-users. From an aesthetic viewpoint, reducing vandalism creates a cleaner, more visually appealing rail environment while instilling a sense of pride in the community. Lowering the frequency and/or salience of vandalism incidents lessens the perception that there is no authority over the network while cultivating a more positive journey experience for commuters. Lastly, by providing reliable and efficient services, the overall satisfaction of commuters, with regards to both the network and its operators, is further improved.

Acknowledgements The author wishes to thank Metro Trains Melbourne for their funding of this research through the provision of an industry scholarship. Special thanks also go to Brendan Connors of MTM, who provided invaluable guidance and support throughout the duration of the project.

References

1. Metro Trains Melbourne: Who We Are. Retrieved from www.metrotrains.com.au/who-we-are/ (2019)
2. VicTrack: Our Role. Retrieved from www.victrack.com.au/our-role (2019)
3. Public Transport Victoria: Customer Compensation Code. Retrieved from www.metrotrains.com.au/wp-content/uploads/2019/04/5130-mtm-CCC-Booklet-Update-Apr-19_v3.pdf. (2019)
4. Metro Trains Melbourne: Inside Metro's Covert Network Security and Surveillance Team. Retrieved from https://www.metrotrains.com.au/inside-metros-covert-network-security-and-surveillance-team/. (2019)
5. Public Transport Victoria: Track Record: Issue 74 Jan–Mar 2018, Victorian Public Transport Services Quarterly Performance Bulletin. Retrieved from https://www.ptv.vic.gov.au/footer/data-and-reporting/track-record/. (2018)
6. Easteal, P., Wilson, P.: Preventing crime on transport: rail, buses, taxis, planes. Australian Institute of Criminology, Canberra (1991)
7. Petterson, G., Stafford, J.: Graffiti and vandalism on and around public transport. In: Paper Presented at the European Transport Conference, Strasbourg, France. Retrieved from https://trid.trb.org/view/840922. (2004, Oct)
8. Wilson, J., Kelling, G.: Broken windows. Atlantic Month **249**(3), 29–38 (1982). Retrieved from https://urbanpolicy.net/wp-content/uploads/2012/11/Kelling+Wilson_1982_BrokenWindows_policing.pdf
9. Mueller, S., Thompson, K., Hirsch, L.: A socio-economic study of carriage and platform crowding in the Australian railway industry: Final Report (R2.104). Retrieved from https://www.researchgate.net/publication/264558372_A_socio-economic_study_of_carriage_and_platform_crowding_in_the_Australian_railway_industry_Final_Report (2012)
10. Taylor, M., Marais, I., Cottman, R.: Patterns of graffiti offending: towards recognition that graffiti offending is more than 'kids messing around.' Policing Soc. **22**(2), 152–168 (2012). https://doi.org/10.1080/10439463.2011.605128
11. Crime Prevention Division NSW Department of Justice and Attorney General: Graffiti Vandalism the Motivations and Modus Operandi of Persons Who Do Graffiti. Retrieved from NSW: www.melbournegraffiti.com/news/The-motivations-and-modus-operandi-of-persons-who-do-graffiti.pdf (2009)
12. Halsey, M., Young, A.: The meanings of graffiti and municipal administration. Australian New Zealand J. Criminol. **35**(2), 165–186 (2002). https://doi.org/10.1375/acri.35.2.165
13. Van Vliet, W.: The cherry question or the role of social science research in designing against vandalism. In: Christensen, H., Johnson, D., Brookes, M. (eds.) Vandalism: Research, Prevention and Social Policy. United States Department of Agriculture Forest Service, Portland (1992)
14. Craw, P., Leland, L., Bussell, M., Munday, S., Walsh, K.: The mural as graffiti deterrence. Environ. Behavior **38**(3), 422–434 (2006). https://doi.org/10.1177/0013916505281580
15. Grant, C.: Graffiti: taking a closer look. FBI Law Enforce. Bullet. **65**(8), 11–15 (1996). Retrieved from https://search.proquest.com/docview/59777578/
16. Halsey, M., Young, A.: 'Our desires are ungovernable' writing graffiti in urban space. Theor. Criminol. **10**(3), 275–306 (2006). https://doi.org/10.1177/1362480606065908
17. Lachmann, R.: Graffiti as career and ideology. Am. J. Sociol. **94**(2), 229–250 (1988). https://doi.org/10.1086/228990
18. Bandaranaike, S.: Graffiti: a culture of aggression or assertion. In: Paper Presented at The Character, Impact and Prevention of Crime in Regional Australia Conference Convened by the Australian Institute of Criminology, Townsville, Australia. Retrieved from https://www.semanticscholar.org/ (2001, Aug)
19. Garner, P., Rashid, O., Coulton, P., Edwards, R.: The mobile phone as a digital spraycan. In: Paper Presented at the International Conference on Advances in Computer Entertainment

Technology, California, USA. Retrieved from https://www.researchgate.net/publication/220982669_The_mobile_phone_as_a_digital_SprayCan (2006)
20. Allen, D.: Fighting Graffiti: an Investigation of Causes and Solutions (NPCR 1252). Retrieved from https://www.cura.umn.edu/search/index.php (2007)
21. Alvi, S., Schwartz, M., DeKeseredy, W., Maume, M.: Women's fear of crime in canadian public housing. Violence Against Women **7**(6), 638–661 (2001). https://doi.org/10.1177/10778010122182640
22. King, P., Setter, M.: Young people, graffiti and the community. In: Paper Presented at the Graffiti and Disorder Conference, Brisbane, Australia. Retrieved from https://webarchive.nla.gov.au/wayback/20040903031913/https://www.aic.gov.au/conferences/2003-graffiti/#papers (2003, Aug)
23. Thompson, K., Offler, N., Hirsch, L., Every, D., Thomas, M., Dawson, D.: From broken windows to a renovated research agenda: a review of the literature on vandalism and graffiti in the rail industry. Transp. Res. Part A Policy Pract. **46**(8), 1280–1290 (2012). https://doi.org/10.1016/j.tra.2012.04.002
24. Growth Management Queensland: Guide for Development in a Railway Environment. Retrieved from https://www.tmr.qld.gov.au/-/media/busind/techstdpubs/Rail/guide-for-dev-in-railway/Guidefordevelopmentinarailwayenvironment.pdf?la=en (2010)
25. Public Transport Victoria: Track record: Issue 74 Jan–Mar 2018, Victorian Public Transport Services Quarterly Performance Bulletin. Retrieved from https://www.ptv.vic.gov.au/footer/data-and-reporting/track-record/ (2018)
26. Metro Trains Melbourne: Application of Exterior Protective Film to Rolling Stock Windscreens. Retrieved from Metro Trains Melbourne (2013)
27. Evolution Rail: Facilities. Retrieved from https://evolutionrail.com/hcmt-project/facilities/ (2018)
28. Hartson, R.: Cognitive, physical, sensory, and functional affordances in interaction design. Behaviour Inf. Technol. **22**(5), 315–338 (2003). https://doi.org/10.1080/01449290310001592587
29. Coaffee, J., O'Hare, P., Hawkesworth, M.: The visibility of (in) security: the aesthetics of planning urban defences against terrorism. Secur. Dialog. **40**(4–5), 489–511 (2009). https://doi.org/10.1177/0967010609343299
30. Briggs, R.: Joining Forces: From National Security to Networked Security. Retrieved from https://www.demos.co.uk/files/JoiningForces.pdf (2005)
31. Public Transport Victoria: Network Performance. Retrieved from https://www.ptv.vic.gov.au/footer/data-and-reporting/network-performance/ (2020)
32. White, K.: Graffiti—looking beyond the symptoms. Graffiti and disorder: local government, law enforcement and community responses. In: Paper Presented at the Graffiti and Disorder Conference, Brisbane, Australia. Retrieved from https://webarchive.nla.gov.au/wayback/20040903031913/http://www.aic.gov.au/conferences/2003-graffiti/#papers (2003)
33. Ferri, E.: Causes of criminal behaviour. In: Muncie, J., McLaughlin, E., Langan, M. (eds.) Criminological Perspectives. Sage Publications, London (1999)

Chapter 14
Creating the Circuit Breakers: An Examination of the Sociotechnical System Factors Which Impede and Enable the Delivery of Safe and Healthy Neighbourhood Street Design in Aotearoa New Zealand

Hamish Mackie, Lily Hirsch, Rebekah Thorne, Karen Witten, and Adrian Field

Abstract Taking a socio-technical systems approach, the aim of this chapter is to describe the barriers and enablers to innovative street projects that promote wellbeing. We explore these barriers and enablers through the lens of five proposed, current, or delivered niche street re-design projects or programmes in Aotearoa New Zealand. Through a thematic analysis of project and programme information, the key themes of leadership, funding, policies and procedures, organizational norms, community and delivery tensions, and social environment emerged. These themes were used to analyse the extent to which the projects and programmes succeeded as niches and influenced the wider system. While there was varying success across the projects and programmes in influencing the wider regime and social landscape, the analysis found that niches need to be supported within government planning systems as a way of managing investment risk and testing future scenarios. The lessons provide direction for those seeking to expedite transport system change so that positive health, safety, environmental, and social outcomes can be realised.

Keywords Sociotechnical system · Street design · Innovation · Self-explaining roads · Active travel

H. Mackie (✉) · L. Hirsch · R. Thorne
Mackie Research and Consulting, Level 2, 2 Princes Street, Auckland Central 1010, New Zealand
e-mail: hamish@mackieresearch.co.nz

K. Witten
SHORE and Whariki Research Centre, Massey University, Level 7, 90 Symonds Street, Auckland Central 1010, New Zealand
e-mail: K.Witten@massey.ac.nz

A. Field
Dovetail Consulting Ltd., PO Box 78-146, Grey Lynn, Auckland 1245, New Zealand
e-mail: adrian@dovetailnz.com

S. Coxon and R. Napper (eds.), *Advancing a Design Approach to Enriching Public Mobility*, Intelligent Systems Reference Library 198,
https://doi.org/10.1007/978-3-030-64722-3_14

14.1 Introduction

14.1.1 The Contemporary Urban Transport Problem

Traffic congestion, traffic injuries, decreased active travel, air pollution, and transport-related greenhouse gas emissions are undermining population wellbeing in many countries around the world [1]. With the highest rate of car ownership in the OECD and over 90% of distances travelled by car, Aotearoa New Zealand (hereafter Aotearoa)[1] is particularly affected by these issues [2]. For example, Aotearoa has high levels of road transport emissions and traffic injuries compared to other developed countries [2, 3]. Obesity-related disease costs the country approximately $1 billion per annum [4], and in Auckland (Tamaki Makaurau), the largest city in Aotearoa, traffic congestion and its associated costs are high and rising [5].

The negative impacts of automobility tend to disproportionately affect marginalised populations—contributing to greater health and social inequities [1]. This is apparent in the higher rates of traffic injuries as well as physical inactivity-related diseases experienced particularly by low socioeconomic, Māori, and Pasifika populations in Aotearoa [6, 7]. As a result, transport systems that prioritise private car use over-active and public transport modes are increasingly seen as undermining public health, environmental, and social equity goals [1, 8], and are incompatible with sustainable approaches to urban planning, or to achieving the targets of the Sustainable Development Goals [9].

An enormous effort is required within the transport sector to respond to the negative externalities it generates, and the systems and ways of working that reinforce this current paradigm are heavily entrenched. Streets have historically been designed to maximise traffic flow, and, while this is changing in some urban centres, the majority of suburban and peri-urban streets remain overwhelmingly car-oriented [10]. These suburban street designs increase people's reliance on private motor vehicles [11, 12] and make the use of active modes unsafe and uncomfortable [13].

14.1.2 Designing Neighbourhood Streets for Community Health and Wellbeing, Cohesion, and Equity

Redesigning streets to improve the uptake and safety of physical modes can contribute to the ecological and social health of cities [14, 15]. Neighbourhood streets in suburbs that are safe and easy for walking and cycling, allow easy access to public transport, and are safe and intuitive to drive around are associated with significant health, social, economic, and environmental benefits [16–18]. New approaches to neighbourhood

[1] The authors have elected to use the Māori name for New Zealand throughout this chapter. Aotearoa is the indigenous name for New Zealand and is gradually becoming accepted as a bilingual name for the country.

street design, such as Complete Streets (USA and Canada) and 20MPH zones (UK), make it feasible to retrofit existing streets and routes in a cost-effective manner [19–21]. In jurisdictions unfamiliar with such innovations they can be seen as costly, but economic evaluations have indicated the societal, health, and environmental benefits can significantly outweigh the costs [22, 23]. Conversely, for large-scale motorway projects that can cost billions, the benefits often struggle to outweigh the costs [24].

Re-design techniques for active, sustainable streets generally have a strong focus on community co-design, placemaking, reducing vehicle speed and/or access, and testing and trialing innovative designs [25–28]. Streets that support all modes for commuting, recreation, and social interaction, can promote stronger and healthier communities and sustain environmentally sustainable travel [29]. For low-income communities in particular, healthy neighbourhood street designs may contribute to injury reduction, health, social, economic, and environmental benefits [30, 31]. Less traffic congestion also has productivity benefits [32, 33].

In Aotearoa, significant road safety benefits from neighbourhood street re-designs have already been demonstrated. In the five years following the Point England Liveable Streets project in Auckland [34], the social cost of crashes reduced by 48% [35], although performance has slipped in recent years through treatments not being maintained Similar safety benefits are now emerging from the Te Ara Mua—Future Streets project in Māngere, Auckland [36]. Wider health benefits in terms of disability adjusted life years (DALYs), have been estimated from the earlier Model Communities Programme [23].

14.1.3 System Solution—Innovation and the Sociotechnical Systems Approach

Despite the potential of new ways of conceptualizing streets, change is difficult. Sociotechnical Systems (STS) Theory contests dominant understandings of service delivery and technological change as purely technical, neutral, and apolitical [37, 38]. Rather, it recognises technical and social aspects are co-constitutive of decision-making and outcomes [39].

STS theory emphasises the causal relationships between different hierarchical levels in complex systems. For example, Rasmussen's work on risk management in workplace settings describes a system hierarchy made up of six levels: government, regulators/associations, company, management, staff, work [40]. This approach highlights how the relationships between decisions, actions, and failures at different system levels lead to particular outcomes, rather than individual people or isolated errors [40].

An STS approach has also been used to understand how innovation or system change is brought about or thwarted. Geels, a seminal contributor to the field, has

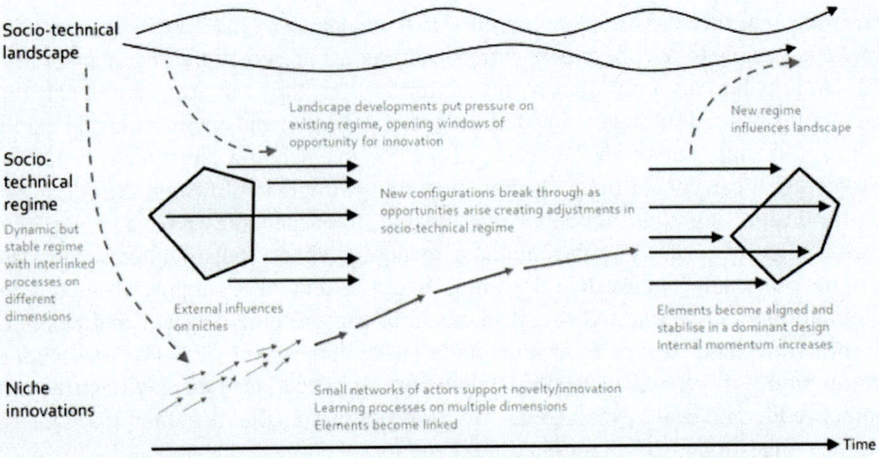

Fig. 14.1 Socio-technical transitions. Adapted from Geels [42]

applied STS theory to investigate the way co-evolution and multi-dimensional interactions between industry, technology, markets, policy, culture, and civil society determine system states [41–43]. Geels posits three analytical levels through which transitions occur: niches, which provide the focal point for emerging innovations; regimes, which bring together established practices and rules that reinforce or stabilise existing systems; and the landscape, the broad demographic and socioeconomic environment influencing niches and regimes (Fig. 14.1) [42, 43]. Coalitions of innovative actors working in niche sociotechnical spaces can activate change by challenging dominant sociotechnical regimes [44]. Changes at the landscape level (e.g. increasing concern over climate change) can also generate pressure on dominant regimes, thereby destabilising them and creating windows of opportunity for niche innovations to take hold [43]. Within this, many different system components and actors impede or allow change to happen [44].

In the transport system, niches may include demonstrations of new ideas by transport personnel external and internal to government agencies who seek to drive change. The regime includes the dominant organisations, rules, and structures that maintain and deliver the transport system, and the landscape is represented by everyday people and the socio-economic context in which they live.

In Aotearoa, the dominant transport landscape and regime of automobility is firmly entrenched [45, 46] and the transport system continues to prioritise the delivery of roading infrastructure catering primarily to private vehicles. This is maintained through a hierarchy of funding programmes from individual councils who create Regional Land Transport Plans (RLTP), which feed into the National Land Transport Programme (NLTP), a system which allocates funding to different 'activity classes'. Within these programmes, infrastructural and system support for active travel are poor and rates of walking, cycling and scooting are generally low. In many (sub)urban neighbourhoods, particularly on the city fringe active travel is not a feasible mode choice [47, 48]. Likewise, variable access to frequent public transport services limits

its utility in many outer city locations [49]. As a result, 'business as usual' streets prevail, thereby perpetuating poor outcomes at a society level.

Referencing Geels' model of socio-technical transitions (Fig. 14.1), niche projects can nudge innovation and challenge system inertia. However, demonstration projects are difficult to initiate and implement, and even successful innovations can face substantial obstacles to translation into everyday practice. When increasing active travel is a desired outcome of niche innovations the complex reasons why automobility remains entrenched are also confronted. In this context, transport policy and decision-makers comprise only one part of a wider transport assemblage within which they have become increasingly constrained by their relationships with other actors [50], including industry, consumers, and civil society [43]. Nonetheless an STS approach provides a way to examine and disentangle existing system attributes that generate, maintain, and (re)produce particular pathways and logics.

In this chapter, an STS analysis approach is used to examine how safe and healthy neighbourhood street innovation in diverse communities in Aotearoa is impeded or facilitated by the current transport system, through the lens of proposed, current, or delivered projects acting as niches. By doing so, this research identifies barriers and solutions to the wider adoption of healthy neighbourhood street design.

14.2 Methods

Five niches are examined. Three proposed or delivered Aotearoa-based projects (or niches) which mostly emerged outside of the regime of the transport delivery system are firstly examined. Through an STS lens we examine the interactions, influences, and effect of each 'safe and healthy neighbourhood street' niche project on the regime and landscape. The projects are then compared with two successfully delivered Aotearoa niche programmes which were generated within the regime of the transport delivery system (by government) to assess the importance of the genesis of niche projects.

The three niche projects which arose outside of the transport delivery system and which are explored throughout this chapter are:

1. **Te Ara Mua—Future Streets (Future Streets)**: A neighbourhood-scale suburban retrofit project in Māngere Central, South Auckland [51].
2. **Safe and Healthy Streets South Auckland (SHSSA)**: A broad multi-suburb initiative in South Auckland, building on the momentum of Future Streets and other projects, to implement safe and healthy streets principles in a way that prioritises community needs and voices.
3. **Aranui Connections**: A scoping exercise to understand the need for, and feasibility of investment in safe and healthy neighbourhood street upgrades in the suburb of Aranui, Christchurch (Ōtautahi), following the establishment of the new Haeata Community Campus.

The background and approach to these projects, and the varying levels of success achieved in the delivery of safe and healthy neighbourhood street innovation are described in Sects. 14.3.1.1–14.3.1.3 below.

In Sect. 14.3.3 the three externally driven niche projects are compared with the two programmes which had a genesis from within the transport delivery system:

4. **Model Communities Programme (MCP)**: Held in two towns in Aotearoa, the programme was designed to improve the safety of walking and cycling by providing fully integrated transport networks for those mode types.
5. **Urban Cycleways Programme (UCP)**: To provide better cycling networks in urban centres in Aotearoa and to improve the rate of urban cycling through the delivery of cycling infrastructure on a national level.

For each of the initial three case study projects, the interactions with the regime, the outcomes from those interactions, and their overall success as niche projects were recorded. This was achieved through project documentation, project reports, and minutes from meetings where the authors chronicled their experiences of participating in each of the projects. Documents from all of these sources were thematically analysed [52].

Initially, six themes for the interactions of each project within the regime were identified. These themes were subsequently used as a framework for assessing each project where a description of the successful (or unsuccessful) characteristics of each project was produced (Table 14.1).

Taking a strengths-based approach, success factors for niches were then identified. This was achieved by reviewing documentation from the two previously successful government driven niche programmes in Aotearoa—The Model Communities Programme (MCP) and the Urban Cycleways Programme (UCP)—and comparing characteristics of these programmes with the three case study projects.

The final step in the analysis was to identify the key factors that are likely to make demonstration projects succeed as niches within the STS, learning from both success and failure.

14.3 Results

14.3.1 Development and Outcomes of 'Safe and Healthy Neighbourhood Street' Innovation Niche Projects

14.3.1.1 Te Ara Mua—Future Streets

Te Ara Mua—Future Streets (Future Streets) is a controlled intervention study in Māngere, a low socioeconomic neighbourhood in a predominantly Pasifika community in South Auckland, Aotearoa [53]. A neighbourhood-scale suburban retrofit was designed and implemented in Māngere Central [51] with an adjacent neighbourhood, Māngere East, serving as a control area.[2]

Future Streets employed a community participatory design approach with the aim of making it safer and easier for people to get around the neighbourhood, especially by walking or cycling. The infrastructure re-design was paired with a research project to understand the effect of the street changes on traffic behaviour, pedestrian and cyclist usability, traffic crashes, mode use, levels of physical activity, and community perceptions of safety and social connection [54]. Future Streets encountered several obstacles in the design and delivery of the intervention including funding uncertainties, conflicts around project governance, regulatory barriers, and rigid project management processes which resulted in delays to implementation [48, 51]. Ultimately, the infrastructure was delivered between 2016 and 2017.

Despite problems in delivering Future Streets, it nevertheless succeeded as a niche project. It was eventually delivered, and outcomes are being monitored and disseminated thanks to a well-funded research programme [54]. Future Streets has influenced other projects/programmes by helping to focus active travel investment to wider Māngere; providing a project design and engagement approach for other area-wide community active travel and safety projects; and influencing policy, such as 'Road to Zero', the national government's Road Safety Strategy 2020–2030 [55].

Future Streets also provided an opportunity for formal reflection on the enablers and barriers to innovation trials. Following the difficulties of delivering Future Streets and other related projects around Auckland, a *'Making Trials Easy'* workshop was held, instigated by the research team. The workshop comprised stakeholders from the city's transport agency, the national transport agency, transport consultancies, and universities. This provided reflection on system barriers and enablers to planning, trialling, and delivering new street designs in Auckland, further informing this analysis.

14.3.1.2 Safe and Healthy Streets South Auckland (SHSSA)

SHSSA is a region-wide initiative at an early scoping stage (as of February 2020). Its aim is to implement safe and healthy street principles in a way that prioritises community needs and voices. Building on the momentum of Future Streets and other neighbourhood-scale projects such as Auckland Transport's Safer Communities [56], it seeks to bring together community, delivery, and political champions to develop and demonstrate new approaches to transport investment, including strong community partnerships, cross-agency collaboration, and innovative solutions. The early project scope included reducing crash deaths and injuries, increasing active transport, and improving access to economic opportunities in South Auckland.

In July 2018, a government-instigated workshop, *'Increasing Safe and Healthy Travel in South Auckland'*, was facilitated by the Future Streets research team. It was

[2]More information about the Te Ara Mua—Future Streets project can be found at https://www.futurestreets.org.nz/

held to gauge support for expanding the Future Streets concept across South Auckland and discuss how stakeholders could work together to make tangible progress in the subsequent two years. Participants from local and national government agencies, community groups, universities, advocacy groups, and health organisations attended, as did local and national politicians [57].

Despite the workshop demonstrating enthusiasm and broad support for the initiative, as well as a need for speedy delivery, for a long period SHSSA failed to move beyond early planning into implementation. Based on the evaluation of SHSSA to date, the project has not yet translated into a successful niche, although some of the concepts and issues that have been identified have been useful for agencies. However, from late 2019 renewed agency-led effort has emerged to deliver on the project's goals. This suggests that there may still be potential for it to have a positive influence on 'business as usual' practices. Note that the analysis carried out here does not cover this most recent period of renewed effort.

14.3.1.3 Aranui Connections

Aranui is a suburb in East Christchurch, Aotearoa which was badly damaged by the earthquakes of 2010 and 2011. Residents in Aranui experience high levels of social deprivation which is compounded by the earthquake-damaged roading network and limited public transport options connecting Aranui to other parts of the city [58].

In 2017 the Haeata Community Campus—a large school complex, catering to all ages—was established in the west of Aranui. It replaced four schools which were closed after the earthquakes. Though the campus is modern, the surrounding local streets do not serve school students well with travel options, particularly active travel [58].

To understand the need for, and feasibility of investment in safe and healthy neighbourhood street upgrades in Aranui, including to and from Haeata Campus, a scoping exercise was conducted. The exercise was led by the national government and delivered by some authors of this chapter. The scoping exercise involved a programme of engagement, coordination with other projects, and primary data collection to triangulate the issues and opportunities for local active travel.

The scoping exercise concluded with a community workshop, *'Future Streets Aranui'*, attended by participants from Haeata Community Campus, local community groups, local and national government agencies, police, health boards, and universities (highlighting the level of local interest). The workshop included group discussions of current transport problems, future scenarios, and possible solutions.

Stakeholders at the workshop confirmed their support for advancing a project to improve streets and transport connections in Aranui. However, although both the community and local government have jointly acknowledged the need for investment in Aranui, progress stalled, with a lack of priority and funding under the regional land transport plan. This is not to say that a project will not happen at some point, and ultimately influence established practices, but as of February 2020, there has been insufficient uptake for the project to be delivered and hence the concepts are not able to have any tangible impact on regime practices.

14.3.2 *System Interactions for the Three Potential Niche Projects*

Analyses of project documentation revealed six key themes that impeded or enabled progress in the three case study projects (Table 14.1). Section 14.3.3 then compares the outcomes from these projects with the two government-generated niche programmes in Aotearoa, the MCP, and the UCP.

14.3.3 *Comparison with Two Previously Successful Government Generated Niches in Aotearoa*

The three niche projects described above are now compared with the previously delivered, and local/national government driven, Model Communities Programme (MCP) and Urban Cycleways Programme (UCP). We examine the relative importance of the genesis of a niche project/programme to their success. The MCP and UCP programmes are introduced below followed by Table 14.2, which highlights the major themes including successes and barriers for each.

14.3.3.1 The Model Communities Programme

Following on from the National Walking and Cycling Strategy in 2005 [59], The MCP was included in the National Land Transport Plan (NLTP) in 2009 and again in 2012. The MCP was designed to improve the safety of urban environments for walking and cycling by providing fully integrated transport networks for those mode types. The MCP aimed to encourage 'novice users' to walk or cycle to school or to work as their main modal choice [23, 60].

Of the 22 cities who bid to be part of the programme, two provincial towns won: New Plymouth (Ngāmotu) and Hastings (Heretaunga). Both towns reportedly had existing walking and cycling programmes and whole-town buy-in from the community to local government level [61].

A total of NZ$13.1 million (NZ$1 ~ USD 0.66 at time of writing) was spread across the two towns with funding provided by the national government and matched by the district councils. Infrastructure changes focussed on the development of direct and practical commuting routes to workplaces, schools, and shops through new shared

Table 14.1 Major themes drawn from each project

Leadership	Funding	Policies and procedures	Organisational Norms	Community and delivery tensions	Social environment
Te Ara Mua—future streets					
• Strong support from science and innovation sector, and Local Board • Committed and passionate project research and delivery individuals determined to see project to completion • Strong researcher-led mandate to improve safety, public health, environmental sustainability, and social equity • Culture of risk aversion, reluctant participation, and reluctant champions from delivery agencies • Incomplete stakeholder leadership	• Insufficient initial infrastructure funding and personnel commitment, as project was 'unplanned'—not accounted for in national funding priorities • Insufficient investment in communication (especially with community) • Local board funding showed community commitment	• No initial 'home' for project—not planned into the system • Innovation in designs and design process stifled by current policies • Lack of clarity around trial/innovation processes preventing unique solutions from being tested • Innovation beyond the usual set of Traffic Control Devices stifled by laws and Road User Rules • Linear project planning process unresponsive to iteration, testing, and early community engagement • Strong research focus	• Unclear project governance exacerbated by unique transport delivery/researcher partnership • Institutions poor at engaging with politicians, and with community needs/aspirations • Different cultures and attitudes towards engagement, design, iteration, and testing new ideas • Locked into 'business as usual' designs • Siloed teams and competing interests (e.g. bus amenity vs raised pedestrian crossings) • Project seen as risky, as opposed to testing future scenarios (risk management)	• Strong initial engagement, but less as project progressed • Mixed local perceptions—some focussing on loss of vehicle amenity, while others advocate for cycling and safe streets	• General reliance on automobility, so some design ideas seen with scepticism • Mixed media narratives—generally positive in published media, but some negative social media • Environmental and urban density issues increasingly reducing viability of automobility

(continued)

Table 14.1 (continued)

Leadership	Funding	Policies and procedures	Organisational Norms	Community and delivery tensions	Social environment
Safe and healthy streets South Auckland					
• Strong national government and local stakeholder support • Strong political leadership but inconsistent institutional champions • Mixed support by transport and related agencies • Strong support from individuals within agencies, but insufficient to unlock project	• Lack of funding to start the project and build business case to obtain more significant funding, despite common underspend within transport delivery agencies and high strategic need • Rigid funding mechanisms, lack of flexibility in pre-programmed investment	• National and regional funding rules and mechanisms do not allow short-notice innovative projects • No 'home' for the project—planning and delivery system unable to accommodate it despite significant political and local demand • Regulatory barriers—unable to progress project without business case	• Agency workplans locked in and difficult to respond to need • Slow response by agencies despite clear identified need • Siloed systems and piecemeal projects with no master plan for infrastructure connectivity (e.g. housing developments not well connected with wider street network projects)	• Communities have limited capacity to participate. A value offering is needed, included social procurement, but no mechanism for this • Strong acknowledgement but a lack of project alignment with community priorities/identities. Institutions focused on individual behavior and project delivery. Communities focused on systemic/socioeconomic change, with authentic engagement	• General reliance on automobility. Not yet a strong enough political agenda to accommodate project, hence seen as relatively unimportant • Insufficient media narratives to inform case for the needed change • Environmental and urban density issues increasingly reducing viability of automobility, but the pace is slow

(continued)

Table 14.1 (continued)

Leadership	Funding	Policies and procedures	Organisational Norms	Community and delivery tensions	Social environment
Aranui connections					
• Lack of leadership on all levels despite agreement about need	• Rigid funding mechanisms • No funding available due to higher council priorities or lack of unallocated funding	• Infrastructure investment based on injury rates, not other metrics e.g. walkability, social cohesion, access to opportunity • Evidence needs to inform design/consent	• Moving away from 'business as usual' difficult • Existing projects are piecemeal, no Master Plan for connectivity • Not well positioned to respond to externally driven project	• Communities have limited capacity • Low institutional capability in engaging with community needs or with politicians • Partnerships and shared control difficult to establish	• Reliance on automobility • Media narratives critical of related work (e.g. cycle lanes) • Environmental and urban density issues increasingly reducing viability of automobility

Table 14.2 Major themes from MCP and UCP [63–65]

Leadership	Funding	Policies and procedures	Organisational norms	Community and delivery tensions	Social environment
Model communities programme					
• Significant local government ownership and initiative in both towns • Key national government personnel worked to make the programmes happen and guided councils through the ingrained rules of their operating system • Strong local leadership in the face of some public backlash (e.g. over loss of parking and delays in infrastructure delivery)	• Some of the biggest contributions to the programme were from infrastructure programmes not part of MCP, but which aligned well with planned routes (e.g. safety, reseals, new roads). In some cases, these programmes/projects were integrated to maximise the benefits of investment. However, due to the set-up of the 2 + 3 year programme it was not always possible to leverage this • Significant national government funding to support existing local work	• Pushed by the national government. Space made for the programme from within the NLTP • The first time a programme of projects was approved by the Transport Agency rather than assessing on a project by project basis • Strong local branding • Little publicly reported independent monitoring and evaluation	• Innovation was in-built in the infrastructure and activation stages • Lack of council readiness for the MCP including working to the national government's programme time frames • The risk averse nature of councils meant that non-traditional changes (i.e. walking and cycling infrastructure) needed to be continually promoted amongst all teams • The culture of each programme continues today, thanks to ongoing council commitment	• The parameters of the programme integrated well with existing local council work • There was very little time between the MCP proposal, being awarded it, and the delivery. Some implications of this included having limited time to plan and resource things properly. Additionally, there was pressure to get quick results, sometimes at the expense of longer-term achievements	• At the programme's commencement in 2009, the social and political acceptance of the value and contribution of walking and cycling to society was less than today • There was an initial underestimation regarding how difficult it would be to encourage adults to walk or cycle rather than drive • Poor public reaction to loss of parking

(continued)

Table 14.2 (continued)

Leadership	Funding	Policies and procedures	Organisational norms	Community and delivery tensions	Social environment
Urban cycleways programme					
• Emerged from National Cycle Trails which was championed by Prime Minister who garnered buy-in from each Region • Also emerged from the MCP • Strong national government leadership	• Complex, untested funding arrangement: Co-funded between the Crown, local government, and NZTA • Some regions found it difficult to access funding due to regulatory barriers and paperwork which led to project delays • Other regions secured funding through the normal mechanisms	• Pushed by the national government but there was no strong national agenda or behaviour change plan behind it • Programme evaluation, but no outcomes or impact evaluation	• A mixed response from councils, some very proactive and ready, others risk averse and slower to come on board	• Capacity and capability at local levels were an issue in some places especially in smaller towns, need to develop a community of practice • Focused on easy wins but not necessarily the most strategic or areas of greatest need. Trips to work and town centres favoured over suburbs and schools • An underlying delivery focus on making progress in building infrastructure to enable the community to choose to bike or walk	• Engagement issues and high-profile problems undermined some efforts • Although the programme was well supported, a degree of national legitimacy was missing

paths, improved links to existing paths, and new cycle lanes [61]. Associated education and promotion was also funded [62]. An evaluation of the outcomes of the programme calculated a benefit–cost ratio of 11:1 [23].

14.3.3.2 The Urban Cycleways Programme

In August 2014, the New Zealand Transport Agency (NZTA) initiated the National Cycling Programme, a multi-faceted programme of work to make urban cycling an easy, safe, and attractive choice [63]. The programme combined the development of urban cycleways with complementary initiatives in legislation, communication, research, and behaviour change. The National Cycling Programme complimented and leveraged change for the UCP—a $333 million infrastructure programme. Under the four-year UCP, changes that focused on cycling infrastructure and shared paths were made in 16 towns and cities across 54 projects. The focus of the national government was on funding and delivery and the programme utilised existing concepts for projects from local government.

For Aotearoa, the UCP represented a significant step-change in the national government's support for cycling. Within the UCP, cycling is viewed as a part of the transport system, building on a growing realisation that there are potentially a range of congestion, health, and environmental benefits to be gained from enabling cycling to flourish. At the time of writing, of the original 54 UCP projects, 17 are still underway, with the whole programme expected to be completed by 2021 [64].

14.3.4 How Do the Five Projects and Programmes Compare?

Both the MCP and UCP were concepts developed by the national government, building on demonstrated readiness or existing progress by councils. This meant that there was strong support and drive for these programmes at the outset, even if expectations between local and the national government were not entirely aligned. Nevertheless, joined up thinking by various aspects of government was sufficient to ensure funding and on-going delivery of each programme.

Meanwhile, the three externally driven projects examined earlier all found it hard to get a foothold in the system and obtain stakeholder buy-in and prioritisation. There was no 'home' for projects emerging from outside the normal planning procedures of the regime, first through council and RLTP and then via the NLTP. Hence it appears that the delivery system was self-reinforcing, with only limited opportunities for innovation by those who planned the NLTP at key points in time.

All projects and programmes—be they government-led or external—struggled to navigate a fast planning and delivery process. Unique programmes that deviated from the norm were difficult to deliver when people (locally or nationally) who represented critical elements of the System, were not active in the project. Rules, regulations, norms, and entrenched ways of working impeded the progress of all niche projects and

Table 14.3 Comparison of regime and externally generated niche projects and programmes

Internally generated	Externally generated
Strongly supported by regime	Sometimes no place in regime
Conceptualised and planned from within	Externally conceptualised and planned
Take longer to develop	Seem to just 'turn up'
Planned into system, therefore fits regime's goals	Struggle to get a 'foothold' in system
Planned into system's existing frameworks	Can chip away at problems with the system
Better funding mechanisms	Struggle for funding
Effective delivery difficult	Effective delivery difficult
Strategy-driven	Challenge a wider range of social issues
Challenges and failures and their associated lessons are more likely to be held internally	More likely to publicly report challenges, failures, and limitations

programmes. This suggests that if innovative projects and programmes are to succeed and avoid floundering through 'business as usual' processes, enabling structures are needed.

While all of the projects and programmes have had elements of design and innovation, externally-driven projects have sometimes identified areas of evidence-based need that are not currently government priorities. For example, Future Streets focused on focused on low socio-economic and Māori/Pasifika needs for safe and active travel, which was not a national government transport focus at the time the project commenced. Thus, more difficult questions may be asked through external projects because they come with the flexibility of being outside of the system.

While all projects and programmes had elements of innovation and risk taking, external projects are more likely to publicly report challenges, failures, and limitations (e.g. within the context of published research papers) and this has the potential to challenge norms. However, government-generated projects and programmes could have a higher chance of success, as they can be planned into the various funding and delivery systems more easily and have the potential to be driven by leadership inside the system. Then again, they may lack innovation as they are generated from within the regime, and therefore reflect the entrenched 'ways of working' within delivery agencies (Table 14.3).

14.4 Discussion

STS Theory offers insights into the role of niches in changing the nature and outcomes of regimes. The five niche projects and programmes presented, all focused on creating opportunities for safer and easier community walking and cycling. Using STS Theory highlights the challenges that these niche projects and programmes typically have in being adopted, delivered, and ultimately in influencing everyday practice. Experiments, demonstrations, or niches to test new ideas must ultimately influence the wider regime—the deep structure of rules, processes, and ideas that reinforce and sustain existing ways of working. These niche projects and programmes may challenge what is considered 'normal' in regimes and help drive system changes [41]. Within this, many different system components, actors, and inter-relationships impede or allow this to happen [44]. Some of the key themes, as experienced by these niches, are discussed below.

14.4.1 Conceptual Acceptance and Leadership

A starting point for all niche projects is that there is some change needed or problem to be addressed. In the cases reported, the need was for new infrastructure to enable a mode shift to safe active travel. Conceptually this problem definition needs to be accepted by those with the power to allow (or disallow) niche demonstrations of safe and healthy streets to occur, and to provide leadership support.

For some time in Aotearoa transport strategies have pointed to the need for safer roads and improved walking and cycling provision, and there was no evidence of overt strategic denial of this need in any of the niches described. However, conceptual denial of the need for niches to demonstrate concepts of safe and healthy streets may be more subtle and there is some evidence across the projects of system actors not buying into the concept, including negative media narratives, resistant organisational (and individual) norms, and risk aversion. Although left unstated in these projects, it may also be that some key actors (individuals or organisations) fundamentally did not believe that these projects were worthy of investment despite a national mandate—resulting in their delivery being undermined in subtle ways. It is difficult to ascertain whether this was the case, or whether individuals agreed with the projects but simply did not see a way to deliver them within system constraints.

The niches described, (with the exception of some of the UCP's higher-profile city centre focussed cycleways), were all community based active travel safety projects and programmes. Within the regime, these appeared to have lower status compared with large infrastructure projects such as new rail systems or motorways. Strong leadership for community transport was harder to develop. An 'A-Team/B-Team' status for projects in organisations can develop, with the A-Team being large budget, high profile, often politically important projects that attract significant resources and are fast-tracked to ensure effective delivery. On the other hand, B-Team projects struggle through the complex planning system, often without strong senior level support, with a high risk of not being delivered. Although the MCP and UCP programmes enjoyed nationwide support, some of the regional projects within these umbrellas, and the three other externally driven projects described all showed evidence of 'B-Team' status. It is clear that safe neighbourhood walking and cycling were not seen as important as other priorities at all levels. One effective way to get buy-in for projects and programmes across all system levels is to highlight the project need in terms of its direct benefit for people—how it can make their daily lives better. This was demonstrated by the Future Streets project, which only succeeded once the potential human-scale benefits to the local community became apparent to the delivery agency and political leaders [66].

Delivery of the UCP and MCP programmes suggest niches have a better chance of being implemented if they are nested within the regime with a level of cross-system buy-in from the outset. Conversely, we have presented examples of potentially very important niche projects, which, due to their lack of cross government support, have, or are still struggling to emerge in any tangible form.

Champions were critical to all projects and programmes described, ranging from proactive delivery officers to the highest levels of government. Yet in some ways these individuals acted to paper the cracks when cross-system leadership was lacking. Champions tended to be passionate individuals who go above and beyond their job descriptions to make things happen—often at personal risk. They need better support from senior organisational leadership.

Although it is difficult to make a definitive link between the underlying social environment and the success of the projects and programmes as niches, it is likely that at a landscape level, social norms have a political influence on the emergence, delivery, and ultimately the impact of the projects and programmes. Certainly, resistance to loss of parking from cycle lanes, and preferences for automobile provision are evident from all projects and programmes.

Overall, the analysis suggests that mostly these projects and programmes were accepted by system actors, but variable levels of leadership across government have challenged the delivery of some of the niches. The greatest challenges related to the regime's structure—the existing planning and delivery mechanisms.

14.4.2 System Delivery

In this section, the planning and delivery issues which were key to the outcomes of the niches are described.

14.4.2.1 Policies, Procedures, and Ways of Working

The policies, processes, and procedures that are established within transport delivery systems, are (for good reason) in place to ensure quality, accountability, consistency, and safety. However, there is little room in this system for experimentation, innovation, and testing disruptive ideas, which is crucial if change is to happen. This paradoxically suggests that the transport delivery regime has confidence that current policies and procedures are fit for purpose, and yet outcomes for many city transport systems are far from optimal and in dire need of change. For Future Streets in particular, there were conflicts between the intention to be innovative and nimble, and the delivery system's requirement for consistency and repeatability. There was also a lack of clarity about the trial system [48], and innovative techniques such as using temporary measures were therefore not possible within the delivery expectations of the project.

Most projects described have suffered from complicated, inconsistent, or onerous business cases, funding applications, or planning approvals. These requirements, designed to make funding processes robust also created delay and uncertainty, and frustrated relationships with the communities awaiting delivery. The extent to which the system itself is configured to enable innovation to flourish (or not) is generally ignored in the general acceptance of 'how things are'. This points to a need for more sophisticated evaluation of projects—not just on short-term outcomes, but also of the implementation process. Evaluative exercises to better understand the process of delivery and the tensions that exist [48], have proved very useful thus far.

14.4.2.2 Funding

The reality is that there are always more project ideas than money and therefore it is critical to have a strategic planning and prioritisation process in place to determine what to invest in, when, where, and why. However, external niche projects often arise outside of the funding planning process and therefore can get stuck in a vicious planning cycle: not being able to start a project because funding is not in place. However, funding requires a scope of work and a scope of work requires quality stakeholder engagement to define it—especially if it is to respond to community needs and aspirations. Engagement, in turn, is not possible because funding is not in place and conversations might raise community expectations with the risk of non-delivery.

The rigid nature of the NLTP and other funding systems often stymied the occurrence of niche projects as they did not fit within funding cycles and did not necessarily meet strategic priorities (because by definition they are suggesting something new). This is particularly the case for active travel-focussed projects, as funding is still relatively limited in this area, despite strategic signals. For example, Aranui Connections and SHSSA both had strong community, and high-level political support, but funding was not forthcoming. This was due to an apparent lack of mechanisms within the national and local government agencies to expedite projects that fell outside the four-yearly NLTP funding scheme. Many of the projects were delivered within the context of constrained funding, yet it could be argued that the issue was of funding allocation rather than total funding available for transport. Funding allocations given to walking and cycling projects in general (let alone niche walking and cycling projects) never matched the strategic aspiration for more walking and cycling.

14.4.2.3 Community and Delivery Tension

Community engagement and empowerment leads to user-friendly investment, but this effort is hard to justify in linear transport planning systems that prioritise swiftness in the design and delivery process, ironically to make promises of efficient delivery to communities. Up-front engagement and data collection to effectively understand community needs and aspirations is valuable but can be time consuming. Progress updates during the delivery phase is also desirable.

Landscapes are diverse and some align more closely with regime values than others. For example, the norms, cultures, and ways of doing things in diverse low-income communities are very different from the procedural-focussed methods of transport delivery agencies. Experimenting with temporary measures also does not fit well with accepted project management processes and community expectations. Sometimes 'just getting on with it' is what communities want, but even that can be difficult. There is a need to design delivery systems that are responsive and efficient, yet still account for community voice at key stages to yield maximum benefit.

14.4.3 Wider Influence

The regime of transport planning and delivery is likely to reflect the social landscape in which it serves. In this chapter, it was demonstrated how the theme of automobility influenced all project and programme aspects, from their existence, to the extent to which lessons are adopted by the regime. Change is difficult and existing narratives are strong. Nevertheless, a hint of how the regime and landscape slowly changes over time in response to niches was demonstrated by the MCP. Within this early context, walking and cycling projects were scarce, with few resources available in planning systems, yet the hunger for their uptake—demonstrated by the 22 competing bids—was evident. By contrast in today's planning system, walking and cycling projects are more routine, although they still suffer from under-investment.

Of the three externally driven projects, Future Streets showed some regime impact by creating more focus on low-socio economic communities, Māori and Pasifika, suburban community area-wide interventions, and, where appropriate, prioritising pedestrians and cyclists over vehicles. However, various attempts to operationalise the Future Streets idea more widely are still struggling to be established.

The impact of the MCP and UCP may also point to the scale of intervention and time needed for niches to ultimately influence the regime. The nationwide focus of both programmes, with the multiple projects sitting under them, has clearly influenced ongoing practice, attitudes, and ways of working and many of today's new projects have emerged from earlier UCP effort, even if (as of February 2020) some UCP projects are still to be delivered.

There was no evidence of a systematic approach to continuous improvement and scaling up as part of the implementation of any of the niches. In theory, any level of success should be followed by refinement and continued investment, but this has not occurred. This reflects the siloed nature of funding packages, and to the inherent difficulties in scaling prototypes or pilots in widespread practice.

14.4.4 Considerations for Regime and Societal Change

Local niche innovations are as important for the systemic barriers that they reveal, as they are for their local impacts. Niche projects cumulatively cause disruption in the socio-technical system, but they need an immense effort both in terms of delivering niches, and the scale of change that is needed. Only through the relentless pushing of boundaries, iteration, and scaling of promising concepts, with internal continuous improvement will the regime change in any substantial way. To develop a culture of niche projects at a magnitude of effort that will eventually change existing regimes, they must be built into planning and delivery systems. It would be mistaken to assume that these isolated projects will lead to widespread system change, when the vast majority of transport effort is committed to status quo activities. More work is needed to conceptualise the mechanisms by which widespread or nationwide change could be achieved. This may include more or larger niches, through programmes that facilitate testing, iteration, and scaling up.

There is already increasing recognition at an institutional level that process change is necessary to facilitate a faster rate of progress. For example, NZTA's *Innovating Streets for People* programme responds to this challenge by focussing on system issues to make it faster and easier to transition streets to safer and more liveable spaces. This is an important start to a culture of doing things differently from within the Transport planning system. Whilst the infrastructure changes from this programme have a micro focus (e.g. streets, or sections of a street), eventually the programme may create sufficient system change to operate at a scale that leads to widespread and national level planning and delivery change, and a focus on programmes with a macro approach (e.g. connected networks, suburb-level).

While niche projects are important disruptors of the status quo, the regime is also constantly working from within to continually improve. The change pathways are different (evolution vs disruption), but both are valuable for overall system improvement. However, the ability for agencies to respond and evolve is often hampered by organisational changes, high staff turnover, and loss of institutional memory. External consultants, researchers, and advisors are often the holders of knowledge as they remain when agency staff are continually leaving and arriving. While this may seem slightly peripheral to the success of niches, it is a crucial system issue that needs to be addressed. The reality is that a community of practice exists around most niches with government officials, researchers, consultants, and community stakeholders all working together. The more that these effective multi-agency teams can continue to work together in ongoing programmes, the more likely that niches will succeed, and regimes and landscapes will be altered.

14.4.5 *Implications for Practice*

The literature suggests that niche projects are important to make change in regimes happen [42]. The analysis carried out in this chapter suggests that there are conditions under which niche projects are more likely to succeed in concept, delivery, and regime impact. These are briefly outlined:

- *Wider system acceptance that things need to change*: Aligning planning and delivery objectives with strategic objectives with processes and ways of working to match.
- *An acceptance that niches can help manage long-term investment risk, rather than represent unnecessary risk to be avoided*: Testing ideas through niches, before substantial nationwide investment is a prudent risk management strategy.
- *A built-in innovation function within national and local government, with niches as accepted practice*: Mechanisms to expedite, fund, manage, and evaluate niches outside of existing planning processes, with a requirement that feedback from lessons learned are formally absorbed into regime processes for continual improvement.
- *Efficient delivery and learning*: Making sure niche projects are delivered quickly so that momentum is maintained, trust in promising concepts is maintained, and evaluation, iteration and scaling up can occur in a timely way.
- *A plan for scaling up and regime influence*: Formal and structured implementation of successful niche outcomes acknowledging that scale, time, and network completeness is needed for full benefits to be realised. A staged approach to implementation may therefore be prudent; single concept test, followed by more demonstrations, and eventually if success is emerging then roll-out and embedding in regime policies and processes. This process has worked well with other road safety projects [67].
- *A community of practice*: More focus is needed on shared learning between city councils and national government agencies, as well as practitioners, policy makers, community stakeholders, and researchers working together, with the goal of innovation and continually improved transport implementation and delivery.

14.5 Conclusion

Neighbourhood streets that support all modes for commuting, recreation, the environment, and social interaction are critical for the health, social, and economic wellbeing of towns and cities. By innovatively redesigning neighbourhood streets, we can promote stronger, healthier, and safer communities, and encourage the uptake of environmentally sustainable travel.

However, automobility, embedded within an STS that buffers against change still structures our daily lives. The social norms, customs, and practices of car use are held in place by the regime: supportive infrastructure; policy logics; and professional practices, with a grip that can stifle innovation.

The magnitude of change that is needed for transport systems to deliver on today's societal and environmental challenges—including urgent road safety, public health, climate change, and urban liveability—is enormous. Despite promising initiatives, overall there is a theme of an inadequate rate of progress.

Currently, barriers maintained inherently within the system obstruct the fulfilment of street improvements that cater to safe and healthy outcomes for all users. For urban street re-design projects, the issue is often to do with national and regional priorities, funding allocation, co-investment models, and entrenched ways of working, rather than absence of need or community buy-in. A significantly increased programme of internal and external niche projects is needed within built-in innovation functions with appropriate resourcing and empowered senior leadership to challenge the current system and ultimately add greater public value.

Funding Acknowledgement This work was supported financially by the Ministry of Business Innovation and Employment Science and Innovation Research Funding, contract ID MRACO1501.

References

1. Giles-Corti, B., et al.: City planning and population health: a global challenge. Lancet **388**(10062), 2912–2924 (2016). https://doi.org/10.1016/S0140-6736(16)30066-6
2. OECD: Environmental Performance Review: New Zealand 2017 Highlights'. Available at https://www.oecd.org/environment/country-reviews/Highlights_OECD_EPR_NewZealand.pdf (2017)
3. Briggs, D., Mason, K., Borman, B.: Rapid assessment of environmental health impacts for policy support: the example of road transport in New Zealand. Int. J. Environ. Res. Public Health **13**(1), 1–23 (2015). https://doi.org/10.3390/ijerph13010061
4. Lal, A., et al.: Health care and lost productivity costs of overweight and obesity in New Zealand . Australian New Zealand J. Publ. Health J. Publ. Health **36**(6), 550–556 (2012). https://doi.org/10.1111/j.1753-6405.2012.00931.x
5. Leung, C. et al.: Benefits from Auckland Road Decongestion, p. 103. Available at https://www.ema.co.nz/resources/EMAReports and https://www.Documents/Advocacy/Submissions/NZIER report on Auckland Benefits of Decongestion.pdf (2017)
6. Hosking, J. et al.: Social and Geographical Differences in Road Traffic Injury in the Auckland Region. Section of Epidemiology and Biostatistics, School of Population Health, The University of Auckland. Report produced for Auckland Transport (2013)
7. Ministry of Health: Understanding Excess Body Weight. Available at https://www.health.govt.nz/system/files/documents/publications/understanding-excess-body-weight-nzhs-apr15.pdf (2015)
8. Shaw, C., et al.: Health consequences of transport patterns in New Zealand's largest cities. N. Z. Med. J. **131**(1472), 64–72 (2018)
9. Macmillan, A. et al.: Suburb-level changes for active transport to meet the SDGs: causal theory and a New Zealand case study. Sci. Total Environ. **714** (2020)https://doi.org/10.1016/j.scitotenv.2020.136678
10. Pocock, M.: The Effects and Efficacy of Large-Scale Motorway Development: A Case Study of Wellington. University of Otago, New Zealand (2018)

11. Hinde, S., Dixon, J.: Changing the obesogenic environment: insights from a cultural economy of car reliance. Transp. Res. Part D Transp. Environ. **10**(1), 31–53 (2005). https://doi.org/10.1016/j.trd.2004.09.003
12. Bean, C.E., Kearns, R., Collins, D.: Exploring social mobilities: narratives of walking and driving in Auckland, New Zealand. Urban Stud. **45**(13), 2829–2848 (2008). https://doi.org/10.1177/0042098008098208
13. Frank, L., Engelke, P.: The built environment and human activity patterns: exploring the impacts of urban form on public health. J. Plann. Liter. **16**, 202–218 (2001). https://doi.org/10.1177/08854120122093339
14. Lovell, S.T., Taylor, J.R.: Supplying urban ecosystem services through multifunctional green infrastructure in the United States. Landscape Ecol. **28**(8), 1447–1463 (2013). https://doi.org/10.1007/s10980-013-9912-y
15. Heath, G.W., et al.: The effectiveness of urban design and land use and transport policies and practices to increase physical activity: a systematic review. J. Phys. Act. Health **3**(s1), S55–S76 (2016). https://doi.org/10.1123/jpah.3.s1.s55
16. Renalds, A., Smith, T.H., Hale, P.J.: A systematic review of built environment and health. Family Community Health **33**(1), 68–78 (2010). https://doi.org/10.1097/FCH.0b013e3181c4e2e5
17. Perk, V. et al.: Capturing the benefits of Complete Streets. Tampa, Florida. Available at https://www.nctr.usf.edu/wp-content/uploads/2016/01/NCTR-977-04-Capturing-the-Benefits-of-Complete-Streets-2015-1.pdf (2015)
18. Sallis, J.F., et al.: Co-benefits of designing communities for active living: an exploration of literature. Int. J. Behavioral Nutr. Phys. Activity **12**(1), 1–10 (2015). https://doi.org/10.1186/s12966-015-0188-2
19. Fraker, H.: The Hidden Potential of Sustainable Neighborhoods: Lessons From Low-Carbon Communities. Island Press, Washington. ISBN: 9781610914086
20. Smart Growth America: What are Complete Streets? Smart Growth America. Available at https://smartgrowthamerica.org/program/national-complete-streets-coalition/publications/what-are-complete-streets/ (2018)
21. Webster, D.C., Layfield, R.E.: Review of 20 mph Zones in London Boroughs. TRL Limited (2003)
22. Sallis, J.F., et al.: Use of science to guide city planning policy and practice: how to achieve healthy and sustainable future cities. Lancet **388**(10062), 2936–2947 (2016). https://doi.org/10.1016/S0140-6736(16)30068-X
23. Chapman, R., et al.: A cost benefit analysis of an active travel intervention with health and carbon emission reduction benefits. Int. J. Environ. Res. Publ. Health **15**(5), 962 (2018). https://doi.org/10.3390/ijerph15050962
24. Pickford, M.: State highway investment in New Zealand: the decline and fall of economic efficiency. Policy Q **9**(3), 28–35 (2013)
25. Gehl, J.: Cities for People. Island Press, Washington (2013). ISBN: 9781597265737
26. Kehagia, F.: Transforming small towns by remedial street design. Transp. Res. Proc. **24**, 507–514 (2017). https://doi.org/10.1016/j.trpro.2017.05.090
27. Kott, J.: Streets of yesterday, today and tomorrow. World Transp. Policy Pract. **21**(4), 42–51 (2016). ISSN: 1352-7614
28. Mallo, D., Parsons, R. and Tardiveau, A.: Participatory design methods in the co-production practice of urban space. In: Challenges and Best Practice in Co-Production, University of Sheffield. Newcastle University (2016)
29. Keall, M.D., et al.: Health Outcomes Projected from Transportation Modelling for the New Zealand Transport Outlook. University of Otago, Wellington (2016)
30. Arseni, O., Racioppi, F.: Making THE (Transport, Health and Environment) link: Transport, Health and Environment Pan-European Programme and the Sustainable Development Goals. World Health Organization Regional Office for Europe (2018)
31. Rachele, J.N., et al.: Automobile dependence: a contributing factor to poorer health among lower-income households. J. Transp. Health **8**, 123–128 (2018). https://doi.org/10.1016/j.jth.2017.11.149

32. Requia, W.J., et al.: The health impacts of weekday traffic: a health risk assessment of PM2. 5 emissions during congested periods. Environ. Int. **111**, 164–176 (2018). https://doi.org/10.1016/j.envint.2017.11.025
33. Simeonova, E. et al.: Congestion Pricing, Air Pollution and Children's Health. Johns Hopkins Carey Business School Research Paper no. 17-05 (2018).https://doi.org/10.2139/ssrn.2955260
34. Mackie, H.W., et al.: Road user behaviour changes following a self-explaining roads intervention. Accid. Anal. Prev. **50**, 742–750 (2013). https://doi.org/10.1016/j.aap.2012.06.026
35. Unpublished data, available upon request
36. Mackie, H. et al.: Te Ara Mua—Future streets: lessons and outcomes from a researcher, community, and practitioner partnership'. In: International Conference on Transport and Health Melbourne, Australia. Available at https://www.tphlink.com/preliminary-programme1.html (2019)
37. Geels, F.W.: Regime resistance against low-carbon transitions: introducing politics and power into the multi-level perspective. Theor. Culture Soc. **31**(5), 21–40 (2014). https://doi.org/10.1177/0263276414531627
38. Maassen, A.: Heterogeneity of lock-in and the role of strategic technological interventions in urban infrastructural transformations. Euro. Plann. Stud. **20**(3), 441–460 (2012). https://doi.org/10.1080/09654313.2012.651807
39. Vreugdenhil, R., Williams, S.: White line fever: a sociotechnical perspective on the contested implementation of an urban bike lane network. Area **45**(3), 283–291 (2013). https://doi.org/10.1111/area.12029
40. Rasmussen, J.: Risk management in a dynamic society: a modelling problem. Safety Sci. **27**(2), 183–213 (1997). https://doi.org/10.1016/S0925-7535(97)00052-0
41. Geels, F.W.: From sectoral systems of innovation to socio-technical systems: insights about dynamics and change from sociology and institutional theory. Res. Policy **33**(6–7), 897–920 (2004). https://doi.org/10.1016/j.respol.2004.01.015
42. Geels, F.W.: The multi-level perspective on sustainability transitions: responses to seven criticisms. Environ. Innov. Soc. Trans. **1**(1), 24–40 (2011). https://doi.org/10.1016/j.eist.2011.02.002
43. Geels, F.W.: A socio-technical analysis of low-carbon transitions: introducing the multi-level perspective into transport studies. J. Transp. Geographyeogr. **24**, 471–482 (2012). https://doi.org/10.1016/j.jtrangeo.2012.01.021
44. Marletto, G.: Car and the city: socio-technical transition pathways to 2030. Technol. Forecast. Soc. Chang. **87**, 164–178 (2014). https://doi.org/10.1016/j.techfore.2013.12.013
45. Hebbert, M.: Engineering, urbanism and the struggle for street design. J. Urban Des. **10**(1), 39–59 (2005). https://doi.org/10.1080/13574800500062361
46. Geels, F.W., et al.: Sociotechnical transitions for deep decarbonization. Science **357**(6357), 1242–1244 (2017). https://doi.org/10.1126/science.aao3760
47. Burdett, B.: Inclusive transport in action: policy drivers that account for the way transport decision-makers think. In: Inclusive Transport in Action. Wellington, New Zealand, 15 Nov 2018. 5th Transport Knowledge Conference. https://doi.org/10.1016/c2016-0-04941-0
48. Opit, S., & Witten, K.: Unlocking transport innovation: a sociotechnical perspective of the logics of transport planning decision-making within the trial of a new type of pedestrian crossing. Building Better Homes Towns and Cities SRA—The Architecture of Decision-making. Auckland, New Zealand: Shore Whariki, Massey University (2018)
49. Mavoa, S., et al.: GIS based destination accessibility via public transit and walking in Auckland, New Zealand. J. Transp. Geogr. **20**(1), 15–22 (2012). https://doi.org/10.1016/j.jtrangeo.2011.10.001
50. Docherty, I., Shaw, J.: The transformation of transport policy in Great Britain? 'New Realism' and New Labour's decade of displacement activity. Environ. Plann. A **43**(1), 224–251 (2011). https://doi.org/10.1068/a43184
51. Mackie, H., et al.: Te Ara Mua-future streets suburban street retrofit: a researcher-community-government co-design process and intervention outcomes. J. Transp. Health **11**, 209–220 (2018). https://doi.org/10.1016/j.jth.2018.08.014

52. Braun, V., Clarke, V.: Using thematic analysis in psychology. Qual. Res. Psychol. **3**(2), 77–101 (2006). https://doi.org/10.1191/1478088706qp063oa
53. Auckland Council: Māngere-Ōtāhuhu Local Board Profile: Initial Results From the 2013 Census (2014)
54. Macmillan, A.K., et al.: Controlled before-after intervention study of suburb-wide street changes to increase walking and cycling: Te Ara Mua-Future Streets study design. BMC Publ. Health **18**(1), 850 (2018). https://doi.org/10.1186/s12889-018-5758-1
55. New Zealand Government: Road to Zero: New Zealand's Road Safety Strategy 2020–2030. Wellington, New Zealand (2019)
56. New Zealand Transport Agency: National Land Transport Programme 2018–2021. Wellington, New Zealand (2018)
57. Te Ara Mua—Future Streets Research Team: Safe and Healthy Streets in South Auckland: Notes from Workshop 6 July'. Auckland, New Zealand: Te Ara Mua—Future Streets Research Team (2018)
58. Newton, J., Hirsch, L., Mackie, H.: Future Streets Aranui: Community insights. Auckland, New Zealand (2019)
59. Minstry of Transport: Getting There—On Foot, by Cycle. Wellington, New Zealand (2005)
60. New Zealand Transport Agency: Model Communities Information Sheet. Available at https://www.nzta.govt.nz/assets/resources/model-communities-infosheet/docs/2011-03.pdf (2011)
61. New Zealand Transport Agency: The Walking and Cycling Model Community Story with Hastings and New Plymouth. Wellington, New Zealand (2013)
62. Dance, G.: Model Communities—Making Walking and Cycling the Easiest Transport Choice. Wellington, New Zealand (2010)
63. Mackie, H., Field, A., Hawley, G.: National Cycling Programme Evaluation. Mackie Research, Dovetail, Auckland (2016)
64. Personal Communication, NZTA (2019)
65. CicloCivica Ltd.: Share Hub: Final Report. Prepared by CicloCivia Ltd. for NZ Transport Agency (2016)
66. Witten, K., et al.: Te Ara Mua-future streets: knowledge exchange and the highs and lows of researcher-practitioner collaboration to design active travel infrastructure. J. Transp. Health **9**, 34–44 (2018). https://doi.org/10.1016/j.jth.2018.03.001
67. Mackie, H. et al.: The signs they are a-changin': development and evaluation of New Zealand's rural intersection active warning system. J. Australasian College Road Safety **28**(3), 11–21. ISSN: 1832-9497 (2017)

Chapter 15
Practising the Art of Wayfinding Design and Finding the Way to Artwork Materials: Designing Los Angeles County's Transit Future, Applying Lessons Learned in New Project Delivery Methods for Metro Signage and Artwork

Susan Gray, Lance Glover, and Allison Porterfield

Abstract This chapter presents select case studies of Los Angeles County Metropolitan Transportation Authority (Metro) Arts & Design through the lens of 'lessons learned.' During nearly 30 years of project delivery amidst tectonic shifts in the construction environment, refined frameworks have been developed using time-tested materials and processes that optimize the customer experience with the aim of creating a world-class transit system. As mediators of that experience, station architecture, environmental graphic design and artwork operate in synergistic harmony. Coordinating the methods of wayfinding and artwork integration allows for maximal visual impact while ensuring adaptability for future growth and ease of maintenance. Through this combined effort it is possible to create navigable landmarks and a unique sense of place within the complexities of a massive urban transit network. This study includes a description of a new subset of typologies in wayfinding and environmental graphic design, as well as the integration of contemporary visual art specific to the requirements of a transit system in transition—where expansion represents design challenges and consequent opportunities. The discussion deals both with the specifics of implementation in design build contracting as well as the broader conceptual framework.

Keywords Architectural glass · Arts-in-Transit · Artwork master planning · Best practices · Branding · Contemporary art and design · Digital customer information · Enamel coated steel · Environmental graphic design · Identification signage · Industrial design · Lessons learned · Life-cycle costs · Maintenance envelope · Modular 'kit of parts' · Mosaic · Overhead Contact System (OCS) · Percent for art · Placemaking · Public art maintenance · Transit system ·

S. Gray (✉) · L. Glover · A. Porterfield
Metro Arts & Design, Los Angeles, USA
e-mail: grays@metro.net

S. Coxon and R. Napper (eds.), *Advancing a Design Approach to Enriching Public Mobility*, Intelligent Systems Reference Library 198,
https://doi.org/10.1007/978-3-030-64722-3_15

Transportation planning · Signage design standards · System legibility · Wayfinding

15.1 Introduction

"*In short, space is a practiced place. Thus, the street geometrically defined by urban planning is transformed into a space by walkers. In the same way, an act of reading is the space produced by the practice of a particular place: a written text, i.e., a place constituted by a system of signs.*" Michele de Certeau examining the ways in which people individualize mass culture and utilitarian systems through ritual in order to make them their own [1].

The Los Angeles County Metropolitan Transportation Authority (Metro) is the largest of 44 transit operators in Los Angeles County, a vast area of 4058 square miles (10,510 km^2) with over 10 million inhabitants [2]. Metro is also one of the largest transportation authorities in the United States, ranking second in the nation in terms of average weekday bus rides for third quarter 2019. In the same data, the agency is third nationwide in total trips across service modes and fifth for average weekday rides [3]. Metro provided service in 2019 for an average of nearly 1.2 million weekday rides, traveling an average of 2.6 million daily miles [4].

Metro operates bus, rail, bike and microtransit services, as well as planning and directing funding for rail and freeway projects across the county. The multimodal transit agency was the product of the 1993 merger between the Southern California Rapid Transit District (RTD) and the Los Angeles County Transportation Commission (LACTC) [5]. These predecessor agencies, separately and in aggregate, have been working to build a complex mass transit system for decades. Arts and design have been principle building tools since the inception of Metro, as it is undeniable that the reintroduction of sustainable public transit must attract ridership in order to compete with a dominant car culture. In 1979, RTD awarded a major contract to the Los Angeles firm of Bass/Yager and Associates (well-known for creating a multitude of diverse and distinctive corporate identity programs, among them United Airlines and AT&T) to develop a Visual Identification Program for fleet graphics and iconic bus information signs [6, 7]. In 1983 RTD adopted an Art-In-Transit policy and, in 1989, LACTC hired its first staff to develop the art program which has grown to fund public art, tours, temporary installations, integrated works, and other cultural programming [8]. For the past 15 years at LA Metro, the Signage and Environmental Graphic Design (SEGD) and Public Art programs have operated under an organizational umbrella that takes a holistic approach to the customer experience. This programmatic marriage provides the opportunity to simultaneously strategize on the pragmatic and poetic, to orchestrate seamless visual integrations and intentional hierarchies. Together, these experience-oriented design values are strategically balanced with the realities of constructing, operating and maintaining public transportation.

As FTA Circular 9400. 1A says, "The visual quality of the nation's mass transit systems has a profound impact on transit patrons and the community at large. Good design and art can improve the appearance and safety of a facility, give vibrancy to its public spaces and make patrons feel welcome" [9, 10]. These are just some possibilities. Well-executed creative expression in transit can have a huge range of benefits, such as creating a cohesive brand identity, easing wayfinding, building community, promoting placemaking, offering beauty, encouraging exploration, improving safety, facilitating stewardship and supporting diverse representation (see [10–13]). Recognizing these myriad benefits for transit riders, this paper will bring arts and design together and spotlight ways to maximize impact through lessons learned. The focus will center on the evolution of a station identifier system and a permanent artwork program for Metro Rail.

15.2 Overview of Metro Arts and Design

Metro Arts & Design is the primary in-house creative arm of an over 10,000-person agency that plans, designs, builds, operates and maintains its transit modes across Los Angeles County [14]. This internal group is tasked with signage and environmental graphic design, public art implementation, art asset management and cultural programming for fleet, stations and facilities for a system that provides over 374 million annual rides on bus and rail across the 4000+ square mile county [15].

Such a comprehensive scope is atypical in modern transit organizations. For example, unlike New York's Metropolitan Transportation Authority (NYC MTA), Chicago Transit Authority (CTA) and Massachusetts Bay Transportation Authority (MBTA), Metro has SEGD staff dedicated to developing and overseeing signage design standards. Of note, while Metro is currently in the midst of an unprecedented system expansion, NYC MTA is predominantly focused on rehabilitation and upgrades of existing stations [16]. Metro's expansive scope directly translates to Metro's Arts & Design department, which connects at numerous stages of transit operations and along with responsibility, comes with great opportunity and risks for artwork and signage (Fig. 15.1).

Metro's creative and cultural programming connects with its diverse ridership in a variety of methods. Running the gamut from architectural to the performative, diagrammatic to sculptural, the program's core work is permanent integrated artwork and wayfinding signage for Metro Rail corridors, Bus Rapid Transit (BRT) and special facilities. Per the Metro Rail Architectural Design Criteria, wayfinding signage is established as an element of design consistency with site responsive placement, while artwork has been characterized as an element of variability.

Fig. 15.1 Station signage alongside mural by Richard Wyatt, part of *City of Dreams/River of History*, a Metro artwork commission with May Sun that also includes integrated paving elements, a salt-water aquarium and a river bench (Union Station East Portal). Image courtesy of Metro

15.3 Local Growth and Planning Shifts Impacts Art and Design Methodologies

The historic passage by LA County voters of two local funding initiatives, Measure R (2008) followed by a 'no sunset' sales tax increase with Measure M (2016) [17], combined with a winning bid for the 2028 Olympics and the resulting "Twenty-eight [projects] by '28" initiative [18], means explosive growth for Metro and a shifting organizational landscape. In this time of expansion, there are exciting opportunities for new projects with growing demands for a range of nimble, responsive project delivery mechanisms.

A resulting significant change in Metro's contracting structure over the past two decades parallels changes in the private sector and lends more readily to the expedited delivery of complex infrastructure as cited above. A method that was once Design-Bid-Build (DBB) between government sector and contractors is now typically Design-Build (DB). With DB, early construction phases can start in parallel with advanced architectural design and are less hindered by bid rounds. The increased speed lends to a more controlled and modular design approach, but with very little room for architectural innovation. The future holds new possibilities with Public-Private Partnership (P3) with the potential for contractor's financial risk to shift from construction into operational phases. In contrast to these approaches, another project delivery model worthy of consideration is Construction Management General

Contractor (CMGC), which is in use by many transit agencies in the United States (see [19–21]).

In planning for the delivery of new transit corridor initiatives under a DB framework, it became clear that multiple new rail and bus rapid transit corridors in various stages of development warranted a dramatic reduction in custom design elements in order to improve the accuracy of cost estimating for project construction, maintenance and operations and to improve the quality of the outcome under an expedited contracting model. Consequently, the need to adopt a consistent architectural vernacular for all Metro stations became paramount. Exploration began around 2010 with local architects Johnson Fain, and the consistent application of signage and artwork was included in planning efforts. Application of the systemwide station design approach was officially adopted by Metro's Board of Directors in January 2018 (see [22, 23]).

While Metro has long pursued a modular approach to the station format [24] the architectural expression of this approach (with the notable exception of wayfinding signage) initially varied erratically from one station to the next. Previously, each station was designed relatively independently through architect and artist collaboration with art and architectural excellence in mind (including winning the prestigious, well-established Progressive Architecture Award for such collaboration [25]), but with varying results in overall maintenance costs. The new standards ensure that all stations have the quality of contemporary design with emphasis on accessibility and navigability, better safety with open sightlines, cleanliness with high performing materials, and more environmentally-sound, sustainable materials and landscaping. This also presents a unified customer experience with a more consistent identity. The architecture (and intuitive placement of signage and artwork) becomes a more consistent part of the Metro experience, providing a recognizable service identity and a seamless transition between system nodes. The adaptable approach also accounts for the future, so that the system maintains the aesthetic standards with additional system growth and surrounding neighborhood integration.

Lastly, Metro has negotiated a major advertising revenue generating contract for its rail environment, which presents potential visual competition with signage and artwork arrangement. The Arts & Design program has been strategic in the utilization of signage and public art assets to ensure prominent locations ensuring minimal compromise to the Metro brand and customer experience.

For Metro Arts & Design, these shifts in the organizational landscape and process have led to a more consistent palette and ordering of spaces within the transit corridors. With architectural materials and components predetermined, the canvas of opportunities for art and signage is known well in advance of a project and the process for incorporating aesthetic elements is streamlined. Logically, it followed to adopt similar design strategies and maintainable materials standards with modular signage elements in stainless steel, painted aluminum and light-emitting diodes (LEDs), and artworks in a refined palette of mosaic ceramic tile, porcelain enamel steel and architectural glass. This planning builds future-proofed projects that allow Metro staff, design consultants and artists to focus energies on refining unique site-responsive attributes.

15.4 Metro Pin: A Family of Station Identifiers

"The only way you can find a good answer is to clearly understand the question."—Saul Bass [26].

LA Metro has undertaken the design of a family of branded rail station identifier elements with the aim of enhancing system legibility in a fluid, rapidly changing urban landscape. These designs take on their own 'kit-of-parts' approach with systemwide station design to address the requirements of high public visibility, conformity with signage and brand design standards, replicability, adaptability, ease of maintenance and phased implementation. Various sizes and configurations have been developed to adapt to the wide range of conditions found on the system, from open plazas to retrofit applications in tightly constrained spaces. Designs have been developed with DB delivery in mind.

15.4.1 Design Challenge

Prior to investigations into a systemwide station design idiom, by 2011, a new range of station identifiers was deemed necessary to replace an aging system of existing pillar identification signs, originally designed in the late 1990s, that suffer from several design deficiencies, particularly an outmoded overall appearance, limited visibility and susceptibility to vandalism.

The existing "Grand Pylon" design for Metro Rail station identification signs featured a large structural volume that unfortunately presented a readily accessible, enticing surface area for vandalism, and attractive planes for bird roosting that warranted equally unattractive mitigating spikes. Regular maintenance and repair of these factory-applied finishes became increasingly problematic as the transit system and number of stations grew over time, factors that led to a limited lifecycle. As well, the existing pylon signs, at 15 ft. high (approximately 4.5 m), proved diminutive in dense and vertical urban environments, and often became visually lost in the overall fabric of the station surroundings (Fig. 15.2).

15.4.2 Design Process

To address these challenges, and cognizant of international best practices in wayfinding systems like those created in London [27], New York [28] and Switzerland [29], Metro Arts & Design's SEGD team established objectives that the new designs be iconic, improve visibility within the urban landscape, increase legibility from a distance, and offer improved vandal-resistance and maintainability. They should, above all, be reflective of a world class transit system with a distinct,

Fig. 15.2 Existing Grand Pylon station identifier. Image by ALAS Media, courtesy of Metro

"designed for LA" quality. They should also be adaptable to a variety of site conditions and construction delivery methods. The design brief also included two sizes of parking identifiers (called "Tack" and "Peg"), and a solution for station identification in congested urban settings where limited space might preclude implementation of free-standing signage elements entirely.

Continuing the practice of enlisting seasoned professional design consultants, Metro engaged the services of Sussman/Prejza & Co. [30], an international leader in the field of environmental graphic design, to produce the conceptual design through an iterative process. Once the conceptual design was developed into a family of sign types, the Metro in-house SEGD team further refined all details to optimize constructability. These designs are memorialized in the Metro Rail Design Criteria (MRDC) and Metro's Design Directive Drawings to ensure consistent application documents that detail, describe and specify methods and means to ensure consistent application of high-quality architectural materials, finishes and other specialties, such as lighting and signage (Fig. 15.3).

In addition to a full set of 1 in. scale models, two first article, full-sized working prototypes were produced and installed, one for the multimodal transit plaza at the

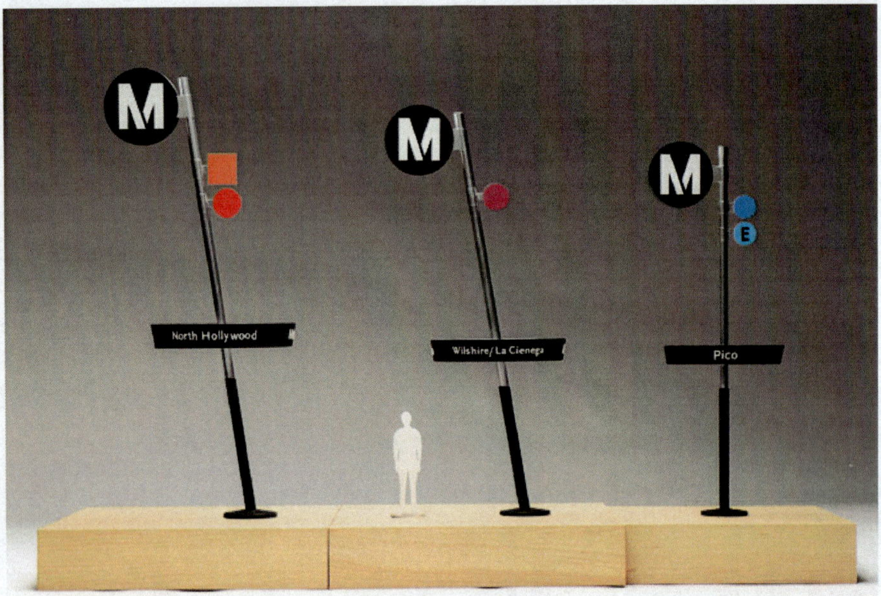

Fig. 15.3 Scale models of the new Metro Pin family. Image courtesy of Metro

Fig. 15.4 New Metro Pin station identifier at Anaheim St Station. Image courtesy of Metro

Metro North Hollywood Station, and the other in an increasingly dense urban setting at the Metro Pico Station near the Los Angeles Convention Center. Both stations areas are undergoing substantial redevelopment and it will doubtless prove instructive to see how the designs translate from their current situations to denser, more heavily trafficked environments [31]. In addition to these two prototypes, a comprehensive signage refurbishment project for Metro's oldest rail line, running between downtown Long Beach and downtown Los Angeles, includes a number of these new signs.

15.4.3 Design Solution

The family of station identifiers is based on the notion of a consistent, uniform and modular 'kit-of-parts' that strongly reinforces the agency's established brand identity. The designs are contemporary, bold and well-scaled for applications and conditions; they improve daytime and nighttime visibility and are easily recognizable within a variety of urban environments. They are designed to be highly constructible within DB contract scenarios and easily maintained, in consideration of the life-cycle costs consistent with the demands of a heavily-used, urban transit system. Reducing the area of sign structure within reach of the public creates a relatively 'tag-free' zone and the reachable area includes a sacrificial aluminum 'skin,' which can be replaced if damaged beyond repair. Internal illumination of graphic elements is accomplished via dimmable LEDs which are energy efficient and feature long life-expectancy. Issues of life-cycle costs and ease of maintenance are paramount for a transit system where even the smallest repair, if proximal to the tracks, requires special allocation and must be accomplished during a 2- to 3 hour window between daily operating cycles. Considerations for safety, both of riders and transit workers, has always been paramount in the design of the Pin. The minimum clearances indicated in Fig. 15.5 assure that any damage to or required maintenance of the unit would have the least impact on the safety and well-being of the riding public as well as to Metro employees. Careful siting of the units has also taken into account potential collision hazards from vehicles, balancing visibility with safety concerns. But perhaps the most critical safety benefit of the new design is that it affords the rail car operator much better sightlines around station entrances—the previous pylon design, being considerably bulkier at the base, has the tendency to create blind spots for operators during platform arrivals and departures (Fig. 15.4).

The full family of station identifiers consists of three variations; two that utilize a slanted tubular steel support, at 22 ft. and 24 ft. heights, and a third, at 20 ft. in height and supported by a vertical tubular steel support. The smallest of the three was adapted from the other designs specifically for confined spaces, such as a retrofit condition adjacent to the rail track bed—this variation can fit within the Overhead Contact System (OCS) maintenance envelope between the tracks, a particularly versatile design in terms of locating for ideal visibility. Additional sizes, smaller or larger than these basic three, may be developed for future locations with specific considerations.

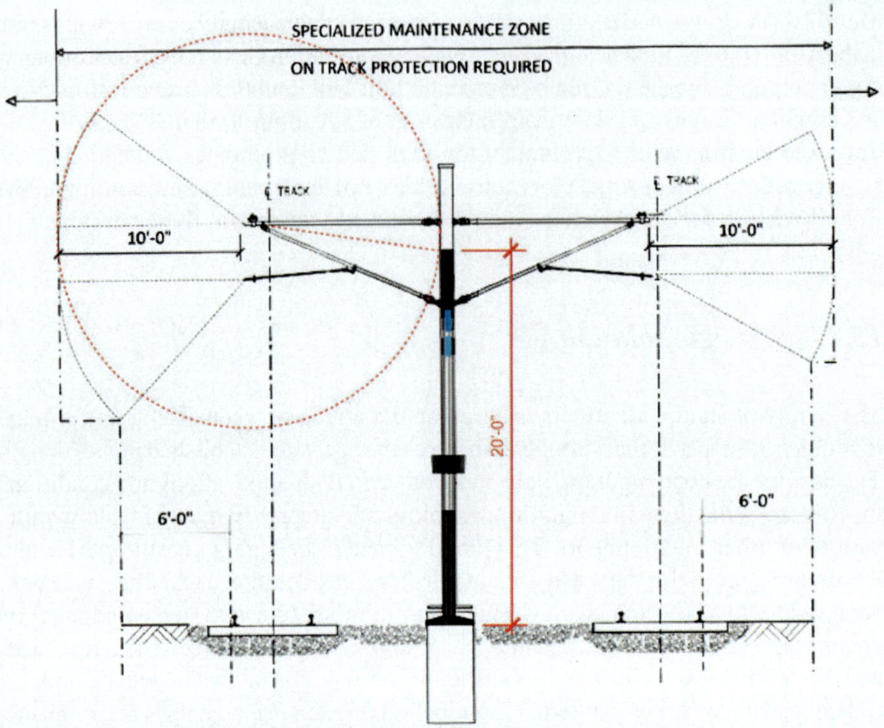

Fig. 15.5 Diagram indicating Overhead Catenary System (OCS) maintenance envelope with Metro Pin deployed between tracks. Image courtesy of Metro

While the basic allocation of components is consistent across the designs, there are two options for the horizontal station identifier element that respond to specific site conditions; the first is an axially-symmetrical, internally-illuminated ring supported from the center tube by a bracket and four angled support arms, more expansive and suited for open locations such as plazas, and the second is an internally-illuminated ellipse, supported in a similar fashion, that is designed for tighter spaces. Each of these options uses standard sign cabinet construction and 'push-thru' translucent acrylic letters. The internally illuminated logo element also utilizes a translucent acrylic graphic, with the addition of a recessed return with a subtle halo lighting effect, so that the full brand is clearly depicted in low light and nighttime conditions. Line identification is accomplished with 'Flag' elements that are constructed of painted aluminum panels that form a sandwich over an integral support bracket and arm. The gusseted steel base is bolted to a robust concrete footing and receives a painted aluminum cover plate (Fig. 15.6).

The aesthetics are a straightforward rendering of industrial materiality, with just enough streamlining of components to signify speed and movement. Proportions have been carefully calibrated to respond to the station environment, with graphics scaled appropriately to be the most visually effective. The design balances its industrial

Fig. 15.6 New Metro Pin "Ellipse" connection detail. Image courtesy of Metro

aesthetic with a degree of colorful playfulness—the graphic elements and the way they attach to the structure are reminiscent of Tinkertoys, the popular children's wooden construction set.

This 'kit-of-parts' approach lends itself to site applications where a stand-alone identifier might not be appropriate, for example immediately adjacent to a building or where a building sits directly atop a station entrance. In these cases, the kit can be deconstructed into individual elements that are arranged around and supported by other structures. The station identifying 'ring' element can be suspended from a soffit or transformed into a linear 'beam' that could run between columns, and the logo element can become a flag or blade sign mounted directly to a façade. In this manner, the new identifiers are flexible for a variety of site conditions, so that consistent brand impact is preserved in less than optimal circumstances while remaining adaptable for future station needs and operational service changes (Fig. 15.7).

15.5 Adoption of Refined Artwork Materials Palette

"The Commission recognizes that the inclusion of art in the design of public spaces creates a more inviting environment, enlivening a functional world and contributing to a positive experience for the system's future riders."—Los Angeles County Transportation Commission, which merged into LA Metro [32].

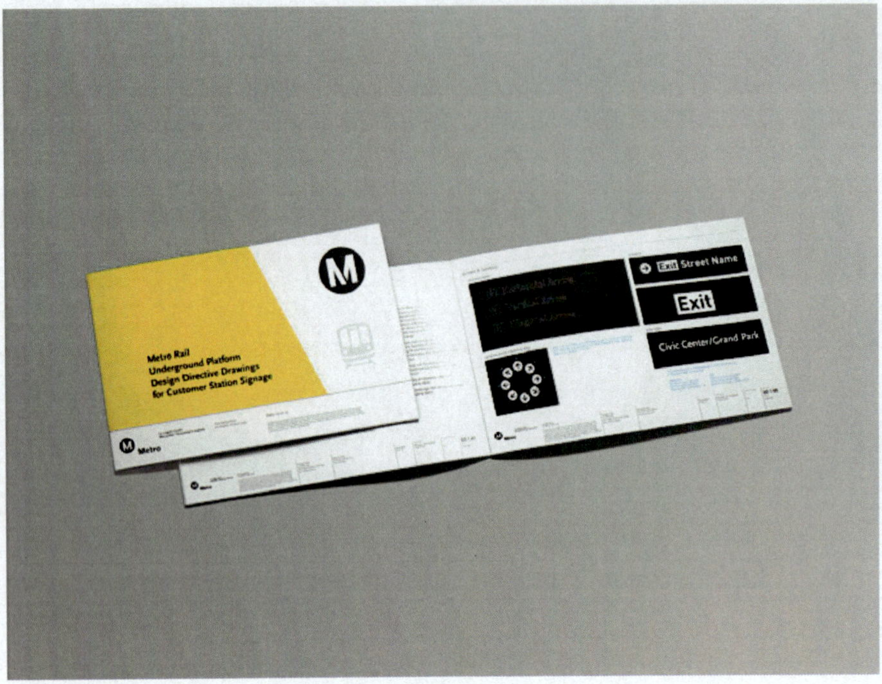

Fig. 15.7 Metro SEGD Design Directive Drawings. Image courtesy of Metro

In tandem with the station identifier elements, Metro reevaluated the approach to permanent artwork commissions on new corridor lines. These new commission scopes address some similar requirements of the identifiers to create a more visible impact in rapidly expanding urban environment and to apply lessons learned to ease maintenance concerns and promote artwork longevity. Additionally, there was the opportunity to respond to changes in project delivery methods.

15.5.1 Design Challenge

LA Metro was an early leader in the field of artist/architect collaboration in the commissioning of art, and historically created rail station designs that blurred the lines of architecture and art into an integrated whole. At first this open process to artworks involved only rail stations; however, coinciding with the expansion of the system it has been more broadly applied to Metro facilities and other components of transit infrastructure. Project delivery was a linear process of Design-Bid-Build which allowed for years, sometimes decades, for a committed partnership to fully blossom into a major artwork investment. However, with the changing building/bidding approach, artwork implementation needed to keep pace with overall design development and contractually fit within expedited execution. An artwork parameter needed to be set before artist selection.

For example, *Hooray for Hollywood* at the Hollywood/Vine Metro Rail Station is one of LA Metro's most popular public art tour destinations. The commission was completed in 1999 and Artist Gilbert "Magu" Luján closely collaborated with the architect Miralles Associates, Inc. to design a unique station that focuses on one of the most famous intersections in the world—Hollywood and Vine. Artist and architect worked together to evoke the history, glamour and excitement of the Hollywood film industry's yesteryears and its great movie palaces.

From the street level plaza down to the subway platform, passengers pass through wall details that mimic theater prosceniums to see 240 hand-glazed art tiles depicting low-rider street culture. The interior handrailing design houses the musical notes to the song "Hooray for Hollywood." Film reels cover the ceiling while two original film projectors from the 1930s, donated by Paramount Pictures, are on exhibit [33]. The resulting station delights and surprises, but like many of the collaboratively-designed Metro Rail B Line (Red) stations, it has provided maintenance challenges over the last twenty years (Figs. 15.8 and 15.9).

Given the location, this busy station receives much use—and abuse. The fiberglass benches, that mimic classic lowrider cars that once frequented the neighborhood, were within direct contact of the public waiting on the rail platform. With heavy use and direct access, the benches fell victim to significant vandalism. As custom pieces, replacements required specialized refurbishment and almost immediately encountered the same fate. It also became apparent that there was a need to fabricate with concept of removability without permanent damage to other station elements and finishes.

Other stations completed around the same time through artist/architect collaboration that have employed tile, porcelain enamel steel and granite in high traffic areas have proven enormously successful from a materials standpoint. In one such example, artist May Sun and Escudero-Fribourg Architects created copper, granite and porcelain enamel panels with text and photographs set in a field of randomly-colored wall tiles that pay homage to the native Mestizo heritage and original European settlement, as well as to the panethnic backgrounds of more recent immigrants who constitute a large portion of Metro Rail users (Fig. 15.10).

Without significant stockpiling of custom parts and detailed artwork fabrication records, losses can lead to altered or compromised creative visions as stations go through necessary modernization, repairs and safety upgrades. Fiberglass was one of several materials that was deemed inappropriate for a location within arm's reach. Staff began to refine the materials palette with an eye towards future station modifications and systemwide expansion.

Fig. 15.8 Metro Hollywood/Vine Station interior in 1999 by Gilbert "Magu" Luján and Miralles Associates. Image courtesy of Metro

15.5.2 Design Process

Before considering new approaches to any design process, it is practice for Metro to do extensive research, looking at case studies and piloting applications. Retrofits to five B and D Line (Red and Purple) stations, where the factory-produced field tile had failed to perform, permitted room to test large-scale mosaic works. While mosaic can be expensive, the assumption was that the artwork life span would be increased exponentially, and with reduced maintenance costs this well worth the

Fig. 15.9 Metro Hollywood/Vine Station during fiberglass artwork removal in 2010. Image courtesy of Metro

investment. For example, at Universal City/Studio City Station the resulting 58 ft long piece (approximately 17.7 m), *Universal Delights*, completed in 2010 by Stephen T. Johnson, is a mesmerizing and colorful attraction for customers waiting at revenue collection [34]. The vast mosaic provides visual engagement for those queuing to pay for fares in what is otherwise a dull gathering space. Standing out as a piece of art, the result was well-received by the artist, customers and staff. In addition to warming and humanizing this entry point to the turnstile area, the location is easily accessible to maintain and the artwork avoids misuse (Fig. 15.11).

Looking at artwork precedents, the mosaics of ancient Rome have proven to be as vibrantly colored now as they were 2000 years ago, with some even designed to be walked upon [35]. Other leaders in the transit field provided valuable mosaic case studies, including NYC Metropolitan Transportation Authority (MTA) and Chicago Transit Authority (CTA) (Figs. 15.12, 15.13).

Further observations were made of how porcelain enamel steel panels had been deployed in LA Metro's signage standards, and new digital printing technologies accommodated more detailed and colorful work by artists than previously imagined.

Prototyping continued at a smaller-scale series along Metro's G Line (Orange), which was a bus rapid transit project (BRT) with DB delivery with artworks rendered in rustic terrazzo in phase one and mosaic floor medallions in phase two. Porcelain enamel coated steel panels were tested in lower and higher elevations with and

Fig. 15.10 Metro Hollywood/Western Station *Untitled* artwork by artist May Sun and Escudero-Fribourg Architects features high-performance ceramic tile and porcelain enamel steel. Image courtesy of Metro

without protective film. It was assessed that the elevated applications worked, but needed to be higher and covered with film, while the floor applications failed because of foot traffic and the required high-pressure cleaning. The G Line (Orange) also tested fixtures where artwork and signage could agreeably share support structures to minimize architectural bulk (Fig. 15.14).

In a similar fashion, digitally printed glass tile was tested in the heart of Boyle Heights, with the artist's goal to create a shared, focal space for the multi-generational communities of the area. Images of birds, alluding to migration and travel, are carried throughout portions of the plaza, escalator walls and mezzanine. Inside the mezzanine around bird silhouettes is a map of the Boyle Heights neighborhood and text that alludes to travel and destination. A nest-like metal wire sculpture with a glowing, cast fiberglass egg is suspended from a vaulted ceiling in the center of the mezzanine. While the sculpture is a mesmerizing focal point, the installation presented some coordination challenges with the DB contractor. Of note, the glass application demonstrated a means to significantly impact future subway stations with minimal DB interface. Of additional benefit, the panels have required no maintenance whatsoever over the past decade (Figs. 15.16, 15.17).

Fig. 15.11 Mosaic artwork *Universal Delights* by artist Stephen T. Johnson, 2010 (Metro Universal City/Studio City Station). Image courtesy of Metro

Fig. 15.12 The New York Metropolitan Transportation Authority Arts & Design's mosaic artworks served as a case study. Pictured here is artist Vik Muniz's *Perfect Strangers* at the Second Avenue-72nd Street Station [37, 38]

Fig. 15.13 The New York Metropolitan Transportation Authority Arts & Design's mosaic artworks served as a case study. Pictured here is artist Stephen T. Johnson's *DeKalb Improvisation* in the DeKalb Avenue Station. Image courtesy of Stephen T. Johnson

15.5.3 Design Solution

Given the shift to a DB delivery model it was apparent that the luxury of an architect and artist collaboration was no longer viable with an expedited scenario. To control costs, the DB architect has a very limited scope and is often the first design discipline to be released on a project.

Based on DB lessons learned on the E and G Line (Expo and Orange) corridors, Metro's artwork opportunities for DB projects over the past decade have been evaluated by staff based on square footage and scale, sightlines, paths of travel, potential future construction, and potentially competing visual elements such as revenue-generating advertising. Early art program involvement continues to be the key to maximizing prospects, and this work has been undertaken from preliminary engineering phases in consultation with the DB architects to ensure accurate artwork description in the construction documents.

Expansive 2D commissions have been generally regarded to be most impactful and most effective to integrate into this contracting scenario. Once staff have defined placement and materials, artworks have been designed in the artist's studio to be fabricated off site by skilled artisans with a proven track record in artwork fabrication and delivered to the DB contractor for installation. In keeping with past projects,

Fig. 15.14 Pilot for porcelain enamel steel and rustic terrazzo, artwork titled *Lucky California*, by artist Phung Huynh. Signage is housed compatibly in same structure (Metro Laurel Canyon Station). Image courtesy of Metro

artists and staff continue to provide construction oversight to ensure quality artwork execution in accordance with the artist's original intent.

Metro Arts & Design staff have identified materials for system expansion and within the realm DB project delivery. Observations were carefully evaluated both within the Metro system and at transit agencies in the United States and abroad. Considerations have included resistance to rusting, pitting, fading, delamination, solvents, pollutants and graffiti, etc. The refined artwork palette of proven materials resulted in a selection of mosaic, porcelain enamel steel, ceramic tile and embellished architectural glass. These applications have been prioritized for placement in elevations within a neutral architectural backdrop of steel and glass, concrete, granite, and terrazzo; well-positioned to create an immersive experience for the customer that establishes memorable connections to the station environment (Figs. 15.19 and 15.20).

Large format mosaics are in the works for nine new concourses along three future corridors. These mosaics are in varying stages of fabrication off site by artisan laboratories that have demonstrated a technically commanding caliber and the ability to support an artist's creative vision. This is an environment where artists may collaborate with a team of skilled craftspeople and where developments in technology

Fig. 15.15 The Brussels Intercommunal Transport Company (STIB-MIVB) porcelain enamel artworks served as a case study. Pictured is artist Jean François Octave's *Le Heysel, reflet du monde au 20ème siècle (et 21e)* at Heizel Station [39]

Fig. 15.16 Concourse artwork by artist Nobuho Nagasawa, *Landings* (Metro Soto Station). Image courtesy of Metro

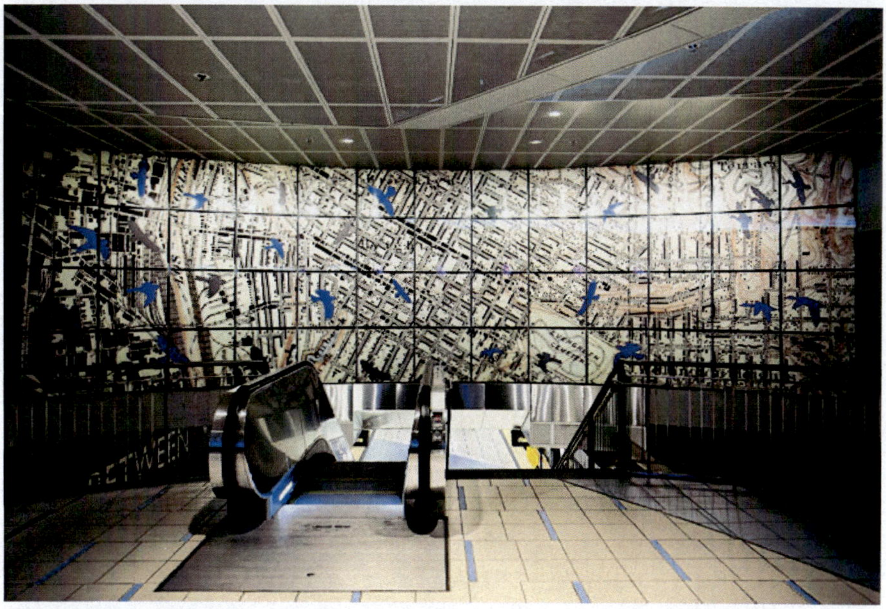

Fig. 15.17 Detail: Metro pilot for digital printing artwork on glass, *Landings*, by artist Nobuho Nagasawa (Metro Soto Station). Image courtesy of Metro

Fig. 15.18 Transport for London glass artworks served as a case study. Pictured is Daniel Buren's *Diamonds and Circles* at Tottenham Court Road Station [36]

Fig. 15.19 Iteration of Metro Systemwide Design Standards including Metro branding and placeholder artwork locations for architectural glass at subway station entrances. Image courtesy of Metro

provide a dizzying array of innovative techniques that surpass the skillset of any DB contractor (Figs. 15.21 and 15.22).

Artwork content also continues to be informed by each artist's own independent research, including community input, and is highly specific to the surrounding neighborhood context. For example: currently in design development, Todd Gray's artwork titled, *Mining the Archive: S. Charles Lee, Architect,* dug into photographic and architectural archives. Inspired by the architectural drawings and photographs of the nearby historic Saban Theater, Gray recontextualizes these images with multicultural textile patterns for a more inclusive narrative of the area. Utilizing digitally

Fig. 15.20 Typical concourse Metro subway station includes opportunities for immersive glass tile and mosaic artworks with integrated customer information and signage. Courtesy of Metro

printed glass covering over three thousand square feet of wall surface area, the artwork will provide a large-scale, immersive experience where riders on the escalators are entering the theater of the Metro system [40].

Artist Eileen Cowin, commissioned for artwork at the future Metro Martin Luther King Jr Station along the Metro Crenshaw/LAX Corridor, captured community members in businesses and locations near the station. Arranged in non-linear narratives on fourteen filmstrip-like panels of porcelain enamel steel, Cowin's juxtaposition of photographs invites the viewer to create their own narratives as they wait on the station platform [41] (Figs. 15.23 and 15.24).

Porcelain enamel steel engineered in a cleat attachment system has proven the premium material for platform walls where materials need to withstand the train's dynamic envelope and where fire/life safety issues are apparent above the energized track. This approach is being implemented in twelve new subway stations. Given the material high performance endurance and residence to fading, a similar methodology is also being applied in five new above grade station platforms.

Metro Arts & Design staff continue to observe material performance and explore new materials and technologies as they evolve. Innovations in cast-in-place concrete for retaining walls and bridges is one such example (Figs. 15.23 and 15.24).

Fig. 15.21 Dry fit of mosaic artwork, *Only Light, Only Love*, by artist Shinique Smith to be installed at the concourse level of the future Metro Martin Luther King Jr Station. Image courtesy of Metro

15.6 Design Harmony Accommodates Elements of Consistency and Variability

Transit systems are for users and meant to be activated, and the system artworks do not retain the same meaning without an active, viewing public. However, a radically different environment than the temporary, controlled space of a pristine gallery, artworks must also meet practical demands and exist in harmony with signage. Art and design must be of a quality and timeless standard that remains meaningful for a range of viewers and retains vibrancy for future generations.

Fig. 15.22 Dry fit of mosaic artwork, *At the Same Time*, by artist Rebeca Méndez to be installed at the concourse level of the future Metro Expo/Crenshaw Station. Image courtesy of Metro

Working within a design thinking process enables designation of large surface areas for artworks in proven materials while still prioritizing wayfinding for patrons. For both signage and artwork, the approach meets the import and practical drivers for increased durability, ease of maintenance access and reduction in the degree of maintenance. For both design disciplines, a modular strategy allows for system growth. Panelized artworks can be removed and reinstalled in case of a major station impact; and pole mounted signage elements can be removed and reconfigured to allow for system growth and any new station identification needs. In some circumstances where there is minimal physical infrastructure or footprint, it has been possible to bring visual wayfinding and commissioned art together under single architectural feature such as a canopy where the elements of variability (art) and continuity (signage) coexist harmoniously without adding unnecessary clutter (Figs. 15.25 and 15.26).

Fig. 15.23 Installation of artwork, *You are heading in the right direction*, by artist Eileen Cowin. These porcelain enamel steel panels are on the train track wall adjacent to the center platform of the Metro Martin Luther King Jr Station under construction. Integrated sign band is mounted flush with the artwork. Image courtesy of Metro

As lessons have been applied from previous projects, placement in active station needs a plan for future access, so that routine and long-term maintenance can be performed on the artwork and signage. Additionally, the materials should be predetermined based on their proven longevity. The transit environment is harsh, with both environmental and human factors, and the best maintenance plan is a prevention plan.

A new prototype on the horizon that holds potential is adopting advertising space in digital station kiosks for rotating digital display of artwork through to wayfinding. The changeability of the display allows Arts & Design to test artwork and wayfinding targeted for more temporal site responsive programming, for example, artwork to celebrate calendar events and signage to help navigate service changes. Further ahead, Metro will introduce a new corridor naming convention, which in addition to colors utilizes letter designations, an approach that is among the international best practices for clarity in transit systems with eight or more lines.

Metro continues to exercise artist and architect collaborations, and unique one of a kind sign types, but these are limited to contracting scenarios (such as DBB) that are

Fig. 15.24 Installation of artwork, *You are heading in the right direction*, by artist Eileen Cowin. These porcelain enamel steel panels are on the train track wall adjacent to the center platform of the Metro Martin Luther King Jr Station under construction. Integrated sign band is mounted flush with the artwork. Image courtesy of Metro

truly supportive of creative partnerships. This contracting format is generally reserved for major infrastructure such as bridges, large-scale, multi-modal transit facilities and customer service centers, as opposed to the station environment. Nonetheless, lessons learned for materials, methods and placement still apply (Fig. 15.27).

15.7 Conclusions

Adopting lessons learned to materials, methods of construction and placement provides vast opportunity to maximize the scale and areas for artwork and signage with a shared system vision, instead of negotiating for space on a station-by-station basis.

Being in harmony within the station design implies a holistic integration of signage and artwork into the transit ecosystem. Wayfinding signs are best located at natural decision points to support the flow of traffic and provide assurance of direction. Locations of artworks should aim to maximize surprise and delight, and skillfully play off and thereby enchance the dynamics of these public spaces.

Fig. 15.25 Ceramic tile and mosaic artwork alongside signage integrated into limited station infrastructure on DB project. Artwork pictured is *Artist Educators* by Shizu Saldamando at Metro Palms Station. Image courtesy of Metro

Fig. 15.26 Porcelain enamel steel panel artwork alongside signage integrated into limited station infrastructure on DB project. Artwork pictured is *Local Color* by Constance Mallinson at Metro 26th St/Bergamot Station. Image courtesy of Metro

Fig. 15.27 Metro Division 13 Bus Maintenance and Operations Facility is a DBB project with integrated art and signage. The luminous artwork, *El Aliso de Los Angeles*, by artist Christine Ulke received two CODA awards and a Public Art Network Year in Review award. Image courtesy of Metro

Additionally, Arts & Design shares equity across stations, with similar materials, placement and scale across various socioeconomic areas in the very diverse county of Los Angeles. Aesthetic consideration is an indicator of societal value in a space, it is visually coded, so maximizing artwork area and signage standards is important to all neighborhoods served by Metro.

This responsive and responsible design approach meets the critical need to realistically plan, so that future generations inherit the legacy of quality art and design. This strategy builds works with increased resilience to environmental conditions and human stressors to survive longer periods when there may be organizational change due to political shifts and economic downturns.

These are just some lessons adopted for rail corridors in a range of Metro Arts & Design programming. As Metro continues to grow and evolve with P3 contracting and digital technologies on the horizon, it will be vital to stay agile and continue to reevaluate approaches to materials, methods and placement. It is a learning process to create dignified transit experiences for all types of users. It is artists and designers that capture the humanness of the journey and elevate the everyday experience. An important role of transit staff is to remove administrative obstacles to ensure that creative visions translate as clearly as possible and last for years to come.

References

1. Certeau Michel, D.: The Practice of Everyday Life. Translated by Steven Rendall. University of California Press. p. 117. Retrieved from: https://monoskop.org/images/2/2a/De_Certeau_Michel_The_Practice_of_Everyday_Life.pdf (1984)
2. United States Census Bureau (n.d).: QuickFacts: Los Angeles County. Retrieved 26 Mar 2020 from: https://www.census.gov/quickfacts/losangelescountycalifornia
3. American Public Transportation Association: Transit Ridership Report: Third Quarter 2019. Retrieved from: https://www.apta.com/wp-content/uploads/2019-Q3-Ridership-APTA.pdf. 22 Nov 2019
4. Los Angeles County Metropolitan Authority (n.d.).: Metro Interactive Estimated Ridership Stats. Retrieved from: http://isotp.metro.net/MetroRidership/Index.aspx
5. Katz, R.: AB152 (Katz). MTA Library Archives. Retrieved from: http://libraryarchives.metro.net/DPGTL/legislation/1992-ab-152-los-angeles-county-metropolitan-transportation-authority-reform-act.pdf. 18 Dec 1990
6. Southern California Rapid Transit District.: Easier public identification: firm developing 'new look' for district. RTD Headway. **6**(9). Retrieved from: http://libraryarchives.metro.net/DPGTL/employeenews/Headway_1979_Sep.pdf (1979, September)
7. Special Collections, Margaret Herrick Library, Academy of Motion Picture Arts and Sciences. (1949–1996). Saul Bass Papers
8. Corwin, M.: Subway Showcase: Public art: Metro Red Line stations will have murals, neon sculptures and other works with distinct designs at each stop on the route. Officials say installations will curb vandalism. Los Angeles Times. Retrieved from: https://www.latimes.com/archives/la-xpm-1992-11-15-me-1124-story.html. 15 Nov 1992
9. Federal Transit Administration.: Design and Art in Transit Projects. *FTA Circular 9400.1.A*. U.S. Department of Transportation, Washington, DC., 9 Jun 1995
10. Federal Transit Administration.: Arts-in-Transit (AIT). U.S. Department of Transportation, Washington, DC. Retrieved from: https://www.transit.dot.gov/regulations-and-guidance/arts-transit. 17 Jan 1995
11. San Francisco International Airport.: The principals of R.E.A.C.H. Retrieved from: http://media.flysfo.com.s3.amazonaws.com/pdf/about/b2b/SFO-principles-of-REACH.pdf (2013)
12. American Public Transportation Association.: Best Practices for Integrating Art into Capital Projects. APTA brochure, APTA SUDS-UD-RP-007-13 Washington, D.C., 28 June 2013
13. American Public Transportation Association.: Why Design Matters for Transit. APTA brochure, APTA SUDS-UD-RP-003-11 Washington, D.C., (31 Ma 2011; revised 26 Mar 2013)
14. Los Angeles County Metropolitan Authority.: Facts at a Glance. Metro Website. Retrieved from: https://www.metro.net/news/facts-glance/. Last updated 24 June 2018
15. Los Angeles County Metropolitan Authority (n.d.).: Interactive Estimated Ridership Stats. Metro Website. Retrieved from: http://isotp.metro.net/MetroRidership/Index.aspx
16. Metropolitan Transportation Authority.: MTA Capital Program 2020–2024: Rebuilding New York's Transportation System. Retrieved from https://new.mta.info/sites/default/files/2019-09/MTA%202020-2024%20Capital%20Program%20-%20Full%20Report.pdf (2019)
17. Puentes, R.: Spotlight on Measure M. Eno Center for Transportation. Retrieved from: https://www.enotrans.org/article/spotlight-on-measure-m/. 26 July 2019
18. Angeles County Metropolitan Authority (n.d.).: Twenty-Eight by '28 Initiative. Retrieved from: https://www.metro.net/projects/resources/28-by-2028/
19. Federal Transit Administration.: "6.7 Artwork." BPPM Procurement Object Types: Special Considerations. U.S. Department of Transportation, Washington, DC. Retrieved from: https://www.transit.dot.gov/funding/procurement/bppm-procurement-object-types-special-considerations. Last updated 25 Jan 2016

20. Federal Transit Administration.: 6.1.1 The Traditional Construction Process—Design/Bid/Build.In: BPPM Procurement Object Types: Special Considerations. U.S. Department of Transportation, Washington, DC. Retrieved from: https://www.transit.dot.gov/funding/procurement/bppm-procurement-object-types-special-considerations. Last updated 25 Jan 2016
21. Federal Transit Administration.: Cost Benefits to Construction Manager/General Contractor Approach. U.S. Department of Transportation, Washington, DC. Retrieved from: https://www.transit.dot.gov/regulations-and-guidance/cost-benefits-construction-managergeneral-contractor-approach. Last updated 8 Feb 2016
22. Welborne, M.: Systemwide Stationwide Design. Metro Interoffice Memo with Attachment: Metro Rail Design Criteria and Architectural Directive Drawings. Los Angeles County Metropolitan Transportation Authority (25 Nov 2013)
23. Los Angeles County Metropolitan Transportation Association.: Systemwide Station Design Standards, Metro Board Report. Retrieved from: https://media.metro.net/projects_studies/tod/images/approved_boardreport_systemwide_station_design_standards_policy.pdf. (17 Jan 2018)
24. Federal Transit Administration.: Modular Station Design. U.S. Department of Transportation, Washington, DC. Retrieved from: https://www.transit.dot.gov/regulations-and-guidance/modular-station-design. Updated 24 May 2016
25. Architect.: The Los Angeles Metro station at Santa Monica and Vermont. Architect Magazine [Online Projects]. Retrieved from: https://www.architectmagazine.com/project-gallery/the-los-angeles-metro-station-at-santa-monica-and-vermont-6460. 27 Jan 2015
26. Bass, J., Kirkham, P.: Saul Bass: A Life in Film & Design. Laurence King Publishing, vol. 384 (9 Nov 2011)
27. London Underground (n.d.).: Station design idiom. Retrieved from: http://content.tfl.gov.uk/station-design-idiom-2.pdf. 25 Oct 2019
28. Vignelli, M., Noorda, B. (Unimark).: New York City Transit Authority Graphic Standards Manual (1970)
29. Müller-Brockmann, J.: Passenger Information System: Design Manual for the Swiss Federal Railways (1992)
30. Sussman, D., Prejza & Company.: Beyond Graphic Design. In: Process Architecture Book, vol. 124 (1 Oct 1995)
31. Khouri, A.: Chinese developer teams up on big North Hollywood mixed-use project. Los Angeles Times (23 June 2016)
32. Los Angeles County Transportation Commission.: Art Program Policy. Los Angeles County Metropolitan Transportation Authority (Metro) Board Archives. Retrieved from: http://boardarchives.metro.net/Items/1993/04_April/items_g_0252.pdf (1989, 1993)
33. Lujan, G.: Hooray for Hollywood. Metro Art Website. Retrieved from: https://www.metro.net/about/art/artworks/hooray-hollywood/
34. Johnson, S.: Universal Delights. Metro Art Website. Retrieved from: https://www.metro.net/about/art/artworks/universal-delights/
35. Stephan, A.: A Brief Introduction to Roman Mosaics. In: The Iris, Behind the Scenes at the Getty. Retrieved October 28, 2019 from: https://blogs.getty.edu/iris/a-brief-introduction-to-roman-mosaics/. 4 April 2016
36. Bal, T.: Daniel Buren, 'Diamonds and Circles, works in situ,' Tottenham Court Road [photograph]. London, Wikimedia. License CC BY-SA 4.0: https://creativecommons.org/licenses/by-sa/4.0/deed.en. Retrieved from: https://commons.wikimedia.org/wiki/File:Daniel_Buren,_%27Diamonds_and_Circles,_works_in_situ%27,_Tottenham_Court_Road,_2016._Photo-_Thierry_Bal,_2016.jpg (1 Dec 2015; published 4 April 2017)
37. Kennedy, R.: Art Underground: A First Look at the Second Avenue Subway. *New York Times*. Retrieved from: https://www.nytimes.com/2016/12/19/arts/design/second-avenue-subway-art.html. 19 Dec 2016

38. Metropolitan Transportation Authority.: 72ndStreet: Vik Muniz, Perfect Strangers [photograph]. Making of: Second Avenue Subway [digital photo album]. Flickr. License CC BY 2.0: https://creativecommons.org/licenses/by/2.0/. Retrieved from: https://flickr.com/photos/mtaphotos/31747106855/in/album-72157677942834956/. 15 Dec 2016
39. Lek, S.: Metrostation Heizel in Brussels [photograph]. Brussels, Wikimedia. License CC BY-SA 4.0: https://creativecommons.org/licenses/by-sa/4.0/deed.en. Retrieved from: https://commons.wki-media.org/wikimedia.org/wki/File:Heizael_metro_station_Brussels.jpg#filelinks. 22 Dec 2017
40. Los Angeles County Metropolitan Transportation Authority (n.d.).: Purple Line Extention Art Program: Wilshire/La Cienega. Metro Arts & Design Webpage. Retrieved 10 July 2020 from: https://www.metro.net/about/art/PLE/
41. Los Angeles County Metropolitan Transportation Authority (n.d.).: Crenshaw/LAX Transit Project Art Program: Martin Luther King Jr." Metro Arts & Design Webpage. Retrieved 10 Jul 2020 from: https://www.metro.net/about/art/crenshawlax/